Protocols for Oligonucleotide Conjugates

Methods in Molecular Biology

John M. Walker, Series Editor

26. **Protocols for Oligonucleotide Conjugates,** edited by ***Sudhir Agrawal,*** *1994*
25. **Computer Analysis of Sequence Data:** *Part II,* edited by ***Annette M. Griffin and Hugh G. Griffin,*** *1994*
24. **Computer Analysis of Sequence Data:** *Part I,* edited by ***Annette M. Griffin and Hugh G. Griffin,*** *1994*
23. **DNA Sequencing Protocols,** edited by ***Hugh G. Griffin and Annette M. Griffin,*** *1993*
22. **Optical Spectroscopy, Microscopy, and Macroscopic Techniques,** edited by ***Christopher Jones, Barbara Mulloy, and Adrian H. Thomas,*** *1994*
21. **Protocols in Molecular Parasitology,** edited by ***John E. Hyde,*** *1993*
20. **Protocols for Oligonucleotides and Analogs,** edited by ***Sudhir Agrawal,*** *1993*
19. **Biomembrane Protocols:** *I. Isolation and Analysis,* edited by ***John M. Graham and Joan A. Higgins,*** *1993*
18. **Transgenesis Techniques,** edited by ***David Murphy and David A. Carter,*** *1993*
17. **Spectroscopic Methods and Analyses,** edited by ***Christopher Jones, Barbara Mulloy, and Adrian H. Thomas,*** *1993*
16. **Enzymes of Molecular Biology,** edited by ***Michael M. Burrell,*** *1993*
15. **PCR Protocols,** edited by ***Bruce A. White,*** *1993*
14. **Glycoprotein Analysis in Biomedicine,** edited by ***Elizabeth F. Hounsell,*** *1993*
13. **Protocols in Molecular Neurobiology,** edited by ***Alan Longstaff and Patricia Revest,*** *1992*
12. **Pulsed-Field Gel Electrophoresis,** edited by ***Margit Burmeister and Levy Ulanovsky,*** *1992*
11. **Practical Protein Chromatography,** edited by ***Andrew Kenney and Susan Fowell,*** *1992*
10. **Immunochemical Protocols,** edited by ***Margaret M. Manson,*** *1992*
9. **Protocols in Human Molecular Genetics,** edited by ***Christopher G. Mathew,*** *1991*
8. **Practical Molecular Virology,** edited by ***Mary K. L. Collins,*** *1991*
7. **Gene Transfer and Expression Protocols,** edited by ***Edward J. Murray,*** *1991*
6. **Plant Cell and Tissue Culture,** edited by ***Jeffrey W. Pollard and John M. Walker,*** *1990*
5. **Animal Cell Culture,** edited by ***Jeffrey W. Pollard and John M. Walker,*** *1990*
4. **New Nucleic Acid Techniques,** edited by ***John M. Walker,*** *1988*
3. **New Protein Techniques,** edited by ***John M. Walker,*** *1988*
2. **Nucleic Acids,** edited by ***John M. Walker,*** *1984*
1. **Proteins,** edited by ***John M. Walker,*** *1984*

Methods in Molecular Biology • 26

Protocols for Oligonucleotide Conjugates

Synthesis and Analytical Techniques

Edited by

Sudhir Agrawal

Worcester Foundation for Experimental Biology, Shrewsbury, Massachusetts

Foreword by

H. Gobind Khorana

Humana Press **Totowa, New Jersey**

999 Riverview Drive, Suite 208
Totowa, New Jersey 07512

Printed in the United States of America. 9 8 7 6 5 4 3 2 1

Library of Congress Cataloging in Publication Data

Main entry under title:
Methods in molecular biology.

Protocols for oligonucleotide conjugates: synthesis and analytical techniques / edited by Sudhir Agrawal.
p. cm.—(Methods in molecular biology; 26)
Includes index.
ISBN 0-89603-252-3
1. Oligonucleotides—Synthesis—Laboratory manuals.
2. Oligonucleotides—Analysis—Laboratory manuals. I. Agrawal, Sudhir. II. Series: Methods in molecular biology (Totowa, NJ): 26.
[DNLM: 1. Oligonucleotides—chemistry. W1 ME9616J v. 26]
QP625.O47P757 1993
547.7'9—dc20
DNLM/DLC
for Library of Congress 93-23127
CIP

Preface

During the past ten years or so, there has been an increasing number of publications describing various aspects of oligonucleotide chemistry and the application of oligonucleotides in basic, as well as applied science. More recently, various oligonucleotide conjugates have been synthesized that are currently being studied for their potential diagnostic and therapeutic applications. With the introduction of new oligonucleotide conjugates, analytical methods for the analysis of these oligonucleotides are also being developed and optimized. The present volume, *Protocols for Oligonucleotide Conjugates: Synthesis and Analytical Techniques*, is designed to bring readers up-to-date on the protocols for the synthesis of various oligonucleotide conjugates and on all the important techniques for the analysis of synthetic oligonucleotides and their analogs. The synthetic chemistry of many of these oligonucleotides and their various analogs has been covered in a companion volume entitled *Protocols for Oligonucleotides and Analogs: Synthesis and Properties.*

The first chapter of *Protocols for Oligonucleotide Conjugates* describes the protecting groups in oligonucleotide synthesis. In this chapter, the author has described all protecting groups that have been used to date in various stages of the development of oligonucleotide synthesis, supported by more than 400 references. The second chapter describes the synthesis and incorporation of modified bases into oligonucleotides to introduce reporter groups. Chapters 3 and 4 describe the derivatization of synthetic oligonucleotides with amino groups and with the thio groups, respectively. In both chapters, protocols have been provided for the introduction of the amino or thio groups at various positions in the oligonucleotides, for the attachment of various reporter groups and for the purification of oligonucleotide conjugtes. The postsynthetic functionalization of oligonucleotides for the introduction of reporter groups is discussed in Chapter 5. Protocols for the use of oligonucleotide conjugates as diagnostic probes are described in Chapter 6.

In Chapter 7, the authors have incorporated two new degenerate bases into oligonucleotides and have discussed their uses. Very recently, the use of oligonucleotides containing degenerate bases has been increasing in the area of polymerase chain reaction. Chapter 8 describes protocols for the incorporation of radio-labeled nucleotides into oligonucleotides. The remaining chapters (Chapters 9–14) describe various analytical techniques and their applications to oligonucleotides, oligonucleotide analogs, and oligonucleotide conjugates. The techniques described in the present volume are high performance liquid chromatography (Chapter 9), sequencing (Chapter 10), gel-capillary electrophoresis (Chapter 11), nuclear magnetic resonance (Chapter 12), mass spectrometry (Chapter 13), and thermodynamic stability (Chapter 14).

I would like to thank all the authors of various chapters of this book and all the credit for the book's teachings goes to them. I am also thankful to Paul Zamecnik, Thoru Pederson, and Dan Brown for fruitful suggestions. I would also like to thank John Walker, *Methods in Molecular Biology* series editor, for his guidance at all stages of editing. Finally, many thanks go to Thomas Lanigan and Fran Lipton of Humana Press, whose hard work and diligence have made this book a reality.

Sudhir Agrawal
Worcester Foundation for Experimental Biology
222 Maple Avenue
Shrewsbury, MA 01545

Present Address:
Hybridon Inc.
One Innovation Drive
Worcester, MA 01605

Foreword

Since the discovery of the nucleic acids in the latter part of the last century, several decades elapsed before the structures of the constituent purines and pyrimidines, the corresponding nucleosides, and the nature of the internucleotide bonds in the polynucleotides were clarified. By 1952, the chemical structures of both classes of nucleic acids, RNA and DNA, had been established. Soon followed the Watson-Crick proposal for the DNA structure and this ushered in a new era in chemical, biochemical, and molecular biological studies.

As with other classes of biological macromolecules, interest in the synthesis of oligo- and polynucleotides increased very rapidly and systematic studies on the synthesis of polynucleotides were undertaken in the mid-1950s. Indeed, strategies were successfully developed for the synthesis of oligonucleotides of defined sequences. These, in conjunction with enzymatic approaches enabled the synthesis of high molecular weight polymers that were successfully used in the elucidation of the genetic code and later in the total laboratory synthesis of genes fully functional in vivo.

In more recent years new developments in synthesis have advanced the synthetic technology to a highly efficient level. In particular, synthesis on polymer supports, a concept first developed by Merrifield in polypeptide synthesis, has resulted in enormous improvements and speed in synthesis of large size polynucleotides.

Dr. Agrawal is to be congratulated on having produced two authoritative books with contributions from experts in different aspects of polynucleotide synthesis. These monographs should be extremely useful to those who are already practitioners in the field and also those who want to enter this field without any particular background.

H. Gobind Khorana
Massachusetts Institute of Technology
Cambridge, MA

Contents

Contents of the companion volume:

Protocols for Oligonucleotides and Analogs

Synthesis and Properties

Contributors

SUDHIR AGRAWAL • *Worcester Foundation for Experimental Biology, Shrewsbury, MA*
ALEX ANDRUS • *Applied Biosystems, Inc., Foster City, CA*
DIANE M. BLACK • *Institut de Chimie Biologique, Faculté de Médecine, Strasbourg, France*
KENNETH J. BRESLAUER • *Department of Chemistry, Rutgers, The State University of New Jersey, Piscataway, NJ*
DANIEL M. BROWN • *Laboratory of Molecular Biology, MRC, Cambridge, UK*
BARBARA C. F. CHU • *The Salk Institute for Biological Sciences, San Diego, CA*
JACK S. COHEN • *Department of Pharmacology, Georgetown University, Rockville, MD*
JACQUELINE A. FIDANZA • *Department of Chemistry, Boston College, Boston, MA*
PETER T. GILHAM • *Department of Biological Sciences, Purdue University, West Lafayette, IN*
JERZY W. JAROSZEWSKI • *Department of Organic Chemistry, Royal Danish School of Pharmacy, Copenhagen, Denmark*
ROGER A. JONES • *Department of Chemistry, Rutgers, The State University of New Jersey, Piscatatway, NJ*
PAUL KONG THOO LIN • *Laboratory of Molecular Biology, MRC, Cambridge, UK*
THOMAS D. MCCLURE • *Department of Pharmaceutical Sciences, The University of Arizona Health Sciences Center, Tucson, AZ*
LARRY W. MCLAUGHLIN • *Department of Chemistry, Boston College, Boston, MA*
RICH B. MEYER, JR. • *MicroProbe Corporation, Bothell, WA*
LESLIE E. ORGEL • *The Salk Institute for Biological Sciences, San Diego, CA*
HIROAKI OZAKI • *Department of Chemistry, Boston College, Boston, MA*
SIDDHARTHA ROY • *Department of Biophysics, Bose Institute, Calcutta, India*
JERRY L. RUTH • *US Fish and Wildlife Forensics Laboratory, Ashland, OR*

KARL H. SCHRAM • *Department of Pharmaceutical Sciences, The University of Arizona Health Sciences Center, Tucson, AZ*
ETIENNE SONVEAUX • *Laboratoire de Biochimie Physique et des Biopolymères, Louvain-La-Neuve, Belgium*
GEORGE VELLA • *Millipore Corporation, Waters Chromatography Division, Milford, MA*
WILLIAM WARREN • *Millipore Corporation, Waters Chromatography Division, Milford, MA*

CHAPTER 1

Protecting Groups in Oligonucleotide Synthesis

Etienne Sonveaux

1. Introduction

A biopolymer is synthesized by assembling monomeric or oligomeric blocks. Each block features at least a nucleophilic and an electrophilic function, i.e., the α-amino and the carboxylic functions for peptides, the 5'-OH and the 3'-function (phosphate, phosphoramidite, or phosphonate), for nucleotides. The nucleophilic and electrophilic sites are linked together at the coupling step. Protection is a necessity. It guarantees the chemoselectivity of coupling and the solubility of synthons in organic solvents.

There are two classes of protecting groups: persistent and transient. The persistent protections remain on the biopolymer during all the synthesis. They are cleaved at the very end. They cap the functions of the aglycone residue of nucleotides, or of the side chains of amino acids in peptide synthesis. They also cap the phosphate oxygen of oligonucleotides. The transient protections block the functions to be coupled at a given time of the synthesis. They are specifically cleaved before each coupling.

When the synthesis is performed on a solid support, the first monomer of a hundred-mer has to survive to a hundred cleavages of a transient protecting group. The yield of successful removal is thus as limiting as the coupling yield. This is also true for the final deprotection. If each monomeric unit is only 90% deprotected, the yield of a dimer of correct structure is 9^2%, of a trimer $9^3/10$%, and of a *n*-mer $9^n/10^{n-2}$%.

From: *Methods in Molecular Biology, Vol. 26: Protocols for Oligonucleotide Conjugates*
Edited by: S. Agrawal Copyright ©1994 Humana Press Inc., Totowa, NJ

Yields drop dramatically with length. Paradoxically, the crucial problem of protection is thus deprotection. Good results obtained with a protection strategy on small sequences do not guarantee that the method is viable. The discriminating test is the success in obtaining high yields of long oligomers.

In oligonucleotide synthesis, the academic research is nowadays moving from DNA synthesis to the synthesis of RNA and modified DNA/RNA structures. As functions and types of aglycone residues entering oligonucleotide synthesis diversify, protection strategies have to be adapted. That is why this review may be useful.

Its content is as follows. The persistent protections of the nucleic bases and of the 2'-OH of ribonucleotides are considered first. Both are usually introduced at the beginning of the synthesis. The transient protection of the 5'-OH is then discussed. This function is indeed capped before the phosphorylation or phosphitylation of the synthons. The last paragraph describes the protections of phosphorus. This last topic is much related to coupling strategies. The reader is thus invited to consult the other parts of this book to embrace the whole field.

2. Protection of the Heterocyclic Bases and Protocols for Oligonucleotides and Analogs

The protecting groups that have been proposed for the aglycone residues are presented in Tables 1–5. The most useful protections are briefly described in the following paragraphs.

2.1. Thymidine and Uridine

It is possible to synthesize oligo DNA or RNA by one of the three classical methods (phosphotriester, phosphoramidite, phosphonate stategies) without protecting these residues. However, the *N*-3 nitrogen being acidic (T, $pK_a = 9.79$) *(1)*, a certain amount of the anionic form is present in basic media. In these conditions, the thymine and uracil residues react with electrophiles like alkylating agents *(2,3)* (*inter alia*, the methyl phosphate of the internucleotidic bond in one of the phosphoramidite strategies *(3–5)*, carbodiimides *(6)*, activated phosphates *(7–14)* and sulfonates *(15,16)*, bis(diisopropylamino) methoxyphosphine *(17)*, trimethylsilyl chloride *(18)*, and acid chlorides.

The instable *O*-4 phosphorylated products undergo a nucleophilic attack on C-4 by nucleophiles usually present in the medium, like 1,2,4-triazole, 3-nitro-1,2,4-triazole, 1-hydroxybenzotriazole, *N*-methylimidazole or pyridine. In the case of acylation, the *N*-3 acylated derivative is usually obtained *(19)*. It is the thermodynamic product. The *O*-4 acylated derivative, accessible by PTC, spontaneously rearranges to the *N*-3 acylated isomer *(16)*.

In the conditions of a normal oligonucleotide synthesis, the contact times with electrophiles are short and these side reactions are limited *(20)*. Thymidine is less sensitive than uridine *(15)*, and is usually not protected *(21)*. Some side reactions are reversed *(17)*, either during synthesis, or by an adequate final deprotection *(15,22)* (e.g., oximate). The side reactions of uracil in phosphotriester synthesis may be a serious nuisance *(22)*. They have been carefully studied by Reese *(7,15,23,24)*.

Two protecting groups are well established: the *N*-3 anisoyl and the 4-(2,4-dimethylphenyloxy) derivatives *1* and *4* (Table 1).

The *N*-3 anisoyl compound is a little more resistant to nucleophilic attack than the *N*-3 benzoyl *(25,26)*. It is introduced, by PTC *(16)* or in homogeneous conditions *(27) (See also* ref. *28)*, by selective *N*-3 acylation of 3',5'-*O*-(1,1,3,3-tetraisopropyldisiloxane-1,3-diyl)uridine, followed by the protection of the 2'-OH and desilylation. The 2'-*O*-(tetrahydropyran-2-yl)-3',5'-*O*-(1,1,3,3-tetraisopropyldisiloxane-1,3-diyl)uridine is also easily anisoylable on *N*-3 *(29)*. The selective *N*-3 acylation is possible by the so called Jones' method *(in situ* protection of the 5', 3', and 2'-*O* by trimethylsilyl chloride during acylation, followed by the hydrolysis of the silyl ethers in the workup) *(26,28,30)*.

Protection *4* is introduced by reaction of 2'-*O*-(4-methoxytetrahydropyran-4-yl)-3',5'-*O*-dimethoxyacetyluridine with diphenylphosphochloridate and 3-nitro-1,2,4-triazole in pyridine, followed by a treatment with a phenol and triethylamine. The labile 3' and 5'-*O* acyl groups are cleaved by 8*M* ammonia in methanol *(23)*. Care is needed because this type of uracil protection may be substituted by ammonia to give cytosine *(31,32)*.

2.2. Cytidine and Deoxycytidine

These nucleosides are usually protected by acylation on *N*-4. Cytosine is the most basic of the aglycone residues (dC, $pK_a = 4.25$). It is also the most nucleophilic. The rate of acylation decreases in the order: C(*N*-4) > OH > A(*N*-6) > G(*N*-2) *(52)*.

Table 1
Protections of Thymine and Uracil Residues[a]

Properties of the tabulated protecting groups (solvent mixtures are expressed in volumic proportions)
1. Resists to K_2CO_3 0.2*M* (short time); TBAF 1*M*/THF; chromatography on silica ($CHCl_3/CH_3OH$); NEt_3/pyridine. It is cleaved by NH_3 cc in CH_3OH/H_2O; *n*-butylamine 0.04*M*/CH_3OH; TBAF/pyridine/H_2O (8:1:1). *See* refs. *25,26,28,29,33–35.*
2. Resists to CH_3COOH 80% (2 h); TBAF/THF; $ZnBr_2$ 1*M*; pyridine/*t*-butylamine/H_2O (8:1:1) (24 h). It is cleaved by conc. NH_3 in CH_3OH (9:1) (T_{r}., 3 h). *See* refs. *36,37.*
3. Resists to *t*-butylamine 0.14*M*/CH_3OH (20 min). It is cleaved by Zn/pyridine. *See* refs. *16,38,39.*
4. Resists to K_2CO_3 0.2*M*; morpholine 0.05*M*; NH_3 8*M*/CH_3OH (<15 min). It is cleaved by oximate. *See* refs. *23,24,40.*
5. Is cleaved by oximate. *See* refs. *31,41.*
6. Resists to NH_3/CH_3OH. It is cleaved by DBU 0.5*M*/pyridine. *See* refs. *42,43,45.* For possible side reactions involving the *O*-4-alkylthymine residues, *see* ref. *44.*
7. Resists to CH_3COOH 80%; conc. NH_3/CH_3OH. It is cleaved by I_2 0.1*M*/THF/collidine/H_2O. *See* refs. *16,46.*
8. Resists to CH_3COOH 80%. It is cleaved by pyridine/H_2O; TBAF 0.2*M*/THF. *See* ref. *47.*
9. Resists to CH_3COOH 80%; HCl pH 2; oximate; NaOH 0.2*M*. It is cleaved by NH_3 cc/pyridine. (50°C) $(C_6H_5)_3C^+ BF_4^-$ Does not cleave well. *See* refs. *28,48,49.*
10. Resists to NH_2-NH_2 1*M* in pyridine/CH_3COOH (3:1). It is cleaved by conc. NH_3. *See* ref. *50.*
11. Resists to conc. NH_3; DBU 0.5*M*/pyridine (1 h); TBAF 1*M*: THF. It is cleaved by conc. NH_3, 50°C (2.5 h) *See* refs. *21,51.*

[a]*See* opposite page for corresponding structure.

It is possible to specifically acylate the exocyclic amino function of cytidine or deoxycytidine, under controlled conditions: activated esters *(53,54)*, acid anhydrides *(55,56)*, some acid chlorides *(57)*, mixed anhydrides *(58)*, 1-alkyloxycarbonylbenzotriazoles *(43,59)*, carboxylic acids activated by EEDQ*(60)* and thioacetic acid *(61,62)* have been used for this purpose.

The rate of *N*-deacylation by ammonia or sodium hydroxyde decreases in the order C > A > G *(63–69)* (acylated adenine and guanine residues of nucleosides may loose a proton in basic media, rendering them more resistant to nucleophilic attack). It is to be noticed that nucleophiles may attack on C-4, displacing the acylated nitrogen. The *N*-4 acylcytosine residue is deaminated to the corresponding uracil by hot acetic acid *(70,71)*. Similarly, *n*-butylamine gives the *N*-4-butyl derivative *(72)*.

Table 1
Structures

1 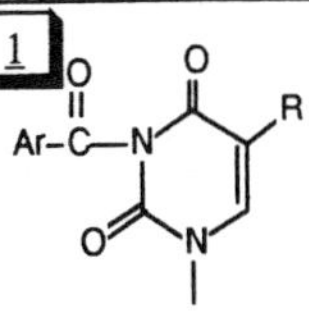 $R = H, CH_3$ 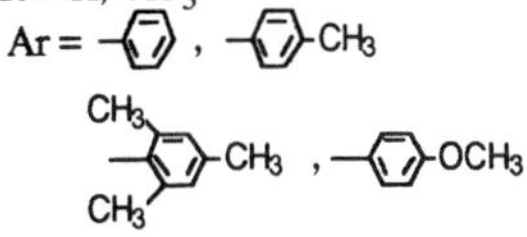	2 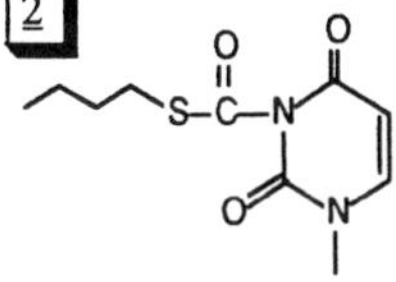	3 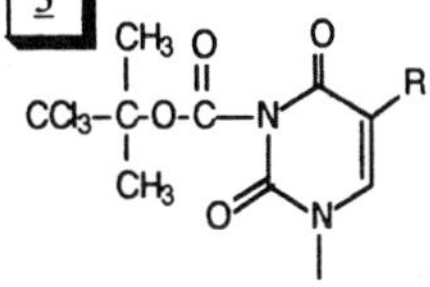 $R = H, CH_3$
4 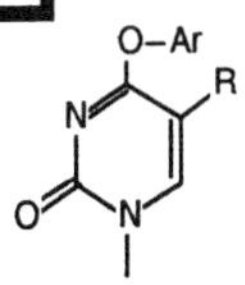 $R = H, CH_3$ 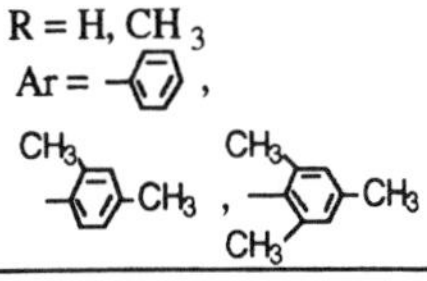	5 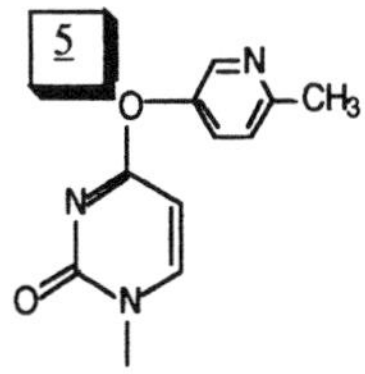	6 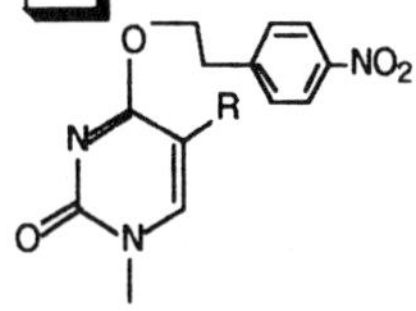 $R = H, CH_3$
7 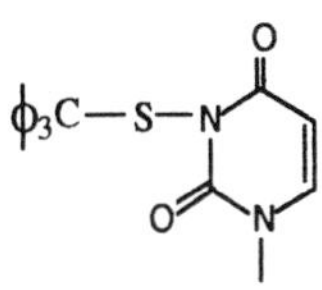	8 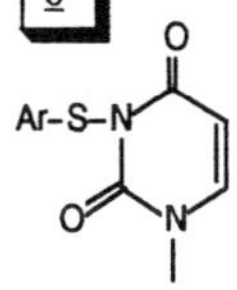 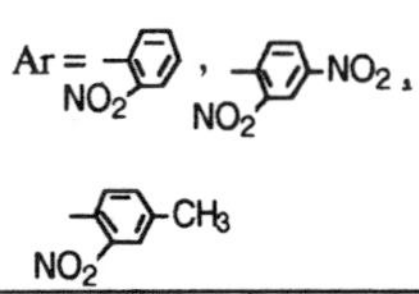	9 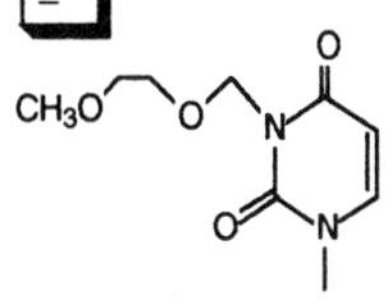
10 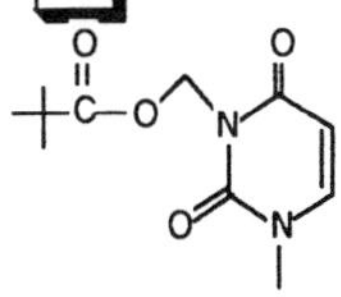	11 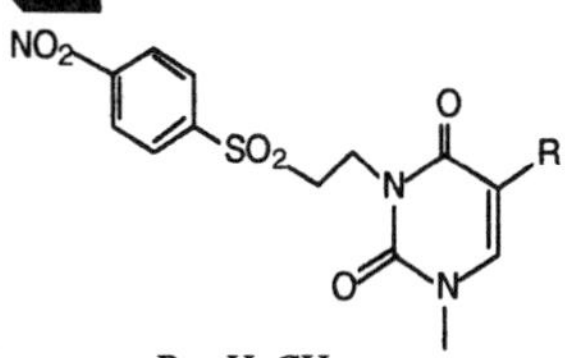 $R = H, CH_3$	

As a result of easy deacylation under acidic or basic catalysis, the protection of cytidine and deoxycytidine need some tuning. The halo-acetyl groups are useless, being to labiles *(73)*. The *N*-4 acetyl protection is fragile*(68,73,74)*, (HCl 0.1*M*, $t_{1/2}$ = 2 h; KOH 0.01*M*, $t_{1/2}$ = 6 h; NaOH 0.2*M*/CH_3OH (1:1), RT, $t_{1/2}$ = 0.2 min; conc. NH_3/C_2H_5OH (1:1), RT, $t_{1/2}$ = 10 min). It is cleaved by boiling ethanol *(61)*. The *p*-methoxybenzoyl group, more resistant to nucleophilic attack, was formerly used with the phosphodiester method *(75)*. The most resistent acyl group is *o,p*-dimethylbenzoyl *(68)*. The benzoyl group is routinely used nowadays, although it is less robust *(70)* (e.g., deacylation of the 2', 3' or 5' *O*-silylated derivatives by methanol; refs. *76,77)*; deacylation by hydrazine in pyridine/acetic acid *(78)*. It is cleaved without problems by ammonia at room temperature *(66)* (NH_3 9*M*, $t_{1/2}$ = 16 min *(70)*; NH_3 5*M*/dioxane (1:1), complete cleavage in 6.5 h*(63)*; NH_3 29%/pyridine (80:20), $t_{1/2}$ = 2 h *(64,65)*.

On a preparative point of view, the acylation method of Jones-Sung-Narang (*in situ* protection of the alcohol functions by trimethylsilyl chloride) *(14,79)* is particularly practical. It is also possible to *N,O*-peracylate the nucleoside and to selectively cleave the ester functions afterwards *(23,71)*. Deoxycytidine has been selectively *N*-4-benzoylated on a multikilogram scale by simply shaking the nucleoside with one equivalent of benzoic anhydride in DMF *(56)*. If a more easily cleavable protection is necessary, one would prefer the *N*-4-isobutyryl group *(64,65)* (cleaved by ammonia 28%, RT, 5 h) *(80,81)*

2.3. Adenosine and Deoxyadenosine

The protection of these nucleotides requires a special comment, because the chemical stability of the nucleoside is altered by the protection of the exocyclic amino function of the nucleic base. The discussion thus starts with an account of the encountered problems.

2.3.1. Stability Toward Acids

Purine nucleosides can loose their purine residue in acidic conditions, leaving an unsubstituted sugar (apurinic site). This reaction is a nuisance because oligonucleotides are repeatedly submitted to acid detritylation during routine synthesis on a solid support.

The mechanism of acid depurination involves a rapid preequilibrium of protonation (or deprotonation) of the purine residue, followed by

Table 2
Protections on *N*-4 of the Cytosine Residue[a]

List of methods of cleavage and relevant references (solvent mixtures are expressed in volumic proportions)	
1.	Conc. NH_3, T_r. Refs. *56,60,74.*
2.	Conc. NH_3, T_r. Refs. *64,65,80.*
3a.	Conc. NH_3, T_r. Refs. *14,56,60,71,79.* This protection is selectively removed by NH_2-NH_2 in pyridine/acetic acid (4:1) Refs. *150–152.*
3b.	Conc. NH_3, T_r. Ref. *23.*
3c.	Conc. NH_3, T_r. Refs. *60,71,82.*
3d.	Conc. NH_3, 50°C. Ref. *68.*
3e.	Conc. NH_3, 50°C. Ref. *57.*
4.	Conc. NH_3, 50°C. Ref. *83.*
5.	TBAF/pyridine/H_2O (1:1) Ref. *84.*
6.	K_2CO_3 0.05*M*/CH_3OH, T_r or conc. NH_3, T_r Ref. *85.*
7.	NH_2-NH_2 0.5*M*/pyridine/CH_3COOH (4:1) Ref. *86.*
8.	Conc. NH_3, T_r. Ref. *78.*
9.	NEt_3/pyridine or oximate. Refs. *63,87–90.*
10.	DBU. Refs. *43,59.*
11.	Co(I)/phthalocyanin anion. Refs. *39,91.*
12.	Pd/C/1,3-cyclohexadiene/EtOAc/EtOH. Ref. *92.*
13.	Pd(0)/$P(C_6H_5)_3$/*n*-butylamine/HCOOH/THF. Ref. *93.*
14.	*p*-Methylthiophenolate/pyridine, T_r. Ref. *94.*
15.	NH_3, 50°C or NH_4OAc/conc. NH_3, 50°C. Refs. *95,96.*
16.	NaOH 0.5*M*/pyridine (1:1) Ref. *97.*
17.	HCl 0.01*M*. Refs. *50,98–100.*

[a]See pages 8 and 9 for corresponding structures.

the cleavage of the glycosidic linkage, that is the rate-determining step *(101–107).* The unstable species are the *N*-7 protonated derivatives (Scheme 1). Both purine nucleosides (or deoxynucleosides) depurinate at about the same rate *(104).* A carbocation being generated at C-1' in the rate-determining step, electron withdrawing substituents on the sugar reduce the tendency to depurinate. Accordingly, adenosine depurinates 1200 times more slowly than deoxyadenosine *(108),* and deoxyadenosine 3',5'-diphosphate (in the middle of an oligonucleotide sequence) depurinates less easily than the nucleotide *(109).* That is why people usually avoid starting a sequence with deoxyadenosine directly attached to the solid support with the 3'-succinate link *(110).*

Depurination is not a problem in the ribo series (except for hypermodified residues like wyosine *[111]).* In the deoxyribo series, the

Table 2
Structures

1

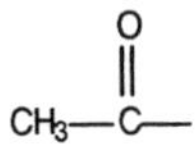

2

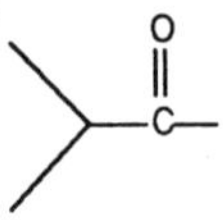

3

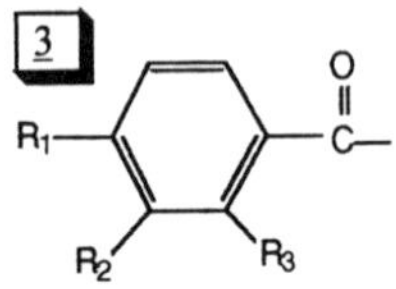

	R_1	R_2	R_3
a	H	H	H
b	+	H	H
c	CH_3O	H	H
d	CH_3	H	CH_3
e	ϕO	CH_3O	H

4

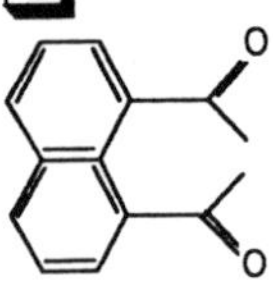

5

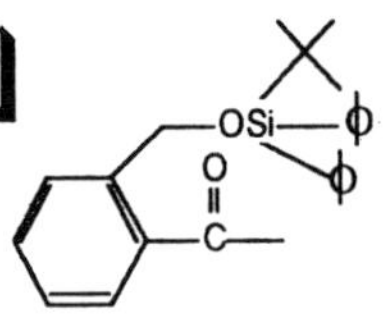

6

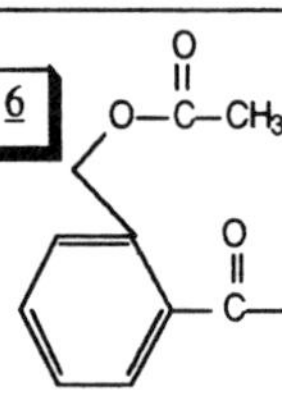

7

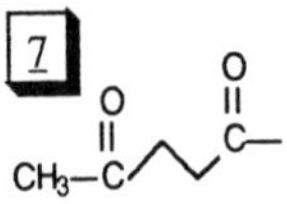

8

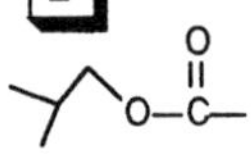

9

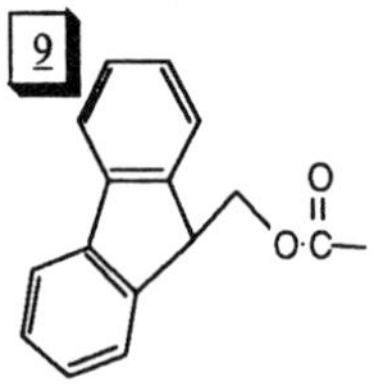

most sensitive residue is *N*-6-acylated deoxyadenosine. The site of first protonation of *N*-6-acylated deoxyadenosine is *N*-7 ($pK_a = 1.75$) *(112,113)*. This quickly depurinating form is thus readily accessible. On the contrary, for deoxyadenosine itself, the site of first protonation is *N*-1*(114)*, and the *N*-1, *N*-7 diprotonated form is accessible only at low pH (pKa1 = -1.48 *(115)* $pK_{a2} = 3.65$ *(1,112)*. The acidic depurination of deoxyadenosine is thus limited by the access to the *N*-7 protonated form, but this is not the case for *N*-6-acylated derivatives. *N*-6 benzoyl

Table 2
Structures *(continued)*

10 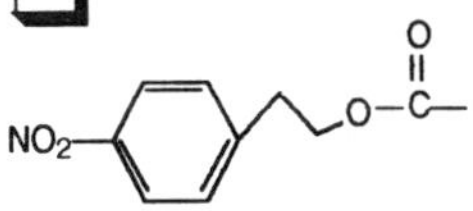	11 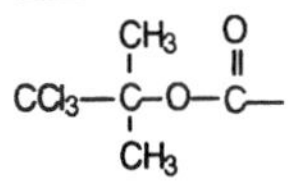	12 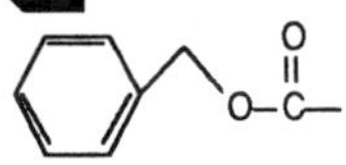
13 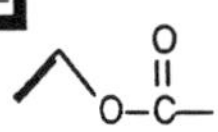	14 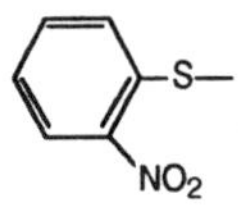	15 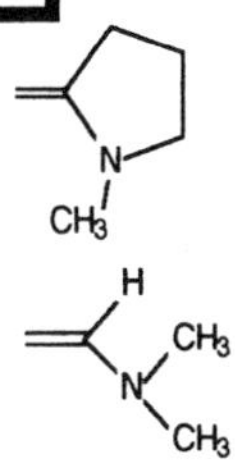
16 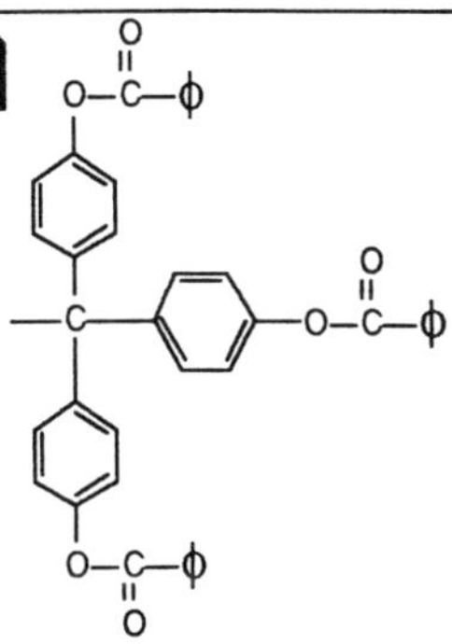	17 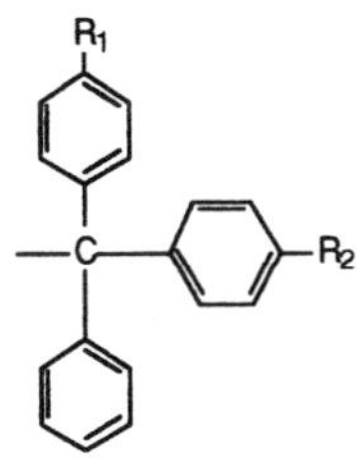 R_1 = H, R_2 = –O-CH_3 R_1 = R_2 = –O-CH_3	

dA depurinates about ten times more rapidly than *N*-2-isobutyryl dG *(109,116)*. A diacylation on *N*-6 corrects this effect by lowering the pK_a (pK_a << 1.4) *(112,117)*. A *N*-6 amidine protection *(95,118,119)* presumably does not shift the first protonation site of deoxyadenosine from *N*-1 to *N*-7, but this protection is cleaved in aqueous acidic conditions at a rate similar to the depurination rate *(112)*.

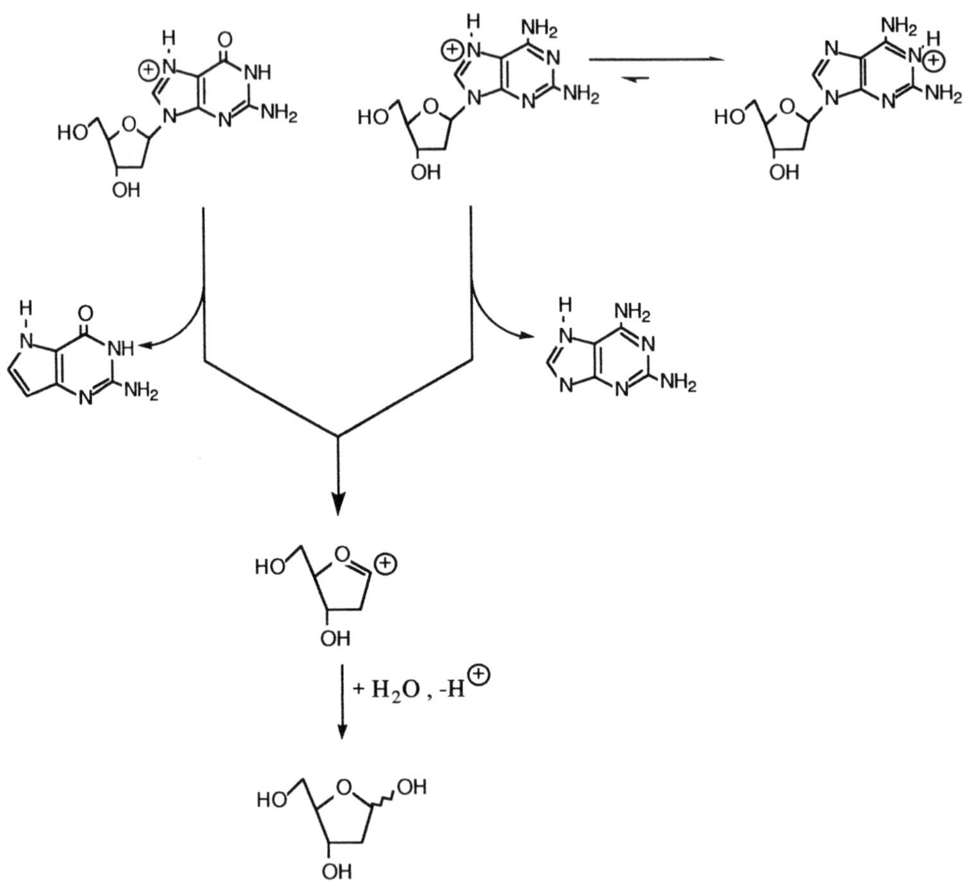

Scheme 1. Mechanism of acidic depurination.

When a partially depurinated oligomer is exposed to the strongly basic conditions of final deprotection, the chain is cleaved at the apurinic site by β-elimination *(120,121)*. The 5'-dimethoxytrityl oligonucleotides usually obtained by an automatic DNA synthesis are thus contaminated by 5'-DMTr truncated sequences difficult to remove by reversed phase chromatography. That this is really the case has been proved by Horn et al. *(122,123)*, who found a method to cleave apurinic sites directly on the solid support, without detaching much of the full length oligonucleotide (1*M* lysine.HCl, pH 9, 60°C, 90 min). Their protocol gave long oligonucleotides of high purity (98–118-mers).

Table 3[a]
Protections on *N*-6 of the Adenine Residue

	List of methods of cleavage and relevant references (solvent mixtures are expressed in volumic proportions)
1a–d.	Hot conc. NH_3. Refs. *23,57,71,79,149,153–155.*
2.	Conc. NH_3, T_r or DBU/CH_3OH (1:9), T_r. Refs. *44b,64-66,69,74,80,81.*
3.	*See* Table 2, entry 5.
4.	*See* Table 2, entry 6.
5.	*See* Table 2, entry 7.
6.	*See* Table 2, entry 9. It is also cleavable by K_2CO_3 0.05*M*/CH_3OH, T_r, 4 h. Ref. *158.*
7.	*See* Table 2, entry 10.
8.	*See* Table 2, entry 11 and ref. *156.*
9.	*See* Table 2, entry 12.
10.	*See* Table 2, entry 13.
11.	Conc. NH_3 40°C, 1 h, or conc. NH_3/pyridine (5:1), T_r, 48 h. Refs. *117,131.*
12.	Conc. NH_3/pyridine (2:3), 54°C, 12 h. Ref. *132.*
13.	Conc. NH_3, 60°C, 5 h. Ref. *130.*
14.	Conc. NH_3, 50°C, 4 h. Ref. *83.*
15.	*See* Table 2, entry 14.
16.	*See* Table 2, entry 16.
17.	*See* Table 2, entry 17 and refs. *113,157.*
18.	*See* Table 2, entry 15.

[a]*See* pp. 12 and 13 for corresponding structures.

The conditions of detritylation in routine synthesis have been carefully optimized to limit depurination to a minimum (weakest possible acid, low concentration of the acid, shortest possible detritylation time) *(124,125)*. It is thus not a serious problem for short oligomers. It remains, however, an intrinsic limitation of the classical methodology, when long oligomers have to be synthesized.

2.3.2. Mono vs Bis Acylation

When deoxyadenosine is reacted with benzoyl chloride in neat pyridine, the *N*-6, *N*-6, *O*-3', *O*-5'-tetrabenzoylated derivative is obtained *(71)*. The second *N*-acylation is thus rapid in pyridine, probably because a certain amount of the ionized secondary amide function (-NH-COPhe) is present ($pK_a = 10.02$) *(113)*. The same type of peracylated products are formed with 4-methoxybenzoyl chloride *(71)*, 2-(*t*-butyldiphenylsilyloxymethyl)benzoyl chloride *(84)*, phenyloxycarbonyl chloride *(126,127)*, *p*-nitrophenylethyloxycarbonyl chloride*(128)* and fluorenyl

Table 3
Structures

1

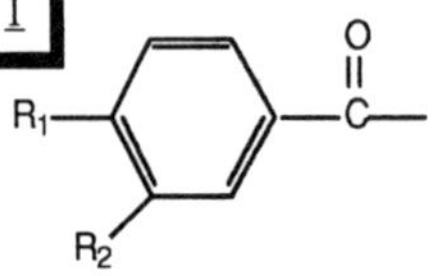

	R_1	R_2
a	H	H
b	H	Cl
c	+	H
d	ϕO–	CH_3O–

2

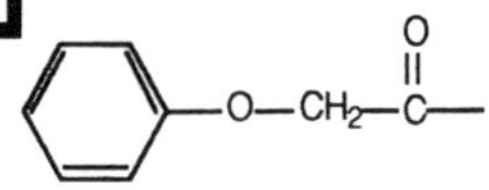

3

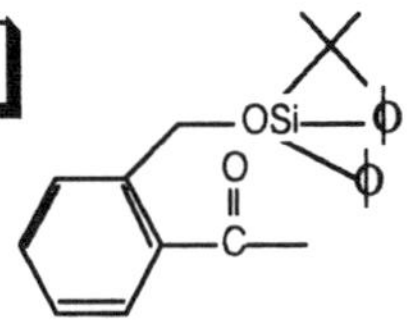

4

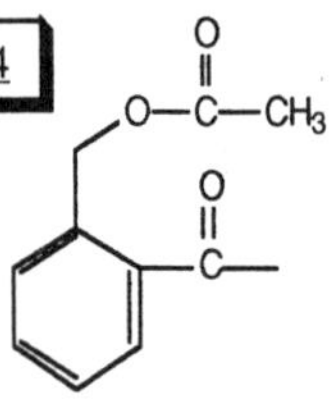

5

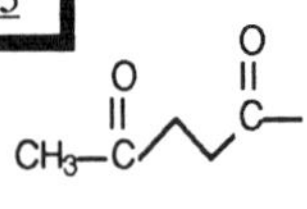

6

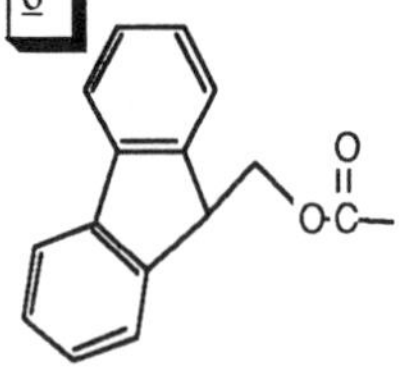

7

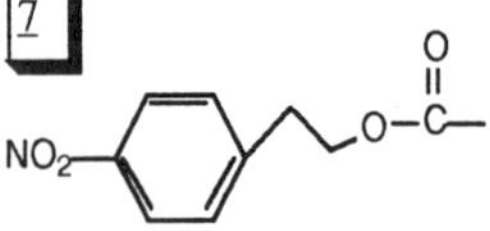

8

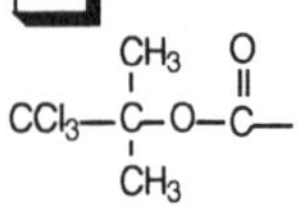

9

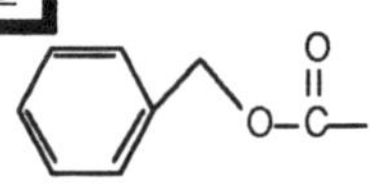

Table 3
Structures *(continued)*

10 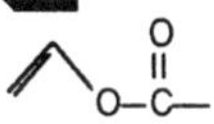	11 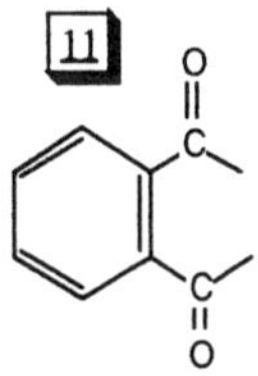	12 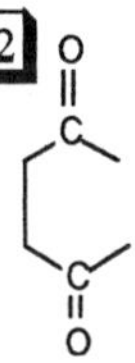
13 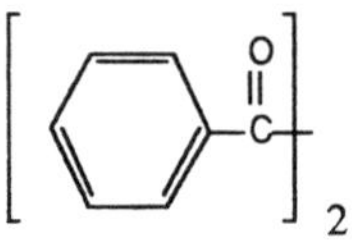	14 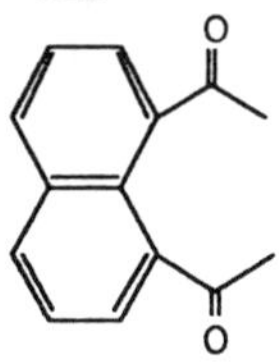	15 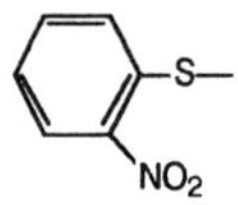
16 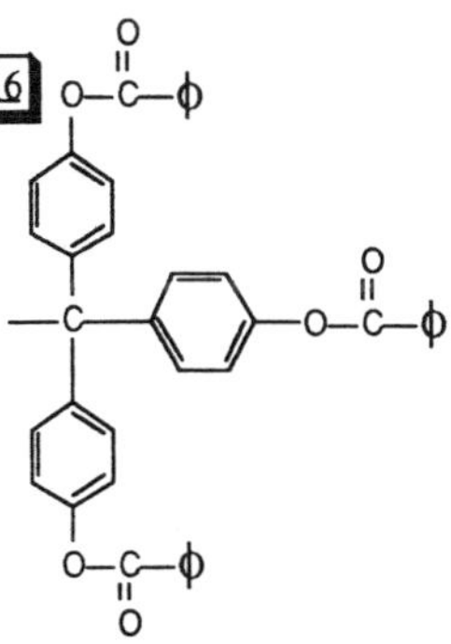	17 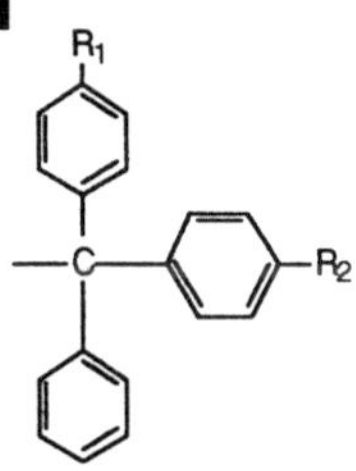 $R_1 = R_2 = H$; $R_1 = H$, $R_2 = -O-CH_3$; $R_1 = R_2 = -O-CH_3$	18 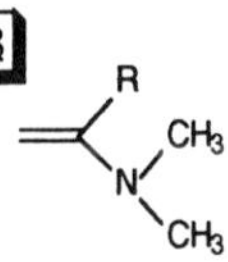 $R = H, -CH_3$

N-6 monoacylation is possible if the reactivity of the acylating agent is reduced and if the medium is not too basic *(43,128)*.

Bis acylated derivatives have been proposed as protecting groups: *N*-6, *N*-6 dibenzoyl dA *(130)*, phthaloyl dA *(117)*, *(131)*, naphthaloyl dA *(83)* and succinyl dA *(132)*. Unfortunately, the phthaloyl group is very easily opened by nucleophiles *(130)* and the imidazole ring of *N*-6, *N*-6 diacylated nucleosides is prone to side reactions *(133)*. The *N*-6, *N*-6, *O*-3', *O*-5'-tetraacyl adenosines quickly depurinate when they are treated with sodium hydroxyde at room temperature (Scheme 2) *(71,92a)*. Adenosine and deoxyadenosine also depurinate in basic medium, but in harsh conditions (NaOH 1*M*, 80–100°C) *(134–136)*. If the purine C-6 subsituent R is less electron releasing than the amino group (R = H, Cl), the depurination by sodium hydroxide happens at room temperature *(137–140)*. This is also the case for the *N*-6 diacylated derivatives of adenosine.

2.3.3. The No Protection Option

Is it necessary to protect the adenine residue? This is an open question because, although the exocyclic amino function is susceptible to react with trityl chloride *(113)*, and the reagents used to phosphorylate nucleosides *(71,141,142)*, it does not react with phosphitylating agents at –78°C *(76,143–145)*. Moreover, it is possible to couple nucleotides by the phosphotriester or phosphoramidite methods without protecting the exocyclic amino function *(145–148,422)*.

2.3.4. The Usual Strategy

A selective *N*-acylation of the adenine residue is usually performed by *O*-silylation before *N*-acylation (e.g., by the Jones' method). It is also possible to 2',3',5'-*O*-triacetylate adenosine by acetic anhydride without concomitant *N*-acetylation, to perform the desired *N*-acylation or bis-acylation with an acid chloride, and finally to rapidly cleave the labile *O*-acetyl functions (basic depurination has to be avoided) *(23)*.

For the sake of simplicity, the *N*-6–benzoyl group is routinely used to protect dA in oligonucleotide synthesis, although it is not entirely satisfactory. Some depurination of *N*-benzoyl-dA was observed, not only in acidic media, but also in the triester coupling conditions *(149)*. Premature *N*-debenzoylation by methanolysis occurs in the case of 2'-*O*-*t*-butyldimethylsilyladenosine *(76,77)*. Hydrazine hydrate 0.5*M* in pyridine/acetic acid (4:1) at room temperature cleaves the benzoyl protection of dA (and of dC) extremely selectively (5'-*O*-MMTr, 5'-*O*-benzoyl esters, cyanoethylphosphates, and the aglycone residues of

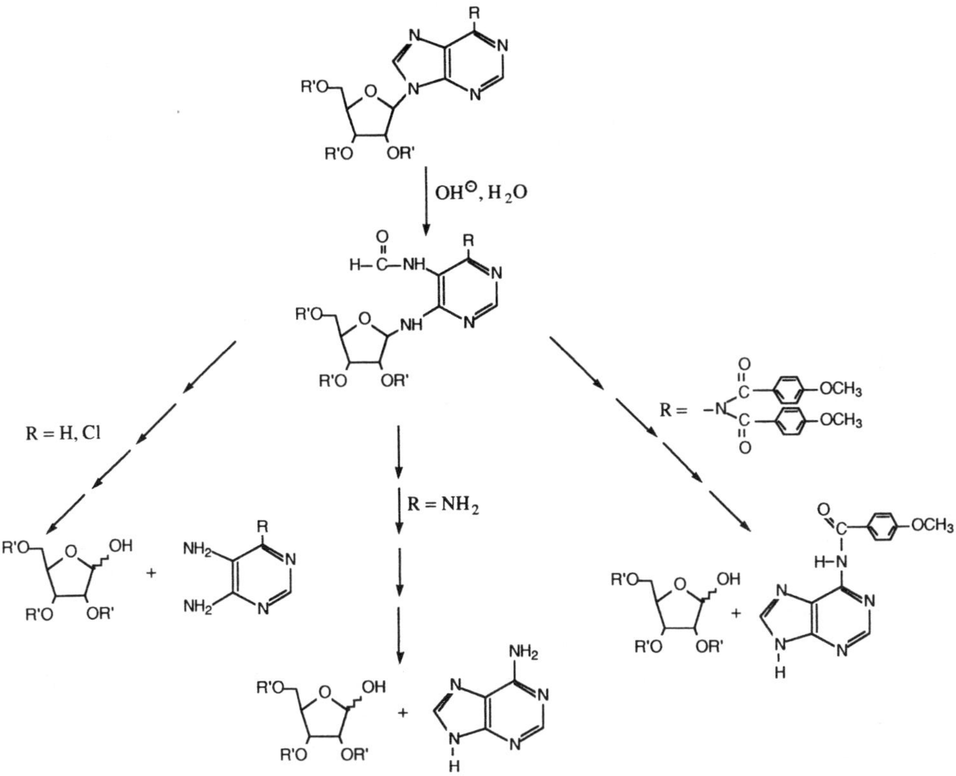

Scheme 2. Mechanism of basic depurination.

pyrimidine nucleosides remain unaffected). The selectivity of hydrazine is probably the result of formation of a hydrogen bond with *N*-1 of dA, concerted with the nucleophilic attack on the *N*-6-carboxyl *(150–152)*. The deprotection of *N*-benzoyl-dA with hot ammonia requires a lengthy treatment. The phenoxyacetyl (PAC) protection, cleaved by ammonia at room temperature, is convenient for the introduction of base-sensitive residues into DNA *(64,65,80,81)*.

2.4. Guanosine and Deoxyguanosine

2.4.1. The Exocyclic Amino Function

This function is neither very basic, nor very nucleophilic (the site of first protonation is *N*-7, with a $pK_a = 1.6$, ref. *1*). The acylation of the alcohol functions of the nucleoside is far more rapid than *N*-acylation

Table 4
Protections on *N*-2 of the Guanine Residue[a]

List of methods of cleavage and relevant references (solvent mixtures are expressed in volumic proportions)
1. *See* Table 2, entry 3e.
2. Conc. NH_3, 65°C, 72 h. Refs. *79,154,163,167.*
3. Conc. NH_3, 60°C. Refs. *157,190.*
4. Conc. NH_3, T_r. Refs. *23,82,164,191.*
5. Conc. NH_3. Refs. *165,192.*
6. *See* Table 3, entry 2.
7. *See* Table 2, entry 7.
8. *See* Table 2, entry 6.
9. *See* Table 2, entry 9 and ref. *189.*
10. *See* Table 2, entry 11.
11. *See* Table 2, entry 13.
12. Conc. NH_3/CH_3OH (9:1), T_r. Ref. *193.*
13. *See* Table 2, entry 14.
14. *See* Table 2, entry 15.
15. HCl 0.01*M*. Refs. *71,76,99,100,162,169,170,194,195.*

[a]*See* opposite page for corresponding structures.

(71,159–161). The coupling step in the triester strategy can be performed without *N*-protection *(146)*. It is also possible to 3'-*O*-phosphitylate the nucleosides in the cold, without protections on the aglycone residue *(76,143,144)*. Direct alkylation of the unprotected nucleoside by 4,4'-dimethoxytrityl chloride gives however the 5'-*O*,*N*-2 dialkylated product *(71,76,79,162)*.

Acyl protections are introduced on *N*-2 by *O*, *N*-peracylation, followed by selective saponification *(57,154,163)*, or, alternatively, by selective *O*-silylation, followed by *N*-acylation (e.g., the Jones' method) *(79,82,87,93,161,164,165)*.

A major role of the *N*-2 acyl group is to enhance the solubility of the oligonucleotides in organic solvents for the block coupling synthesis, and to facilitate their elution in the chromatographic purifications *(23)*. The acetyl group was disregarded by Khorana *(72)*, because it was not lipophilic enough for the block coupling diester method and a little too labile in the chromatographic steps. The benzoyl group was too difficult to remove *(72)*. The isobutyryl group was a good compromise at that time *(163)*, but the more labile phenoxyacetyl group is better

Table 4
Structures

1 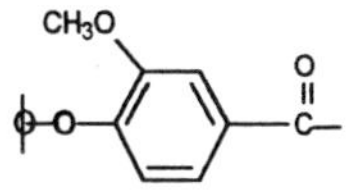	2 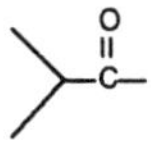	3 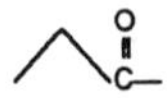
4 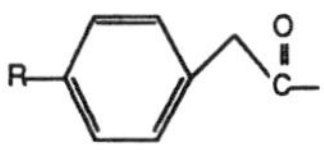 R = H, +	5 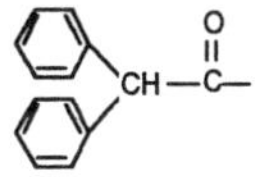	6 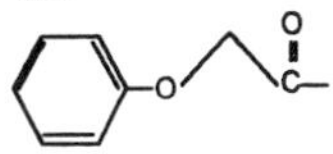
7 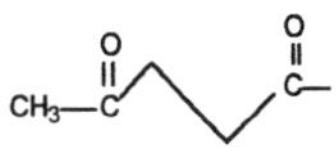	8 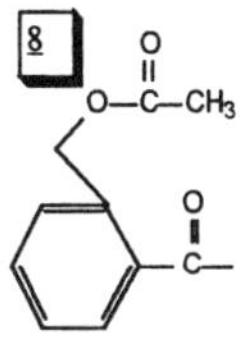	9 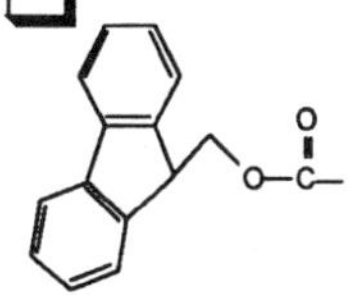
10 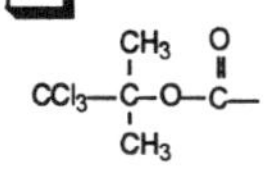	11 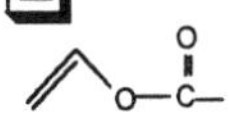	12 
13 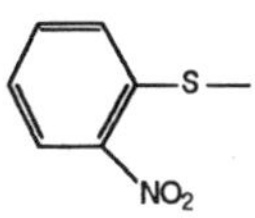	14 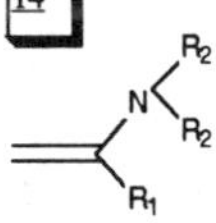 $R_1 = H, R_2 = CH_3$, n C_4H_9 $R_1 = CH_3, R_2 = CH_3$	15 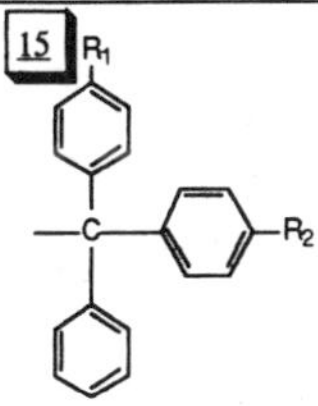 $R_1 = R_2 = H$ $R_1 = H, R_2 = -O-CH_3$ $R_1 = R_2 = -O-CH_3$

Table 5
Protections of the Lactame Function of the Guanine Residue[a]

List of methods of cleavage and relevant references (solvent mixtures are expressed in volumic proportions)
1. Cleaved by oximate. Refs. *23,24,40,187.*
2. Cleaved by oximate. Ref. *41.*
3. Cleaved by conc. NH_3/pyridine (2:1), T_r, or oximate. Refs. *157,190,196.*
4. Cleaved by oximate or conc. NH_3. Refs. *174,184,189,197.*
5. Cleaved by DBU 1*M*/pyridine, 55°C, 24 h, or TBAF 1*M*/THF, 5 h. Refs. *21,43, 178,188,197,198–201.*
6. Cleaved by 1. $NaIO_4$, 2. conc. NH_3, 55°C, 24 h. Ref. *202.*
7. Cleaved by $(Ph)_3C^+$ BF_4^-/CH_3CN/H_2O (4:1) ref. *166.*
8. Cleaved by conc. NH_3/pyridine, 50°C. Refs. *203,204.*

[a]*See* opposite page for corresponding structures.

suited for the solid-supported phosphoramidite strategy *(44b,64–66,74,80,81,167)*. However, the phenoxyacetyl group may be inadvertently replaced by the acetyl group in the acetic anhydride/DMAP mixture, used for capping in automated synthesis *(74)*. The half-life times of the various G protecting groups in NaOH 0.2*M*/CH_3OH (1:1), RT, are as follows: benzoyl, 647; isobutyryl, 271; 4-*t*-butylphenylacetyl, 32; 4-*t*-butylphenoxyacetyl, 8 min *(68)*.

It seems that the acylation of the guanine residue on *N*-2 does not much change the sensitivity of the nucleoside to acid depurination: although the rate of depurination by 80% acetic acid is enhanced by acylation, it remains unchanged if HCl 0.2*M*/dioxane (1:1) is used *(168)*. Amidine and isobutyryl derivatives of dG have comparable stability toward depurination by dichloroacetic acid/CH_2Cl_2 (1:50) *(95)*.

2.4.2. The Lactame Function

The major site of unwanted reactions of the guanine residue is its enolizable lactame function. Indeed, the facile ionization of the hydrogen attached to *N*-1 (pK_a = 9.33) *(1)* gives a strongly nucleophilic anion. Deoxyguanosine is thus modified by carbodiimide at pH 8–8.5 *(6a,b)*; the nucleoside is *N*-1 alkylated under basic catalysis by the Michaël acceptor, 4-nitrophenylvinylsulfone *(21)*; 2,6-dichlorobenzoyl chloride in pyridine acylates *O*-6 instead of *N*-2 *(159)*. As in the case of uridine, the lactame function can be sulfonylated and phosphorylated on *O*-6 by the usual reagents and activated nucleotides of the

Table 5
Structures

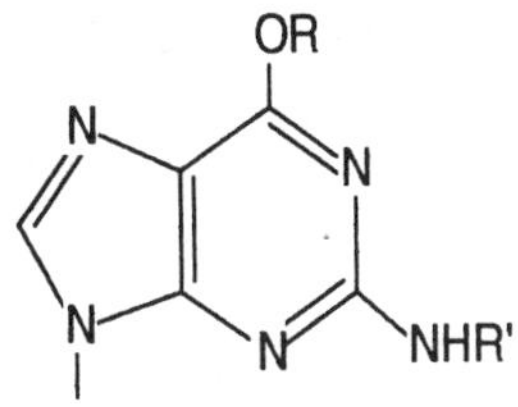

NATURE OF R		
1 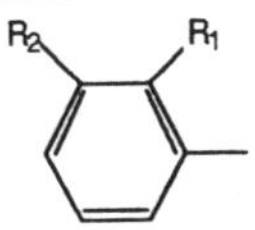 $R_1 = H$, $R_2 = Cl$; $R_1 = NO_2$, $R_2 = H$	2 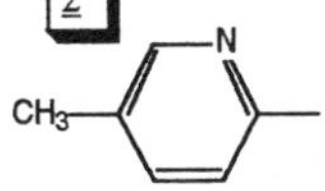	3 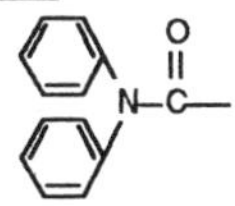
4 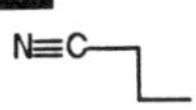	5 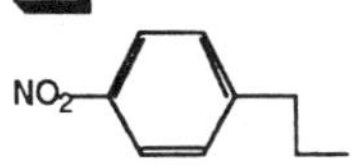	6 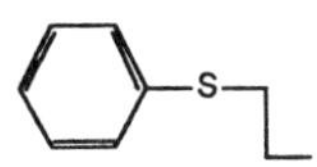
7 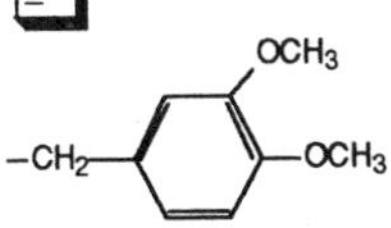	8 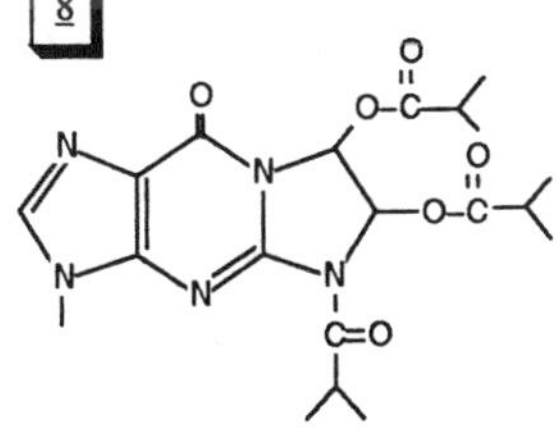	

triester method *(7,15,24,159,169–175)*. A subsequent nucleophilic substitution on C-6 is then possible, with nucleophiles like the hydroxyl anion *(159)*, secondary amines *(159)*, triethylamine *(173,174)*, the azide anion *(176)*, thiols *(176)*, alcohols (with a tertiary amine catalysis) *(24,167, 173, 174,176)*, pyridine *(11,12,172,176–178)*, 3-nitro-1,2,4-triazole *(12,15)*, *N*-methylimidazole *(12)*. The *O*-6 function is also phosphitylated by the phosphitylating agents or the activated nucleotides used in the phosphoramidite method *(116,179–182)*. The *O*-6 phosphitylated products easily revert to G by the attack of nucleophiles on phosphorus [P(III)], but this is no more the case if an oxidation is performed *(182–184)*. The thus obtained *O*-6 phosphorylated intermediates [P(V)] simply depurinate *(183)*, or are further substituted by attack on C-6 of nucleophiles like ammonia (to give 2,6-diaminopurine), DMAP (to give a fluorescent dimethylaminopyridinium salt) *(185)*, and possibly thiophenol *(186)*.

A proper protection of the lactame function of G suppress all these side reactions (*N*-1 acylation is not a good strategy *(187)*). The necessity of protection is however a matter of debate, as the modifications of G are reversible. In the triester method, *O*-6 modified G reverts to G by the oximate treatment *(12,24)*, or even sodium bicarbonate/pyridine *(175)*. In the phosphoramidite method, *O*-6 phosphitylated G [P-(III)] reverts to G by treatment with water or acetate ions *(181,184, 185)*. A proper adjustment of the synthetic protocol is thus in principle able to balance the inconvenience of working with underprotected G *(21,181)*. When full protection is needed, the *O*-6-(β-cyanoethyl), diphenylcarbamoyl, or 3-chlorophenyl groups are usually advocated. The *O*-6 *p*-nitrophenylethyl group is difficult to remove in the case of long oligomers *(21,184,188)*. The *O*-6-(3-chlorophenyl) and related groups are removed by oximate (the deprotection does not work if *N*-2 is not acylated *(189)*.

3. Protection of the 2'-OH Function

It is a persistent protection. It has to be removed at the ultimate step of the synthesis, because oligoribonucleotides featuring a free 2'-OH function are easily cleaved, either chemically, under acidic or basic catalysis (with the neighboring group participation of the 2'-OH) *(205–210)*, or enzymatically (ribonucleases). It is recommended to stock the oligoribonucleotide in the 2'-OH protected form.

The 2'-OH protecting group has to be resistant to the conditions of cleavage of the phosphate and aglycone protections. It also has to withstand the conditions used to remove the 5'-OH protection before each coupling. No isomerization of the phosphate linkage (3'–5' to 2'–5') may occur during the 2'-OH deprotection. Moreover, the chosen protecting group may not easily migrate from the 2'-*O* to the 3'-*O* position of the nucleosides, because the 3'-*O* phosphorylated or phosphitylated synthons must be uncontaminated by their 2'-*O* isomers.

All these requirements lead to the elimination of the acyl-type groups, that migrate too easily from 2' to 3' *(211–214)*, would be prematurely cleaved, or whose cleavage conditions would degrade RNA. Three types of protections are proposed:

1. The acetal/ketal type protections, cleaved by HCl 0.01*M*;
2. The *O*-nitrobenzyl group, cleaved by photolysis; and
3. The *t*-butyldimethylsilyl group, cleaved by the fluoride anion.

These protections enhance the steric crowding of the nucleotides. The coupling is thus slower in the RNA series than in the DNA series.

The proposed protecting groups are presented in Table 6.

3.1. The Acetal/Ketal Type Protections

HCl 0.01*M* is recommended for their cleavage: it does not induce breaking or migration of the phosphodiester internucleotidic linkage *(215)*. A certain amount of 3'–5' to 2'–5' bond isomerization is obtained with acetic acid 80% *(205,206)*.

The mechanism of the acidic cleavage involves the protonation of the 2'-*O* engaged in the acetal/ketal function, followed by the breaking of the C-O bond with formation of a carbocation. This is the rate determining step. It is possible to precisely tune the acid stability of the protection by modifying its structure: Ketals are cleaved about a thousand times more rapidly than acetals (the most substituted cations are the most stable), but those can be stabilized by introducing heteroatoms into their structure (I effect) *(216,217)*. The resistance to acids of the most used protecting groups increases in the order: tetrahydrofuranyl 20 *(218)* < 4-methoxytetrahyropyran-4-yl (MTHP) 3 *(216)* <tetrahydropyranyl (THP) 1 (the relative rates of cleavage are indicated). The 3'-*O* substitution also influences the stability of these protections in acidic medium; the stability increases in the order: 3'-phosphodiester < 3'-OH < 3'-phosphotriester *(191,206,219)*.

The 5'-*O*-DMTr and 5'-*O*-Px groups can be cleaved by zinc bromide without affecting the acetal/ketal protections on the 2'-*O* *(205,220–222)*. The rate of cleavage of 5'-*O*-DMTr and 5'-*O*-Px by protic acids at 0°C is higher than that of 2'-*O*-THP or 2'-*O*-MTHP *(25,100,191, 205,223,224)*. It is thus possible to synthesize medium-sized oligoribonucleotides (8–33-mers), in solution, by using the couple of protections 5'-*O*-DMTr (Px or Mox)/2'-*O*-THP (or MTHP) *(157,191,220,225,226)*. It is also possible to synthesize oligoribonucleotides on a solid support by this strategy (selective cleavage of 5'-*O*-DMTr, Px or Mox by dichloroacetic acid in CH_2Cl_2, at RT) *(34,227–229)*, but the 2'-*O*-acetal/ketal group is partially lost during the synthesis *(230)*. The treatment with concentrated ammonia, that precedes the HCl digestion at the end of the synthesis, cleaves the oligoribonucleotides at the sites where the 2'-OH has been prematurely uncapped. Low yields of the desired sequences are thus obtained *(231)*. All the phosphodiester links are however 5'-3' (i.e., all the sites of possible isomerization have, in fact, been cleaved by the harsh ammonia treatment) *(232,233)*.

Reese proposes the 1-[(2-fluoro)phenyl]-4-methoxypiperidin-4-yl group (Table 6, entry *4b*) as a 2'-*O* protection fully compatible with the repetitive acid cleavages of the 5'-*O*-Px group *(234)*. At high proton activity (strong acid, organic solvent), the nitrogen atom of the piperidine ring is protonated, and the cleavage of the 2'-*O*-ketal function is inhibited. At a lower proton activity (aqueous solution, pH 2), the nitrogen atom is largely unprotonated, and the cleavage of the 2'-*O*-ketal function is easier.

There has been use of orthogonal 5'-*O*/2'-*O* protections, like 5'-*O*-Fmoc/2'-*O*-MTHP *(235–237)*, 5'-*O*-levulinyl/2'-*O*-THP (or MTHP) *(20,152,192,238–243)*, 5'-*O*-trityloxyacetyle/2'-*O*-THP *(244)*.

On the point of view of preparative chemistry, the acetal/ketal protections are specifically introduced on 2'-*O* by a preliminary capping of both 5' and 3'-OH functions by 1,3-dichlororo-1,1,3,3-tetraisopropyldisiloxane *(23,219,226,245–248)* (or the tetra-*t*-butyl analog; ref. *249)*. This capping is removed afterward, by treatment with the fluoride anion *(189)* (Markiewicz's strategy; ref. *250)*. The nucleosides bearing an acetal-type 2'-*O* protecting group are mixtures of two diastereoisomers. The interpretation of NMR spectra and the purification of fully protected oligomers obtained by solution synthesis are all the more difficult. The symmetric MTHP ketal group is a better choice.

3.2. The o-Nitrobenzyl Group

This group is cleaved at pH 3.5, in the absence of oxygen, by UV light filtered through Pyrex *(251–254)*. One oxygen of the nitro func-

Table 6
Protections of the 2'-OH Function of the Ribonucleosides[a]

List of methods of cleavage and relevant references (solvent mixtures are expressed in volumic proportions)
1. Cleaved by HCl, 0.01*M*, T_r. Refs. *29,206,212,218,223,228,244,246,247,250, 297,298,299.*
2. Cleaved by HCl, 0.01*M*, T_r. Refs. *23,89,189,191,216,217,300,301.*
3. Cleaved by CH_3COOH 80%, T_r. Ref. *219* (this reagent may induce some 3'-*O* to 2'-*O* phosphoryl migration, *See* ref. *206*).
4. Cleaved by HCl 0.01*M*, T_r. Refs. *234,302,303.*
5. Cleaved at pH 4, T_r. Refs. *36,217.*
6a. Cleaved by HCl 0.01*M*, T_r. Refs. *2,304.*
6b. Cleaved by $AgNO_3$/Et(i-propyl)$_2$N/DMF/H_2O (9:1), T_r. Ref. *305.*
7. Cleaved by HCl 0.01*M*, T_r (HCl 0.1*M*, T_r is necessary for [A] *n* sequences). Refs. *218,220,221,238,284,306,307.*
8. Cleaved by HCl 0.01*M*, T_r. Refs. *308,309.*
9. Cleaved by $(Ph)_3C^+$ BF4$^-$ CH_3CN/H_2O (4:1) Ref. *49.*
10. Cleaved by CH_3COOH 80%. Ref. *310* (same remark as for entry 3).
11. Cleaved by CH_3COOH 80%. Ref. *311* (same remark as for entry 3).
12. Cleaved by H_2/Pd/C. Ref. *312.*
13. Cleaved by $(Ph)_3C^+$ $BF_4^-/CH_3CN/H_2O$ (4:1) Refs. *166,286.*
14. Cleaved by light (>280 nm), $HCOONH_4$ 0.1*M* (pH 3.5), 1 h, under N_2. Refs. *251, 252,254,256,262,263.*
15. Cleaved by $(But)_4N^+$ F^-/THF or pyridine. Refs. *14,66,69,76,184,266–268, 274–277,280,281,283,295.*
16. Cleaved by DBU. Ref. *313.*

[a]*See* pages 24 and 25 for corresponding structures.

tion is transfered on the methylene group, and the protection is finally liberated on the form of 2-nitrosobenzaldehyde *(255,256).* The protecting group has been largely used by Ikehara's team in the three classical strategies: triester *(253,255,257–260),* phosphoramidite *(261),* and phosphonate *(251).* The protected nucleosides are rather difficult to synthesize *(256,262–265).* The conditions of deprotection have to be carefully controlled, in order to avoid side reactions *(252).*

3.3. The 2'-O-t-Butyldimethylsilyl Group

When 5'-*O*-DMTr and aglycone protected nucleosides are treated with *t*-butyldimethylsilyl (TBDMS) chloride in pyridine, or in DMF/imidazole, a mixture, where 2'-*O* and 3'-*O* monosilylated derivatives dominate, is obtained. They are isolated by chromatography *(266–269),*

Table 6
Structures

1 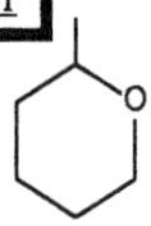	2 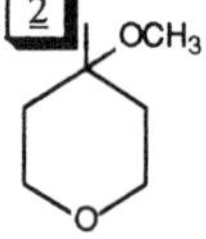	3 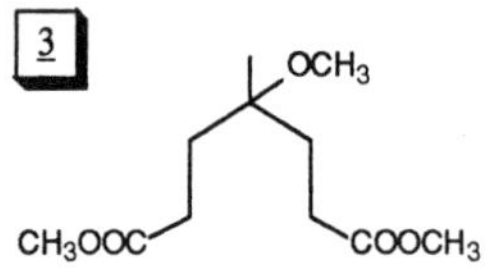
4 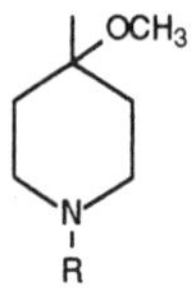 4 a, 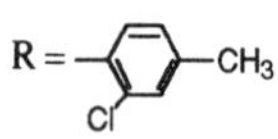 4 b, 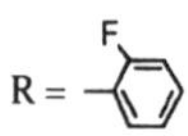	5 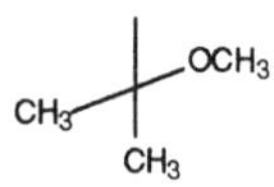	6 6a 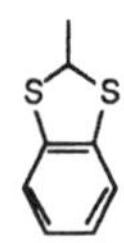 6b 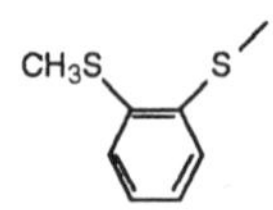
7 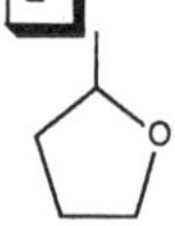	8 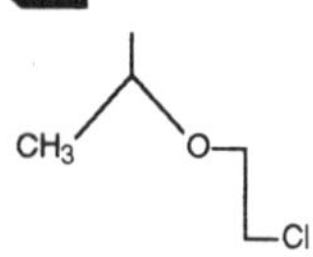	9 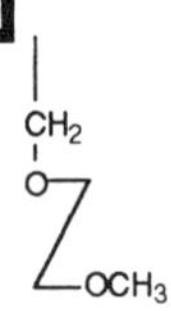

Table 6
Structures *(continued)*

10 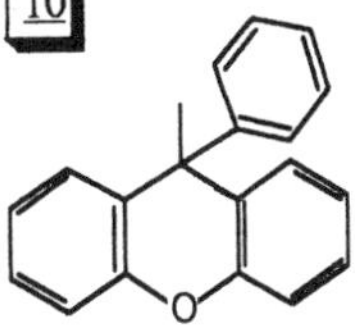	11 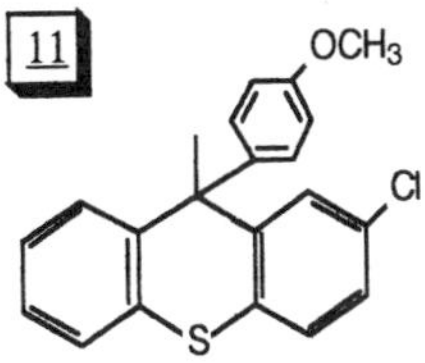	12 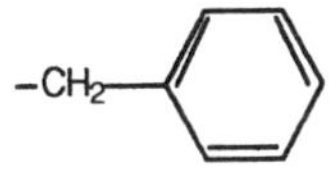
13 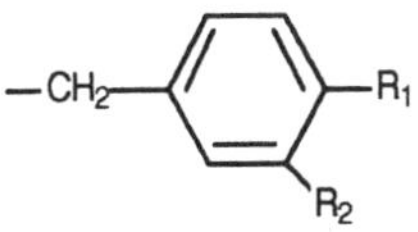 $R_1 = CH_3O$-, $R_2 = H$ $R_1 = R_2 = CH_3O$-	14 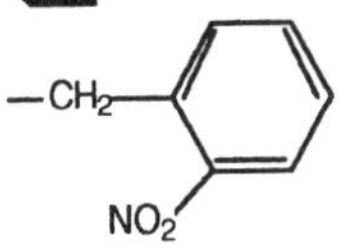	15 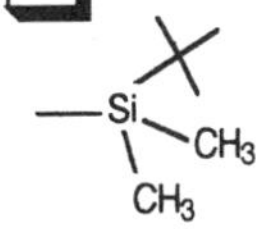
16 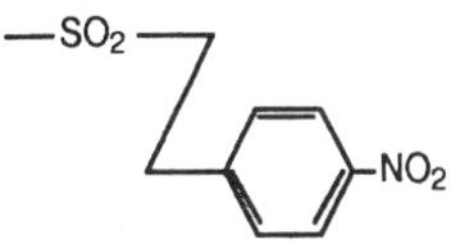		

or selective precipitation *(270)*. The relative amount of the 2'-*O* and 3'-*O* silylated isomers depends on the conditions. For 5'-*O*-DMTr ribonucleosides, the best available degree of selectivity in favor of the 2'-*O* isomer is 70:15 *(271)*. The two isomers equilibrate on standing in alcohols (particularly methanol), and wet solvents *(76,77,266,268,272–274)* (another side reaction is the debenzoylation of 2'-*O*-silylated *N*-benzoylcytidine and *N*-benzoyladenosine in methanol; refs. *76,77)*. However, a clean 3'-*O* phosporylation or phosphitylation of the isolated 5'-*O*-DMTr, 2'-*O*-silylated isomer, giving a single compound, is possible under controlled conditions *(270,272,275–278)*. Phosphitylation with *bis-(N,N*-diisopropylamino) (2-cyanoethoxy)phosphine in acetonitrile, in the presence of diisopropylmammonium tetrazolide as an activator, is to be avoided because it leads to 4–5% 3'-*O*/2'-*O* isomerization *(279)*.

Oligoribonucleotides have been synthesized by using the 2'-*O*-TBDMS group with three current strategies: Pfleiderer's triester method (*p*-nitrophenylethyl protection on phosphates) *(280–282)*, phosphoramidite method *(66,69,188,231,270,283,285)*, and phosphonate method *(287,288)*. The TBDMS group is cleaved by tetrabutylammonium fluoride in THF or pyridine. The huge amount of salt is then efficiently removed by the use of an anion exchange cartridge *(277)*. The TBDMS protection does not resist to an oximate teatment. It is thus not compatible with Reese's triester method (*o*-chlorophenyl protection on phosphates; refs. *282,289)*. Oligoribonucleotides in their diester form may be first treated with fluoride ions and then with ammonia at room temperature (to cleave the aglycone protections; refs. *280,281)*, but the reverse sequence is more reasonable *(66,76,270,283,290–293)*, because concentrated ammonia cleaves RNA *(214)*. However, the TBDMS group is lost to an appreciable extent in hot aqueous ammonia, and chain cleavage results. A mixture of aqueous ammonia and ethanol (3:1) is less aggressive, but still chain cleavage occurs *(69,294)*. There is no evidence of isomerization to 2'–5' internucleotidic linkages *(277,278,295)*. *N*-protections more labile than the standard benzoyl and isobutyryl groups are recommended for A and G (PAC amidites) *(69,285)*. *N*-Deacylation of PAC amidites in anhydrous methanolic ammonia at room temperature avoids the cleavage of the ribonucleotide chain *(285)*.

Automatic RNA synthesis kits, based on the 2'-*O* silylated building blocks are now commercially available *(296)* and have been discussed in detail in Chapter 5 of vol. 20 in this series.

4. Protection of the 5'-OH Function

They are illustrated in Table 7. This protection is removed before each coupling. The removal has thus to be not only quantitative, but also very rapid. It is when this group is introduced on nucleosides that the 5'-*O* is definitely differentiated from the 3'-*O*. The nucleosides protected on the

5'-*O* must not contain a trace of the isomer, protected on the 3'-*O*, because this contamination will enter into an improper orientation in the synthesized DNA sequences.

The protecting groups of the trityl family (e.g., DMTr, 9-phenylxanthen-9-yl=Px, 9-*p*-methoxyphenylxanthen-9-yl=Mox, Table 7, *9,* and *10*) are, without discussion, the best for the routine synthesis of oligodeoxyribonucleotides incorporating the four natural bases only. Trityl chloride, being sterically crowded, alkylates the primary 5'-OH far more readily than the secondary 3'-OH. The introduction of heteroatoms (e.g., the two methoxy groups of DMTr) was necessary to tune the rate of cleavage by acids. The Px and Mox groups are perhaps a little more acid-labile than DMTr; refs. *314,315).*

DMTrCl is recrystallized in a mixture of cyclohexane and acetyl chloride *(223).* The tritylation of nucleosides is usually performed in pyridine *(71,316).* DMAP *(79,317),* the ion-pair loosener perchlorate anion *(318),* and silver nitrate *(271)* are catalysts. It is important not to evaporate the crude mixture after tritylation, because pyridinium hydrochloride is acidic enough to detritylate the product. The crude mixture is instead directly poured into $NaHCO_3$ 5%, and the solution extracted with CH_2Cl_2. Tritylated nucleosides are sensitive to traces of acids in the solvents ($CHCl_3$ has to be avoided). One may add a small amount of pyridine to eluents, to avoid acid-detritylation on silica *(319).*

Suspended or solubilized zinc bromide is extremely selective in cleaving 5'-*O*-DMTr *(174,205,220,221,320,321).* Methanol has to be avoided as a cosolvent because it induces the *N*-deacylation of the aglycone residues *(320).* The longer the sequence, the slower the deprotection by zinc bromide *(322).* That is why the standard deprotection agents are protic acids (typically, dichloroacetic acid in methylene chloride; refs. *246,323–325).* The complex of boron trifluoride with methanol has also been considered *(326).* The choice of dichloroacetic acid comes from a long search for rapid deprotection with a minimum of depurination. Stronger acids are necessary when the deprotection is performed on oligonucleotides immobilized on a basic support (e.g., poly-*N,N*-dimethylacrylamid; ref. *334).* The intense coloration of the waste getting out of the synthesis column, when the acid deprotection is performed, allows us to visualize the progress of the synthesis. A quantitative spectrophotometric determination of the released dimethoxytrityl cation is possible *(71,334).*

A medium-sized oligonucleotide bearing a 5'-*O* terminal dimethoxytrityl group has a great affinity for silanized silica. Untritylated sequences, on the contrary, are less retained. This property allows a rapid purification of crude tritylated oligonucleotides on reversed-phase C_{18} silica cartridges. The pixyl group is not so good in this respect. 5'-*O*-dimethoxytritylated sequences cannot be kept for a long time in aqueous solution when ammonia is not present, because they rapidly detritylate.

The major impetus to develop other 5'-*O* protecting groups than DMTr (and related ones) is the search for a good RNA synthesis, with a ketal/acetal protecting group on the 2'-*O*. The MTHP group is ideal as a 2'-*O* protection (ease of specific introduction on the 2'-*O*, orthogonality to the phosphate and aglycone protections, ease of cleavage). Unfortunately, it is partially cleaved in the detritylation conditions. One thus has very early looked for a 5'-*O* protection cleavable by bases, like alkyl or aryloxy acetyl *(1)*, levulinyl *(2)*, 9-fluorenylmethyloxycarbonyl (Fmoc) *(7)* (Table 7). The synthesis of oligoribonucleotides by this type of strategy has indeed be performed on numerous occasions *(20,152,192,238,235–237,239–244)*. Possible side reactions owing to the conditions of repetitive cleavage of the 5'-*O* protection in basic conditions have to be investigated with great care: Loss of the aglycone and phosphotriester linkage protections, detachment from the polymeric support. Another possible side reaction originates in the fact that oligonucleotides triesters having a free 5'-OH function may cyclize under the action of bases, to generate a terminal 5',3'-cyclic residue *(207,239,327)*. The Fmoc group seems the best candidate as a 5'-*O* base-labile protection. Contrary to other groups (as levulinyl), its 5'-*O*/3'-*O* selectivity is excellent *(235–237,328)*.

Silyl groups are not a pertinent choice for 5'-*O* protection, as fluoride ions, used for their rapid cleavage, also cleave phosphotriester bonds (at a rate depending on the nature of the substituents) *(207, 267, 329,330)*.

Trityl-type protections have been modified to be cleaved by a cascade of more or less specific reagents (use of the protected protecting group or "safety catch" concept) (Table 7; *3,4*).

5. The Phosphate Protection

5.1. The Persistent Phosphate Protection (Scheme 3)

Except for the H-phosphonate method, where there is, strictly speaking, no phosphate protecting group, an oligonucleotide at the end of a synthesis may be pictured as **1** in Scheme 3.

We have discussed at length the protections of the aglycone residues Ri and of the alcohol functions R_1, R_2. The terminal R_3 protection may

Table 7
Protections of the 5'-OH Function of Nucleosides[a]

List of methods of cleavage and relevant references (solvent mixtures are expressed in volumic proportions)
1. Cleaved by dilute ammonia, T_r. Refs. *143,244,299,300,306,331–333,335–337.* For a discussion of the intramolecular attack of the 5'-OH on the proximal phosphotriester function in some deprotection conditions, *See* ref. *239.*
2. Cleaved by NH_2-NH_2/pyridine/CH_3COOH. Refs. *20,152,192,238,239–243, 327,338.* For a discussion of the selectivity of the hydrazine reagent, *See* ref. *50.*
3. Cleaved by NH_2-NH_2/pyridine/CH_3COOH, followed by pyridine/CH_3COOH (1:2), 50°C. Ref. *339.*
4. Cleaved by NH_2-NH_2/pyridine/CH_3COOH, followed by pyridine/CH_3COOH (1:3). Refs. *50,340.*
5a. Cleaved by Ag^+/H_2O/acetone/2,4,6-collidine, followed by morpholine. Refs. *155, 301,341.* It does not work for long oligomers: *See* ref. *23.*
5b. Cleaved by Hg(II)/H_2O/THF/2,4,6-collidine, followed by K_2CO_3 or NEt_3/H_2O/ THF. Refs. *23,342,343.*
6. Cleaved by NEt_3/pyridine. Refs. *310,344,345.*
7. Cleaved by Et_3N/pyridine, or piperidine/CH_3CN, or DBU/CH_3CN. Refs. *235–237,311,328,346.*
8. Cleaved by β-cyanoethanol/NEt_3/H_2O (1:1:1), followed by DBU 0.01*M*/pyridine. Ref. *347.*
9. Cleaved by $ZnBr_2$ or $CHCl_2COOH$/CH_2Cl_2. *See* the text for leading references.
10. Cleaved by $ZnBr_2$ or $CHCl_2COOH$/CH_2Cl_2. Refs. *41b,191,314,315.*
11. Cleaved as DMTr. Ref. *348.*
12. Cleaved by $ZnBr_2$ or a protic acid. Refs. *349–351.*

[a]*See* pages 30 and 31 for corresponding structures.

be an acyl group (usually benzoyl; refs. *22,352,353),* when R_1 is cleaved by acids, a ketal/acetal/orthoester function *(301,332,343),* when R_1 is cleaved by bases, or a silyl group *(225,280,281).* It is the polymer support when the oligonucleotide is synthesized in the solid phase. At the end of the synthesis, R4 has to be removed before all the other groups, because a triester function is very sensitive to nucleophiles and to intramolecular attack by the terminal 5'-OH *(207,239,327),* 3'-OH and internal 2'-OH functions. Two general principles of cleavage are used.

5.1.1. The Persistent Phosphate Protection is Removed by Nucleophilic Attack on Phosphorus

The nucleophile displaces the XR_4- group in **1**.

When X is oxygen, the pK_a of the alcohol HOR_4 has to be as low as practically tractable, in order to be the best leaving group on the phosphorus. It is thus typically a phenol or an acidic alcohol like 1,1,1,3,

Table 7
Structures

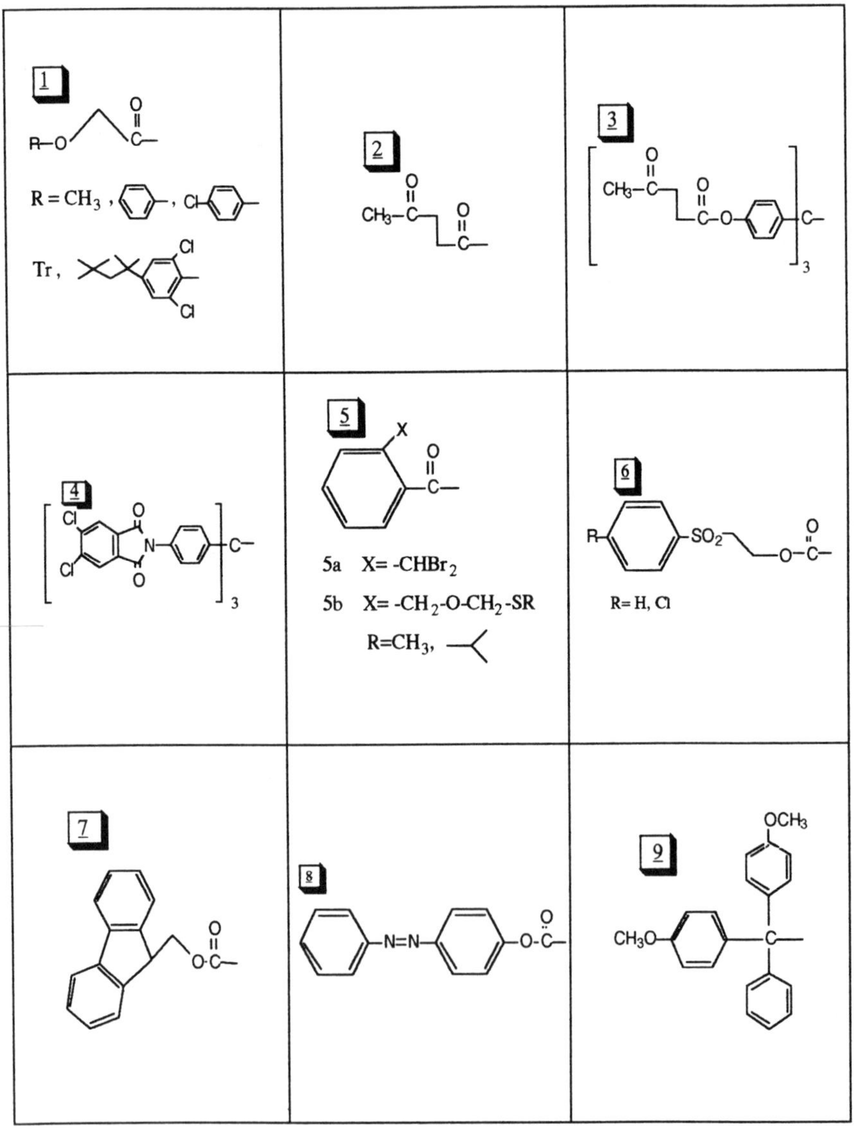

Table 7
Structures *(continued)*

10 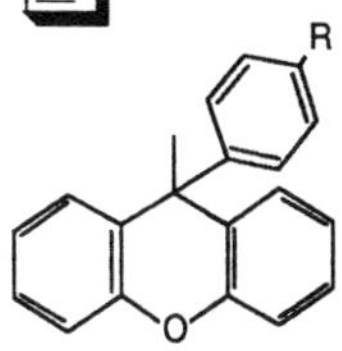 R= H, CH_3O-, $C_{18}H_{37}$-O-	11 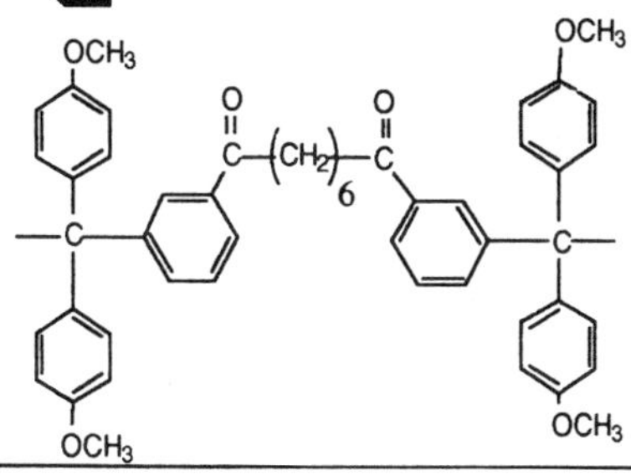
12 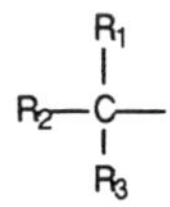 R1, R2, R3 = one of de groups : 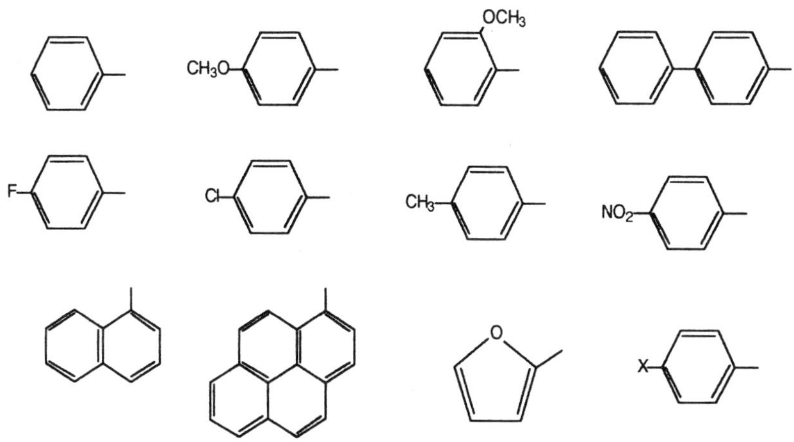 (X= C_8H_{17}- to $C_{16}H_{33}$)	

3,3,3-hexafluoroisopropan-2-ol (pK_a = 9.3) *(354)*. Moreover, the nucleophile has to attack phosphorus, not the 5'-carbon. *o*- or *p*-Chlorophenolate emerged from a long search as the best leaving groups *(23,343,355)*, and the oximate anion as a phosphorus-selective nucleophile (N^1, N^1, N^3, N^3-tetramethylguanidinium syn-2-nitrobenzaldoxi-

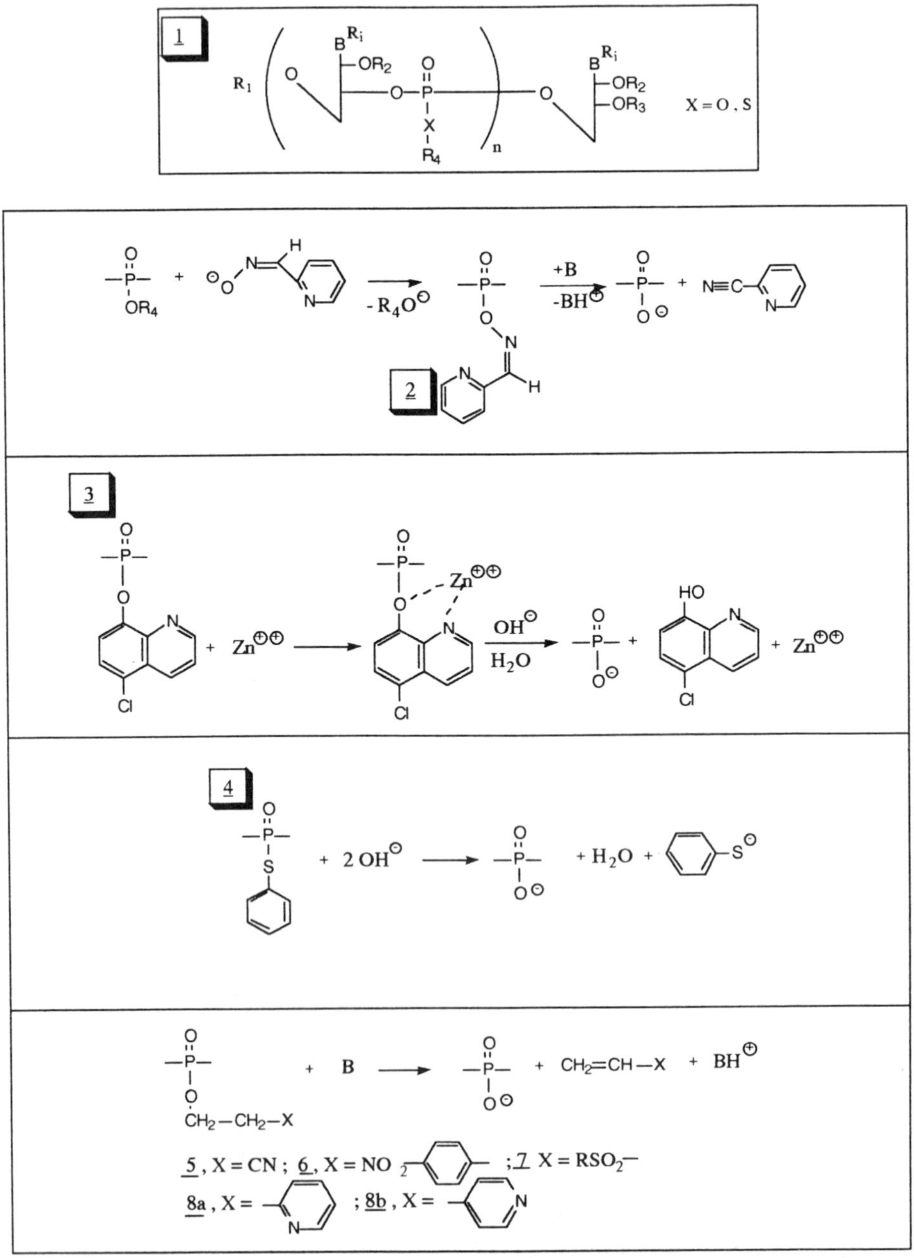

Scheme 3. Persistent phosphate protections and their modes of cleavage.

Scheme 3 *(continued)*

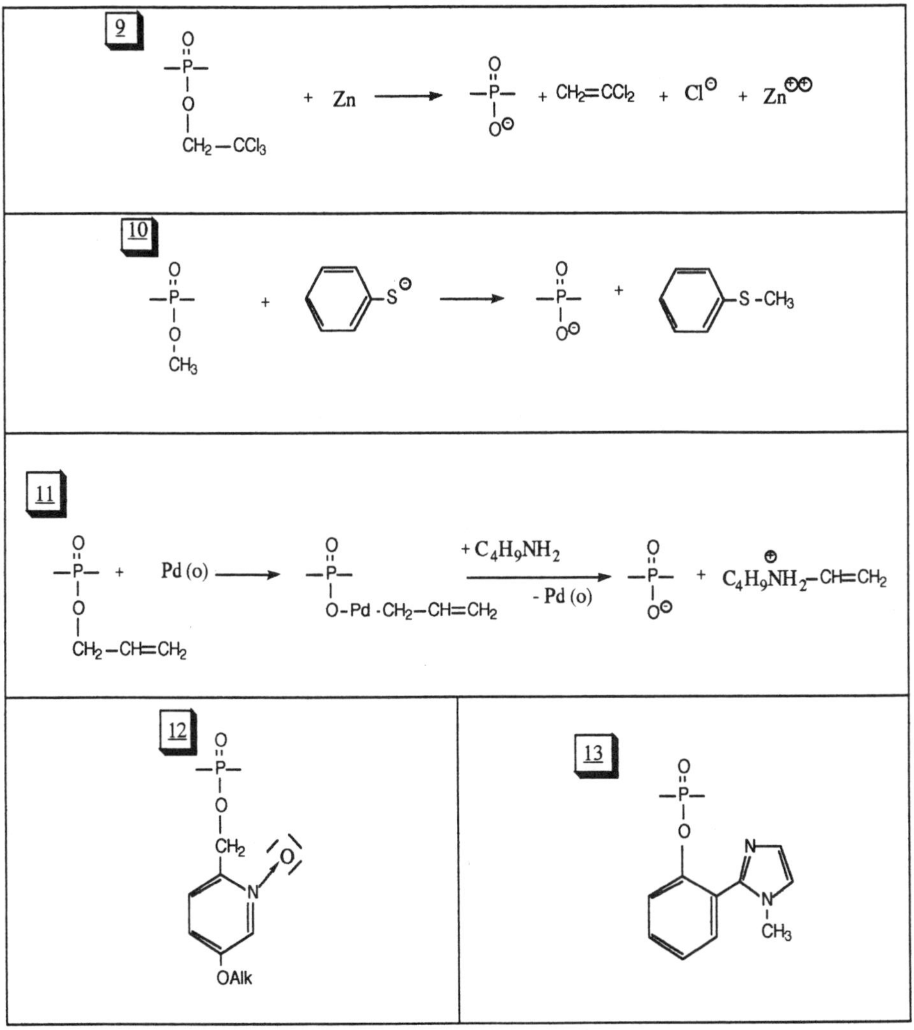

mate, or syn-2-pyridinecarboxaldoximate) *(220,356)*. The first obtained intermediate triester **2** is fragmented by β-elimination *(357)*. The oximate treatment is agressive: It removes the oligonucleotide from the solid support (succinic ester linkage). On prolonged treatment, the acyl protections of the aglycone residues are cleaved too *(358)*.

O-Silyl groups do not resist either *(282)*. The benzohydroxamate anion seems more chemoselective than the oximate anion *(362)*.

The leaving group OR_4 may be engineered to complex cations, in order to enhance its nucleofugacity. The esters of 5-chloro-8-hydroxy quinoline **3** are cleaved by treatment with zinc chloride (or zinc acetate) in pyridine/water, at room temperature *(49,359–361)*.

A thiolate SR_4 (X = S in *1*, structure *4*) is also a good leaving group. It is cleaved by oximate *(204,363)*, the hydroxyl anion *(175,364)* (silver acetate catalysis *(50)*), or bis(tributyltin) oxide *(365,366)*.

5.1.2. The Persistent Phosphate Protection is Removed by C-O Bond Fission

This is usually realized by a base-induced β-elimination, using ethyl esters bearing an electron withdrawing substituent at the 2 position. 2-Cyanoethyl esters **5** are cleaved by triethylamine *(367,368)* or *t*-butylamine *(369,370)* in pyridine, or, at the final deprotection step, by ammonia at room temperature *(371–373)*. It is one of the most classical phosphate protection. It is sensitive to DMAP *(374,375)*, a catalyst used for the 5'-OH capping in automatic synthesis. 2-Cyano-1,1-dimethylethyl esters have also been described *(400)*. 2-*p*-Nitrophenylethyl esters **6** are cleaved by DBU in pyridine *(10,281, 376, 377)*. 2-Ar(alk)ylsulfonyl esters **7** are cleaved by sodium hydroxide *(378,379)*. 2-[2-(or 4-)Pyridyl]ethyl esters **8a,b** need a somewhat tricky *N*-quaternization before being removable by β-elimination *(380,381)*.

The trichloroethyl esters **9** were formerly used in DNA synthesis. The yield of the deprotection by reductive elimination was however too low *(224,382)* and this strategy is now disregarded.

Methyl esters **10** are attacked on carbon by thiolates *(5,188,231, 270,283,383–385)*. *t*-Butylamine has also been advocated *(386)*. Thiophenolate does not cleave the succinate linkage with the solid support. The methyl protection is also a classic in DNA chemistry. However, the methylphosphotriester linkage behaves as an alkylating agent of the thymine residue. Tiny amounts of *N*-3 methylated derivatives are observed *(3)*.

The allyl group of **11** is removed by oxidative addition of Pd(0), to give a σ-bonded Pd(II) complex. The Pd(0) catalyst is regenerated by the attack of butylamine *(93b,420)*.

5.1.3. Persistent Phosphate Protection Allowing an Intramolecular Catalysis at the Coupling Step

The coupling in the phosphotriester method is slower than in the phosphoramidite strategy. Nucleophilic catalysts like *N*-methylimidazole and pyridine-*N*-oxides greatly enhance the coupling rate. They displace bulky leaving groups on phosphorus and are themselves excellent nucleofiles. These nucleophilic catalysts were attached to the persistent phosphate protection, in order to get an intramolecular catalytic effect (compounds **12** and **13**; refs. *387–389)*. The so engineered neighboring group participation indeed enhanced the rate of coupling. The catalysis was chemoselective: It did not enhance the rate of 5'-*O* sulfonylation, the yield-limiting side reaction of the phosphotriester method.

The catalytically active groups were cleaved at the end of the synthesis by nucleophilic attack on carbon by piperidine or thiophenolate for **12** *(387),* and by nucleophilic attack on phosphorus by oximate at 60°C, for **13** *(389).*

5.2. The Transient Phosphate Protection (Table 8)

The fully protected block **13,** used in the Catlin-Cramer synthesis *(390),* is pictured in Table 8. When R_1 is cleaved, the block is engaged in the coupling as the 3'-end partner. When R_5 is cleaved, the block is the 5'-end partner. R_5 has to be orthogonal to R_i, R_1, R_4, and, of course R_2.

The usual strategy is R_1 = DMTr (or Px), R_4 = 2-(or 4-) chlorophenyloxy and R_5 = 2-cyanoethyloxy *(14,22,225,352,369,370,391–396),* or 9-fluorenylmethyloxy *(41b,191,219,397,398),* both being easily removable by β-elimination. There are a number of documentated alternatives illustrated in Table 8. Chattopadhyaya rapidly screened several other possibilities *(311,344,416–419).*

The starting monomeric block **14** is accessible by three methods (Scheme 4):

1. Quenching the phosphorylation mixture obtained according to Itakura, Narang, et al. *(355,392,399)* by the appropriate alcohol HOR_5 *(374);*
2. Coupling the purified phosphodiester **15** with the alcohol HOR_5 in an extra step *(41b,191,219,319,343,396);*
3. Using the presynthesized phosphochloridate **18** as a phosphorylating agent (typically, R_4 = *p*-chlorophenyl, R_5 = β-cyanoethyl) *(394,395,401).*

In the first procedure, a chromatographic step is necessary to remove the byproduct **17** *(178).* In the second procedure, a simple solvent

Table 8
Building Blocks for the Catlin-Cramer Synthesis in Solution[a]

List of methods of selective cleavage of R_5 and relevant references (solvent mixtures are expressed in volumic proportions)
13a,b. *See* text.
13c. Cleaved by phosphorous acid in pyridine. Refs. *99,100,175,403.* The selectivity is however not fully satisfactory: *See* ref. *157.*
13d. Cleaved by oximate. Refs. *10,280,281,376.*
13e. Cleaved by nucleophilic attack on the benzylic carbon by the 2-thiocresolate anion. Refs. *343,404.* This anion may however also attack 5'-C in the sequence: *See* ref. *405.*
13f. Cleaved by oximate. Refs. *406,407.* The amount of internucleotidic triester cleaved by this oximate treatment is unknown: *See* refs. *226,408.*
13g. Cleaved by isopentylnitrite in pyridine/acetic acid (5:4) Refs. *220,221,258,259, 306,409–411.*
13h. Cleaved by zinc/triisopropylbenzenesulfonic acid/pyridine. Refs. *20,192,241, 242,412–415.*

[a]*See* opposite page for corresponding structures.

extraction separates **15** from the byproduct **16** *(402)* (R_1 has to be lipophilic enough, e.g., R_1 = DMTr). The precipitation of the barium salt of the nucleotide **15** also allows to get a pure product *(319).*

6. Conclusion

Although DNA synthesis is now routine, and rapid RNA synthesis will soon be accessible, questions remain about the fidelity of oligonucleotide synthesis. Synthetic oligomers that are homogeneous in anion exchange HPLC, or that feature a single band in PAGE are not necessarily pure products. Minute defects randomly spreaded along the sequence may exist, inducing resistance to restriction enzymes and hypersensitivity to piperidine *(421).* In other words, beyond a certain length, there is not even one molecule whose structure is entirely correct. The quality of the protection strategy is of paramount importance for fidelity. Several side reactions, avoidable by protection or, on the contrary, induced by protection (e.g., depurination of dA), have been studied in detail. Others certainly exist. The harsh treatment by ammonia at the end of the synthesis probably cleaves a great portion of the modified sequences (apurinic sites are an example). Moving to more easily cleavable aglycone protections may unmask some yet undetected DNA modifications.

Table 8
Structures

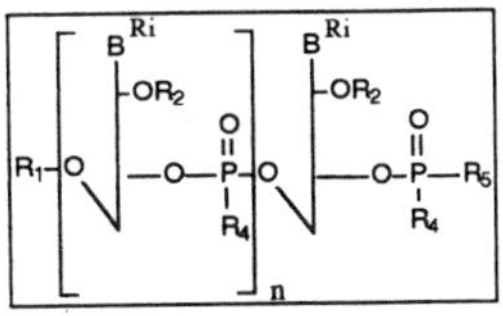

	R1	R4	R5
13 a	DmTr or Px	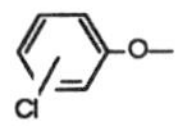	NC–CH_2–CH_2–O–
13 b	DmTr or Px	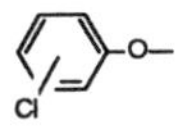	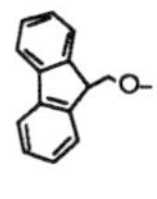
13 c	DmTr	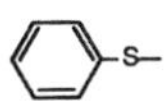	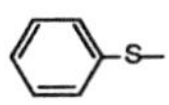
13 d	DmTr	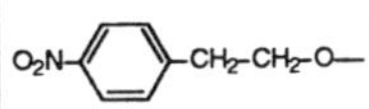	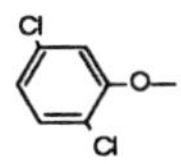
13 e	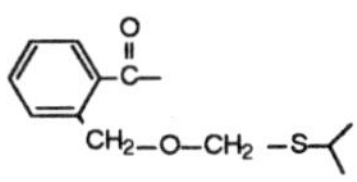	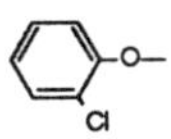	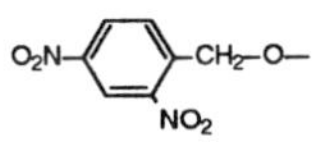
13 f	DmTr	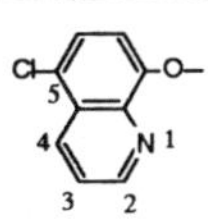	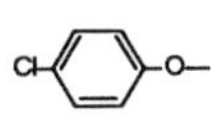
13 g	DmTr	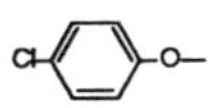	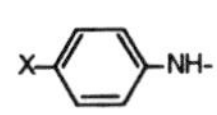 X= -H, -OCH_3
13 h	Acyl	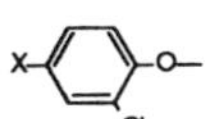 X= -H, -$C(CH_3)_3$	CX_3–CH_2–O– X= -Cl, -Br

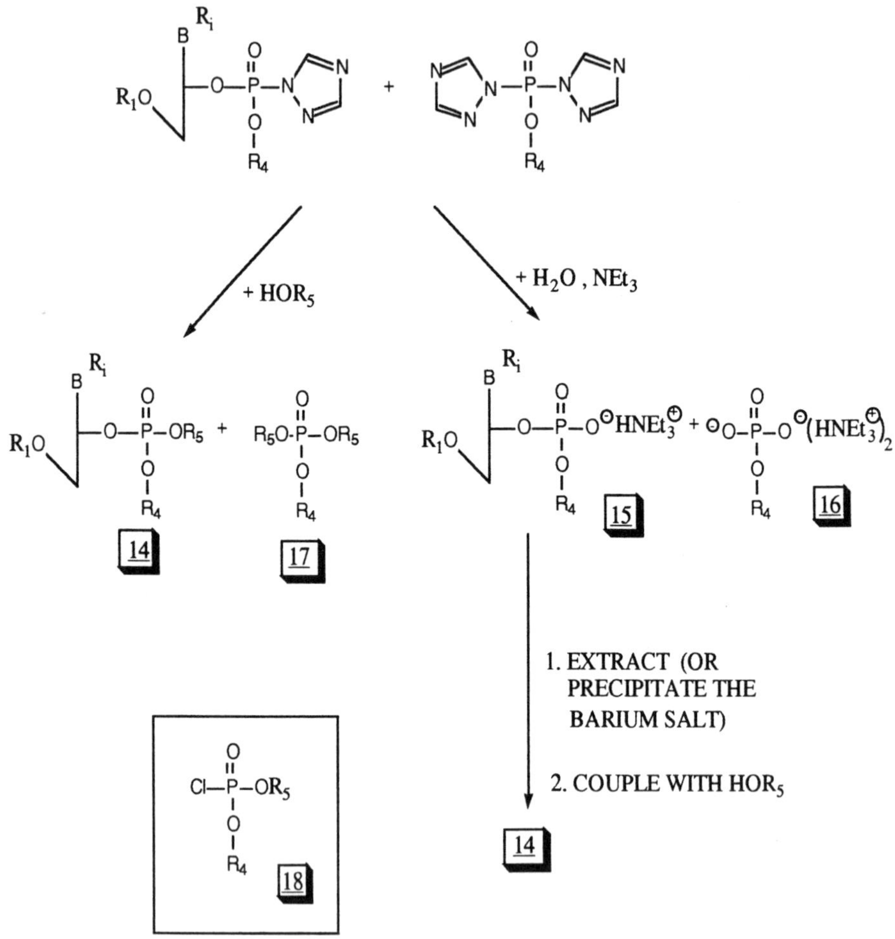

Scheme 4. Synthesis of monomeric Catlin-Cramer's building blocks.

The quest for the optimal synthesis of a potentially endless sequence is by nature endless.

In the next ten years, hypermodified residues will be included in RNA synthesis for tRNA engineering. The introduction of unnatural residues in DNA will also become routine, because antisense DNA has to be unnatural, in order to resist to nucleases and to penetrate into cells. There are thus many opportunities to develop a new chemistry in this field.

Acknowledgments

E. Sonveaux was a research associate of the "Fonds National Belge de la Recherche Scientifique" (FNRS). A. Bidaine is much thanked for the artwork.

Abbreviations

DBU, 1,8-diazabicyclo[5.4.0]undec-7-ene; DMAP, 4-dimethylamino-pyridine; DMF, dimethylformamide; DMSO, dimethylsulfoxide; DMTr, 4,4'-dimethoxytriphenylmethyl (dimethoxytrityl); EEDQ, *N*-ethoxy-carbonyl-2-ethoxy-1,2-dihydroquinoline; Fmoc, 9-fluorenylmethyl-oxycarbonyl; Fpmp, 1-(2-fluorophenyl)-4-methoxypiperidin-4-yl; Mox, 9-(*p*-methoxyphenyl)xanthen-9-yl; MTHP, 4-methoxytetrahydro-pyran-4-yl; NPE, 2-(*p*-nitrophenyl)ethyl; Oximate, N^1,N^1,N^3,N^3-tetra-methylguanidinium syn-2-nitrobenzaldoximate or syn-2-pyridine carboxaldoximate; PAC, phenyloxyacetyl; PAGE, polyacrylamide gel electrophoresis; PTC, phase transfer catalysis; PTSA, *p*-toluenesulfonic acid; Px, 9-phenylxanthen-9-yl (pixyl); RT, room temperature; TBAF, tetrabutylammonium fluoride; TBDMS, *t*-butyldimethylsilyl; THF, tetrahydrofuran; THP, tetrahydropyran-2-yl; Tr, triphenylmethyl (trityl).

References

1. Albert, A. (1973) Ionization constants of pyrimidines and purines. *Synthetic Procedures in Nucleic Acid Chemistry, vol* **2,** *Physical and Physicochemical Aids in Characterization and in Determination of Structure* (Zorbach, W. W. and Tipson, R. S., eds.), Wiley-Interscience, New York, pp. 1–46.
2. Hata, T. and Sekine, M. (1983) Synthesis of 2'-*O*-(1,3-benzodithiol-2-yl)uridine and related compounds as key intermediates in oligoribonucleotide synthesis. *J. Org. Chem.* **48,** 3112–3114.
3. Gao, X., Gaffney, B. L., Senior, M., Riddle, R. R., and Jones, R. A. (1985) Methylation of thymine residues during oligonucleotide synthesis. *Nucleic Acids Res.* **13,** 573–584.
4. Urdea, M. S., Ku, L., Horn, T., Gee, Y. G., and Warner, B. D. (1985) Base modification and cloning efficiency of oligodeoxyribonucleotides synthesized by the phosphoramidite method: methyl versus cyanoethyl phosphorus protection. *Nucl. Acids Symp. Ser.* **16,** 257–260.
5. Andrus, A. and Beaucage, S. L. (1988) 2-Mercaptobenzothiazole: an improved reagent for the removal of methyl phosphate protecting groups from oligodeoxynucleotide phosphotriesters. *Tetrahedron Lett.* **29,** 5479–5482.

6a. Ho, N. W. Y. and Gilham, P. T. (1967) The reversible chemical modification of uracil, thymine, and guanine nucleotides and the modification of the action of ribonuclease on ribonucleic acid. *Biochemistry* **6,** 3632–3639.

6b. Astell, C. R. and Smith, M. (1972) Synthesis and properties of oligonucleotide-cellulose columns. *Biochemistry* **11,** 4114–4120.

6c. Schaller, H. and Khorana, H. G. (1963) Studies on polynucleotides. XXV. The stepwise synthesis of specific deoxyribopolynucleotides (5) Further studies on the synthesis of internucleotide bond by the carbodiimide method. The synthesis of suitably protected dinucleotides as intermediates in the synthesis of higher oligonucleotides. *J. Am. Chem. Soc.* **85,** 3828–3835.

7. Reese, C. B. and Richards, K. H. (1985) Reaction between nucleoside base residues and the phosphorylating agent derived from 1-hydroxybenzotriazole and 2-chlorophenyl phosphodichloridate. *Tetrahedron Lett.* **26,** 2245–2248.

8a. Sung, W. L. (1981) Chemical conversion of thymidine into 5-methyl-2'-deoxycytidine. *J. Chem. Soc. Chem. Commun.* **1981,** 1089.

8b. Sung, W. L. (1981) Synthesis of 4-triazolopyrimidone nucleotide and its application in synthesis of 5-methylcytosine containing oligodeoxyribonucleotides. *Nucleic Acids Res.* **9,** 6139–6151.

8c. Sung, W. L. (1982) Synthesis of 4-(1,2,4-triazol-1-yl)pyrimidin-2(1H)-one ribonucleotide and its application in synthesis of oligoribonucleotides. *J. Org. Chem.* **47,** 3623–3628.

9. Pless, R. C. and Letsinger, R. L. (1975) Solid support synthesis of oligothymidylates using phosphochloridates and 1-alkylimidazoles. *Nucleic Acids Res.* **2,** 773–786.

10. Uhlmann, E. and Pfleiderer, W. (1981) Nucleotide. XIV Substituierte β-Phenyläthyl-Gruppen. Neue Schutzgruppen für Oligonucleotid-Synthesen nach dem Phosphosäuretriester-Verfahren. *Helv. Chim. Acta* **64,** 1688–1703.

11. Adamiak, R. W., Biala, E., Gdaniec, Z., and Mielewczyk, S. (1986) Reactions of nucleoside heteroatomic lactam systems with 4-chlorophenylphosphorodichloridate and 1,2,4-triazole in pyridine. *Chem. Scr.* **26,** 3–6.

12. Adamiak, R. W., Biala, E., Gdaniec, Z., and Mielewczyk, S. (1986) New observations on nucleophile assisted phosphorylations of nucleoside heteroatomic lactam systems during oligonucleotide synthesis. *Chem. Scr.* **26,** 7–11.

13. Divakar, K. J. and Reese, C. B. (1982) 4-(1,2,4-triazol-1-yl)- and 4-(3-nitro-1,2,4-triazol-1-yl)-1-(β-D-2,3,5-tri-*O*-acetylarabinofuranosyl)pyrimidin-2(1H)-ones. Valuable intermediates in the synthesis of derivatives of 1-(β-D-arabinofuranosyl)cytosine (Ara-C) *J. Chem. Soc. Perkin Trans.* **1,** 1171–1176.

14. Sung, W. L., and Narang, S. A. (1982) Modified phosphotriester method for chemical synthesis of riboologonucleotides. Part I. Synthesis of riboundecaadenylate and two fragments constituting the sequence of R-17 translation control signal. *Can. J. Chem.* **60,** 111–120.

15. Reese, C. B., and Ubasawa, A. (1980) Reaction between 1-arenesulphonyl-3-nitro-1,2,4-triazoles and nucleoside base residues. Elucidation of the nature of side reactions during oligonucleotide synthesis. *Tetrahedron Lett.* **21,** 2265–2268.

16. Sekine, M. (1989) General method for the preparation of N^3-and O^4-substituted uridine derivatives by phase-transfer reactions. *J. Org. Chem.* **54,** 2321–2326.

17. Barone, A. D., Tang, J.-Y., and Caruthers, M. H. (1984) *In situ* activation of bis-dialkylaminophosphines. A new method for synthesizing deoxyoligonucleotides on polymer supports. *Nucleic Acids Res.* **12,** 4051–4061.
18. Hata, T. and Sekine, M. (1974) Silyl phosphites. I. The reaction of silyl phosphites with diphenyl disulfide. Synthesis of S-phenyl nucleoside phosphorothioates. *J. Am. Chem. Soc.* **96,** 7363–7364.
19. Kamimura, T., Masegi, T., Sekine, M., and Hata, T. (1984) Structural assignment of N^3-acylated uridine derivatives by means of ^{13}C NMR spectroscopy. *Tetrahedron Lett.* **25,** 4241–4244.
20. den Hartog, J. A. J., Wille, G., Scheublin, R. A., and van Boom, J. H. (1982) Chemical synthesis of a messenger ribonucleic acid fragment: AUGUUCUU-CUUCUUCUUC. *Biochemistry* **21,** 1009–1018.
21. Mag, M. and Engels, J. W. (1988) Synthesis and structure assignments of amide protected nucleosides and their use as phosphoramidites in deoxyoligonucleotide synthesis. *Nucleic Acids Res.* **16,** 3525–3543.
22. Schott, H., v. Biedersee, H., von Sonntag, C., and Schulte-Frohlinde, D. (1986) Synthesis of homologues of deoxyribouridylic acid in preparative amounts using the triester method. *Makromol. Chem.* **187,** 809–827.
23. Brown, J. M., Christodoulou, C., Jones, S. S., Modak, A. S., Reese, C. B., Sibanda, S., and Ubasawa, A. (1989) Synthesis of the 3'-terminal half of yeast alanine transfer ribonucleic acid ($tRNA^{Ala}$) by the phosphotriester approach in solution. Part 1. Preparation of the nucleoside building blocks. *J. Chem. Soc. Perkin Trans.* **1,** 1735–1750.
24. Reese, C. B. and Skone, P. A. (1984) The protection of thymine and guanine residues in oligoribonucleotide synthesis. *J. Chem. Soc. Perkin Trans.* **1,** 1263–1270.
25. Kamimura, T., Masegi, T., Urakami, K., Honda, S., Sekine, M., and Hata, T. (1983) A new protecting tactics for the uracil residue in oligoribonucleotide synthesis. *Chem. Lett.* **25,** 1051–1054.
26. Welch, C. J. and Chattopadhyaya, J. (1983) 3-*N*-acyl uridines: preparation and properties of a new class of uracil protecting group. *Acta Chem. Scand.* **B37,** 147–150.
27. Inoue, H., Hayase, Y., Imura, A., Iwai, S., Miura, K., and Ohtsuka, E. (1987) Synthesis and hybridization studies on two complementary nona(2'-*O*-methyl)ribonucleotides. *Nucleic Acids Res.* **15,** 6131–6148.
28. Sochacka, E. and Malkiewicz, A. (1990) The tRNA "wobble position" uridines. IV. The synthesis of 2'-*O*-methyl-5-methoxycarbonylmethyluridine and its derivatives. *Nucleosides, Nucleotides* **9,** 793–802.
29. Sekine, M., Nishiyama, S., Kamimura, T., Osaki, Y., and Hata, T. (1985) Chemical synthesis of capped oligoribonucleotides, $m^7G^{5'}$pppAUG and $m^7G^{5'}$pppAUGACC. *Bull. Chem. Soc. Jpn.* **58,** 850–860.
30. Sekine, M., Fujii, M., Nagai, H., and Hata, T. (1987) An improved synthesis of N^3-benzoylthymidine. *Synthesis* **30,** 1119–1121.
31. Zhou, X.-X. and Chattopadhyaya, J. (1986) Site-specific modification of the pyrimidine residue during the deprotection of the fully-protected diuridylic acid. *Tetrahedron* **42,** 5149–5156.

32. Nyilas, A. and Chattopadhyaya, J. (1986) Synthesis of $O^{2'}$-methyluridine, $O^{2'}$-methylcytidine, N^4, $O^{2'}$-dimethylcytidine and N^4, N^4, $O^{2'}$-trimethylcytidine from a common intermediate. *Acta Chem. Scand.* **B40,** 826–830.
33. Cruickshank, K. A., Jiricny, J., and Reese, C. B. (1984) The benzoylation of uracil and thymine. *Tetrahedron Lett.* **25,** 681–684.
34. Tanimura, H., Fukazawa, T., Sekine, M., Hata, T., Efcavitch, J. W., and Zon, G. (1988) The practical synthesis of RNA fragments in the solid phase approach. *Tetrahedron Lett.* **29,** 577–578.
35. Sund, C. and Chattopadhyaya, J. (1989) Intra- and intermolecular nucleophilic phosphorus-sulfur bond cleavage. The reaction of fluoride ion with *O*-aryl-0,S-dialkylphosphorothioates, and the degradation of phosphorothioate linkage in di-ribonucleotides by the vicinal 2'-hydroxyl group. *Tetrahedron* **23,** 7523–7544.
36. Takaku, H., Imai, K., and Nakayama, K. (1987) Synthesis of oligoribonucleotides by using 2'-*O*-(1-methyl-1-methoxy)ethyl nucleosides. *Chem. Lett.* **1987,** 1787–1790.
37. Fujii, M., Horinouchi, Y., and Takaku, H. (1987) Use of the (butylthio)carbonyl group to protect uracil and guanine residues in oligoribonucleotide synthesis. *Chem. Pharm. Bull.* **35,** 3066–3069.
38. Kamimura, T., Masegi, T., and Hata, T. (1982) Protection of imide group of uracil moiety by means of 2,2,2-trichloro-tert-butyloxycarbonyl chloride: a selective synthesis of 2'-*O*-methyluridine. *Chem. Lett.* **1982,** 965–968.
39. Zhou, X.-X., Ugi, I., and Chattopadhyaya, J. (1985) A convenient preparation of *N*-protected nucleosides with the 2,2,2-trichloro-*t*-butyloxycarbonyl (TCBOC) group. Structural assignment of *N,N*-bis-TCBOC guanoside and its 2'-deoxy analog. *Acta Chem. Scand.* **B39,** 761–765.
40. Jones, S. S., Reese, C. B., Sibanda, S., and Ubasawa, A. (1981) The protection of uracil and guanine residues in oligonucleotide synthesis. *Tetrahedron Lett.* **22,** 4755–4758.
41a. Zhou, X.-X., Welch, C. J., and Chattopadhyaya, J. (1986) Pyridyl groups for protection of the imide functions of uridine and guanosine. Exploration of their displacement reactions for site-specific modifications of uracil and guanine bases. *Acta Chem.* Scand. **B40,** 806–816.
41b. Welch, C. J., Zhou, X.-X., and Chattopadhyaya, J. (1986) Synthesis of an mRNA fragment of Alanyl-tRNA synthetase gene in *Escherichia coli* using the 6-methyl-3-pyridyl group for protection of the imide functions of uridine and guanosine. *Acta Chem. Scand.* **B40,** 817–825.
42. Schulz, B. S. and Pfleiderer, W. (1983) Synthesis of O^4-*p*-nitrophenylethyl thymidine and uridine derivatives. *Tetrahedron Lett.* **24,** 3587–3590.
43. Himmelsbach, F., Schulz, B. S., Trichtinger, T., Charubala, R., and Pfleiderer, W. (1984) The *p*-nitrophenylethyl (NPE) group. A versatile new blocking group for phosphate and aglycone protection in nucleosides and nucleotides. *Tetrahedron* **40,** 59–72.
44a. Borowy-Borowski, H. and Chambers, R. W. (1989) Solid-phase synthesis and side reactions of oligonucleotides containing *O*-alkylthymine residues. *Biochemistry* **28,** 1471–1477.

44b. Xu, Y.-Z. and Swann, P. F. (1990) A simple method for the solid phase synthesis of oligodeoxynucleotides containing O^4-alkylthymine. *Nucleic Acids Res.* **18,** 4061–4065.

45. This group may be difficult to remove for long, poorly soluble oligonucleotides. *See* the case of dG, refs. *183,184.*

46. Takaku, H., Imai, K., and Nagai, M. (1988) Triphenylmethanesulfenyl group. A new protecting group for the uracil residue in oligoribonucleotide synthesis. *Chem. Lett.* **1988,** 857–860.

47. Welch, C. J., Bazin, H., Heikkilä, J., and Chattopadhyaya, J. (1985) Synthesis of C-5 and *N*-3 arenesulfenyl uridines. Preparation and properties of a new class of uracil protecting group. *Acta Chem. Scand.* **B39,** 203–212.

48. Takaku, H., Ueda, S., and Ito, T. (1983) Methoxyethoxymethyl group for the protection of uracil residue in oligoribonucleotide synthesis. *Tetrahedron Lett.* **24,** 5363–5366.

49. Ito, T., Ueda, S., and Takaku, H. (1986) (Methoxyethoxy)methyl group: new amide and hydroxyl protecting groups of uridine in oligonucleotide synthesis. *J. Org. Chem.* **51,** 931–933.

50. Sekine, M. and Hata, T. (1986) Synthesis of short oligoribonucleotides bearing a 3'- or 5'-terminal phosphate by use of 4,4',4"-tris(4,5-dichlorophtalimido)trityl as a new 5'-hydroxyl protecting group. *J. Am. Chem. Soc.* **108,** 4581–4586.

51. Claesen, C. A. A., Pistorius, A. M. A., and Tesser, G. I. (1985) One-step protection of the nucleoside base in thymine and uridine. *Tetrahedron Lett.* **26,** 3859–3862. The structure proposed by these authors is incorrect. *See* ref. *21.*

52. Sonveaux, E. (1986) The organic chemistry underlying DNA synthesis. *Bioorg. Chem.* **14,** 274–325.

53a. Gait, M. J., Matthes, H. W. D., Singh, M., Sproat, B. S., and Titmas, R. C. (1982) Synthesis of oligodeoxyribonucleotides by a continuous flow, phosphotriester method on a kieselguhr/polyamide support. *Chemical and Enzymatic Synthesis of Gene Fragments, a Laboratory Manual* (Gassen, H. G. and Lang, A., eds.), Verlag Chemie, Weinheim, pp. 1–42.

53b. Steinfeld, A. S., Naider, F., and Becker, J. M. (1979) A simple method for selective acylation of cytidines and cytosines under mild reaction conditions. *J. Chem. Res. Synop.* **1979,** 129.

53c. Igolen, J. and Morin, C. (1980) Rapid synthesis of protected 2'-deoxycytidine derivatives. *J. Org. Chem.* **45,** 4802–4804.

53d. Finlay, M., Debiard, J. P., Guy, A., Molko, D., and Teoule, R. (1983) An efficient one-pot synthesis of a fully protected 2'-deoxycytidine 3'-monophosphate. *Synthesis* **1983,** 303,304.

54. Hata, T. and Kurihara, T. (1973) The *N*-4-benzoylation of deoxycytidylic and cytidylic acids by means of 2-chloromethyl-4-nitrophenyl benzoate. *Chem. Lett.* **1973,** 859–862.

55a. Watanabe, K. A. and Fox, J. J. (1966) A simple method for selective acylation of cytidine on the 4-amino group. *Angew Chem. Internat. Edit.* **5,** 579–580.

55b. Otter, B. A. and Fox, J. J. (1968) *N*-Acyl derivatives of 2'-deoxycytidine. The selective acylation of the 4-amino group of a cytosine nucleoside. *Synthetic Procedures in Nucleic Acid Chemistry, vol 1, Preparation of Purines, Pyrim-*

idines, Nucleosides and Nucleotides (Zorbach, W. W. and Tipson, R. S., eds.), Wiley-Interscience, New York, pp. 285–287.

56. Bhat, V., Ugarkar, B. G., Sayeed, V. A., Grimm, K., Kosora, N., Domenico, P. A., and Stocker, E. (1989) A simple and convenient method for the selective *N*-acylations of cytosine nucleosides. *Nucleosides, Nucleotides* **8,** 179–183.
57. Mishra, R. K. and Misra, K. (1986) Improved synthesis of oligodeoxyribonucleotide using 3-methoxy-4-phenoxybenzoyl group for amino protection. *Nucleic Acids Res.* **14,** 6197–6213.
58. Takaku, H., Shimada, Y., Morita, Y., and Hata, T. (1976) A convenient method for N^4-benzoylation of cytidylic and deoxycytidylic acids by means of *O*-ethyl S-benzoyldithiocarbonate. *Chem. Lett.* **1976,** 19–22.
59. Himmelsbach, F. and Pfleiderer, W. (1983) The use of the *p*-nitrophenylethoxycarbonyl group for amino protection in cytidine and adenosine chemistry. *Tetrahedron Lett.* **24,** 3583–3586.
60. Mercer, J. F. B. and Symons, R. H. (1971) The use of *N*-ethoxycarbonyl-2-ethoxy-1,2-dihydroquinoline in the selective N^4-acylation of cytidine and its derivatives. *Biochim. Biophys. Acta* **238,** 27–30.
61. Van Montagu, M., Molemans, F., and Stockx, J. (1968) Preparation of cytidine, cytidylic acids and ribonucleic acid specifically acetylated in the exocyclic amino group of cytosine. *Bull. Soc. Chim. Belg.* **77,** 171–180.
62. Anteunis, M. and Van Montagu, M. (1965) Locked acetyl group in N(6)-acetylcytidine. *Bull. Chem. Soc. Belg.* **74,** 481–487.
63. Heikkilä, J. and Chattopadhyaya, J. (1983) The 9-fluorenylmethoxycarbonyl (Fmoc) group for the protection of amino functions of cytidine, adenosine, guanosine and their 2'-deoxysugar derivatives. *Acta Chem. Scand.* **B37,** 263–265.
64. Schulhof, J. C., Molko, D., and Teoule, R. (1987) Facile removal of new base protecting groups useful in oligonucleotide synthesis. *Tetrahedron Lett.* **28,** 51–54.
65. Schulhof, J. C., Molko, D., and Teoule, R. (1987) The final deprotection step in oligonucleotide synthesis is reduced to a mild and rapid ammonia treatment by using labile base-protecting groups. *Nucleic Acids Res.* **15,** 397–416.
66. Wu, T. and Ogilvie, K. K. (1988) *N*-phenoxyacetylated guanosine and adenosine phosphoramidites in the solid phase synthesis of oligoribonucleotides: synthesis of a ribozyme sequence. *Tetrahedron Lett.* **29,** 4249–4252.
67. Ralph, R. K. and Khorana, H. G. (1960) Studies on polynucleotides. XI. Chemical polymerization of mononucleotides. The synthesis and characterization of deoxyadenosine polynucleotides. *J. Am. Chem. Soc.* **83,** 2926–2934.
68. Köster, H., Kulikowski, K., Liese, T., Heikens, W., and Kohli, V. (1981) *N*-acyl protecting groups for deoxynucleosides. A quantitative and comparative study. *Tetrahedron* **37,** 363–369.
69. Wu, T., Ogilvie, K. K., and Pon, R. T. (1989) Prevention of chain cleavage in the chemical synthesis of 2'-silylated oligoribonucleotides. *Nucleic Acids Res.* **17,** 3501–3517.

70. Khorana, H. G., Turner, A. F., and Vizsolyi, J. P. (1961) Studies on polynucleotides. IX. Experiments on the polymerization of mononucleotides. Certain protected derivatives of deoxycytidine-5' phosphate and the synthesis of deoxycytidine polynucleotides. *J. Am. Chem. Soc.* **83,** 686–698.
71. Schaller, H., Weimann, G., Lerch, W. B., and Khorana, H. G. (1963) Studies on polynucleotides. XXIV. The stepwise synthesis of specific deoxyribopolynucleotides (4) Protected derivatives of deoxyribonucleosides and new syntheses of deoxyribonucleoside-3' phosphates. *J. Am. Chem. Soc.* **85,** 3821–3827.
72. Weber, H. and Khorana, H. G. (1972) Total synthesis of the structural gene for an alanine transfer ribonucleic acid from yeast. Chemical synthesis of an icosadeoxynucleotide corresponding to the nucleotide sequence 21 to 40. *J. Mol. Biol.* **72,** 219–249.

73a. Goody, R. S. and Walker, R. T. (1967) Some *N*-6 acylated cytosines. *Tetrahedron Lett.* **8,** 289–291.

73b. Goody, R. S. and Walker, R. T. (1971) The preparation and properties of some cytosine derivatives. *J. Org. Chem.* **36,** 727–730.

74. Chaix, C., Molko, D., and Teoule, R. (1989) The use of labile base protecting groups in oligoribonucleotide synthesis. *Tetrahedron Lett.* **30,** 71–74.
75. Agarwal, K. L., Yamazaki, A., Cashion, P. J., and Khorana, H. G. (1972) Chemical synthesis of polynucleotides. *Angew Chem. Internat. Edit.* **11,** 451–550.
76. Ogilvie, K. K., Schifman, A. L., and Penney, C. L. (1979) The synthesis of oligoribonucleotides. III. The use of silyl protecting groups in nucleoside and nucleotide chemistry. VIII. *Can. J. Chem.* **57,** 2230–2238.
77. Ogilvie, K. K. and Entwistle, D. W. (1981) Isomerization of tert-butyldimethylsilyl protecting groups in ribonucleosides. *Carbohydr. Res.* **89,** 203–210.
78. Letsinger, R. L. and Miller, P. S. (1969) Protecting groups for nucleosides used in synthesizing oligonucleotides. *J. Am. Chem. Soc.* **91,** 3356–3359.

79a. Ti, G. S., Gaffney, B. L., and Jones, R. A. (1982) Transient protection: efficient one-flask synthesis of protected deoxynucleosides. *J. Am. Chem. Soc.* **104,** 1316–1319.

79b. Kierzek, R. (1985) The synthesis of 5'-*O*-dimethoxytrityl-*N*-acyl-2'-deoxynucleosides. Improved "transient protection" approach. *Nucleosides Nucleotides* **4,** 641–649.

80. Guy, A., Ahmad, S., and Teoule, R. (1990) Insertion of the fragile 2'-deoxyribosylurea residue into oligodeoxynucleotides. *Tetrahedron Lett.* **31,** 5745–5748.
81. The corresponding phosphoramidites are commercialized by Pharmacia (PAC amidites) *See also* ref. *122* for possible limitations due to the perhaps incomplete cleavage of apurinic sites when smooth conditions are used for the final deprotections.
82. Piel, N., Benseler, F., Graeser, E., and McLaughlin, L. W. (1985) Synthesis of the oligodeoxyribonucleotide, d(CpTpGpGpApTpCpCpApG), and its substrate activity with the restriction endonuclease, BamHI. *Bioorg. Chem.* **13,** 323–334.

83. Dikshit, A., Chaddha, M., Singh, R. K., and Misra, K. (1988) Naphthaloyl group: a new selective amino protecting group for deoxynucleosides in oligonucleotide synthesis. *Can. J. Chem.* **66,** 2989–2994.
84. Dreef-Tromp, C. M., Hoogerhout, G. A., van der Marel, G. A., and van Boom, J. H. (1990) A new protecting group for exocyclic amino functions of nucleobases. *Tetrahedron Lett.* **31,** 427–430.
85. Kuijpers, W. H. A., Huskens, J., and van Boeckel, C. A. A. (1990) The 2-(acetoxymethyl)benzoyl (AMB) group as a new base-protecting group, designed for the protection of (phosphate) modified oligonucleotides. *Tetrahedron Lett.* **31,** 6729–6732.
86. Ogilvie, K. K., Nemer, M. J., Hakimelahi, G. H., Proba, Z. A., and Lucas, M. (1982) *N*-levunylation of nucleosides. *Tetrahedron Lett.* **23,** 2615–2618.
87. Koole, L. H., Moody, H. M., Broeders, N. L. H. L., Qaedflieg, P. J. L. M., Kuijpers, W. H. A., van Genderen, M. H. P., Coenen, A. J. J. M., van der Wal, S., and Buck, H. M. (1989) Synthesis of phosphate-methylated DNA fragments using 9-fluorenylmethoxycarbonyl as transient base protecting group. *J. Org. Chem.* **54,** 1657–1664.
88a. Ueki, M. and Amemiya, M. (1987) Removal of 9-fluorenylmethyloxycarbonyl (Fmoc) group with tetrabutylammonium fluoride. *Tetrahedron Lett.* **28,** 6617–6620.
88b. Atherton, E., Logan, C. J., and Sheppard, R. C. (1981) Peptide synthesis. Part 2. Procedures for solid-phase synthesis using Nα-fluorenylmethoxycarbonylamino-acids on polyamide supports. Synthesis of substance P and of acyl carrier protein 65-74 decapeptide. *J. Chem. Soc. Perkin Trans.* **1.** 538–546.
88c. Sabatier, J.-M., Tessier-Rochat, M., Granier, C., Van Rietschoten, J., Pedroso, E., Grandas, A., Albericio, F., and Giralt, E. (1987) Convergent solid phase peptide synthesis VI: synthesis by the FMOC procedure with a modified protocol of two protected segments, sequence 5-17 and 18-31 of the neurotoxin II of the scorpion Androctonus australis Hector. *Tetrahedron* **43,** 5973–5980.
89. Happ, E., Scalfi-Happ, C., and Chladek, S. (1987) New approach to the synthesis of 2'(3')-*O*-aminoacyl oligoribonucleotides. *J. Org. Chem.* **52,** 5387–5391.
90. Webb, T. and Matteuci, M. D. (1986) Hybridization triggered cross-linking of deoxyoligonucleotides. *Nucleic Acids Res.* **14,** 7661–7674.
91. Schneiderwind, R. G. K. and Ugi, I. (1983) Die 2,2,2-Trichlor-*t*-butyloxycarbonyl-Gruppe, eine neue *N*-Schutzgruppe für Oligonucleotidsynthesen. *Tetrahedron* **39,** 2207–2210.
92a. Watkins, B. E. and Rapoport, H. (1982) Synthesis of benzyl and benzyloxycarbonyl base-blocked 2'-deoxyribonucleosides. *J. Org. Chem.* **47,** 4471–4477.
92b. Watkins, B. E., Kiely, J. S., and Rapoport, H. (1982) Synthesis of oligodeoxyribonucleotides using *N*-benzyloxycarbonyl-blocked nucleosides. *J. Am. Chem. Soc.* **104,** 5702–5708.
93a. Hayakawa, Y., Kato, H., Uchiyama, M., Kajino, H., and Noyori, R. (1986) Allyloxycarbonyl group: a versatile blocking group for nucleotide synthesis. *J. Org. Chem.* **51,** 2400–2402.

93b. Hayakawa, Y., Wakabayashi, S., Kato, H., and Noyori, R. (1990) The allylic protection method in solid-phase oligonucleotide synthesis. An efficient preparation of solid-anchored DNA oligomers. *J. Am. Chem. Soc.* **112,** 1691–1696.

94. Heikkilä, J., Balgobin, N., and Chattopadhyaya, J. (1983) The 2-nitrophenylsulfenyl (Nps) group for the protection of amino functions of cytidine, adenosine, guanosine and their 2'-deoxysugar derivatives. *Acta Chem. Scand.* **B37,** 857–864.

95. McBride, L. J., Kierzek, R., Beaucage, S. L., and Caruthers, M. H. (1986) Amidine protecting groups for oligonucleotide synthesis. *J. Am. Chem. Soc.* **108,** 2040–2048.

96. Vu, H., McCollum, C., Jacobson, K., Theisen, P., Vinayak, R., Spiess, E., and Andrus, A. (1990) Fast oligonucleotide deprotection phosphoramidite chemistry for DNA synthesis. *Tetrahedron Lett.* **31,** 7269–7272.

97a. Sekine, M., Masuda, N., and Hata, T. (1985) Introduction of the 4,4',4"-tris(benzoyloxy)trityl group into the exo amino groups of deoxyribonucleosides and its properties. *Tetrahedron* **41,** 5445–5453.

97b. Sekine, M., Masuda, N., and Hata, T. (1986) Synthesis of oligodeoxyribonucleotides involving a rapid procedure for removal of base-protecting groups by use of the 4,4',4"-tris(benzoyloxy)trityl (TBTr) group. *Bull. Chem. Soc. Jpn.* **59,** 1781–1789.

98. The *N*-4 monomethoxytrityl derivative of cytidine is too resistant to acids. *See* ref. *50.*

99. Honda, S., Terada, K., Sato, Y., Sekine, M., and Hata, T. (1982) New type of prefabricated fully protected ribonucleotide monomer units as useful synthetic intermediates in rapid oligoribonucleotide synthesis. *Chem. Lett.* **1982,** 15–18.

100. Honda, S., Urakami, K., Koura, K., Terada, K., Sato, Y., Kohno, K., Sekine, M., and Hata, T. (1984) Synthesis of oligoribonucleotides by use of S,S-diphenyl *N*-monomethoxytrityl ribonucleoside 3'-phosphorodithioates. *Tetrahedron* **40,** 153–163.

101. Zoltewicz, J. A., Clark, D. F., Sharpless, T. W., and Grahe, G. (1970) Kinetics and mechanism of the acid-catalyzed hydrolysis of some purine nucleosides. *J. Am. Chem. Soc.* **92,** 1741–1750.

102. Zoltewicz, J. A. and Clark, D. F. (1972) Kinetics and mechanism of the hydrolysis of guanosine and 7-methylguanosine nucleosides in perchloric acid. *J. Org. Chem.* **37,** 1193–1197.

103. Panzica, R. P., Rousseau, R. J., Robins, R. K., and Townsend, L. B. (1972) A study on the relative stability and a quantitative approach to the reaction mechanism of the acid-catalyzed hydrolysis of certain 7- and 9-β-D-ribofuranosylpurines. *J. Am. Chem. Soc.* **94,** 4708–4714.

104. Hevesi, L., Wolfson-Davidson, E., Nagy, J. B., Nagy, O. B., and Bruylants, A. (1972) Contribution to the mechanism of the acid-catalyzed hydrolysis of purine nucleosides. *J. Am. Chem. Soc.* **94,** 4715–4720.

105. Jordan, F. and Niv, H. (1977) Glycosyl conformational and inductive effects on the acid catalysed hydrolysis of purine nucleosides. *Nucleic Acids Res.* **4,** 697–709.

106. Lönnberg, H. and Heikkinen, E. (1984) Mechanisms for the solvolytic decompositions of nucleoside analogs. XI. Competitive pathways for the acidic hydrolysis of 9-(β-D-ribofuranosyl)purine. *Acta Chem. Scand.* **B38,** 673–677.
107. For a review, *see* Zielonacka-Lis, E. (1989) The acidic hydrolysis of nucleosides and nucleotides. *Nucleosides, Nucleotides* **8,** 383–405.
108. York, J. L. (1981) Effect of the structure of the glycon on the acid-catalyzed hydrolysis of adenine nucleosides. *J. Org. Chem.* **46,** 2171–2173.
109. Tanaka, T. and Letsinger, R. L. (1982) Syringe method for stepwise chemical synthesis of oligonucleotides. *Nucleic Acids Res.* **10,** 3249–3260.
110. Miyoshi, K., Huang, T., and Itakura, K. (1980) Solid-phase synthesis of polynucleotides. III. Synthesis of polynucleotides with defined sequences by the block coupling phosphotriester method. *Nucleic Acids Res.* **8,** 5491–5505.
111. Itaya, T. and Harada, T. (1984) Hydrolysis of nucleosides related to tyrosine. *J. Chem. Soc. Chem. Commun.* **1984,** 858–859.
112. Remaud, G., Zhou, X.-X., Chattopadhyaya, J., Oivanen, M., and Lönnberg, H. (1987) The effect of protecting groups of the nucleobase and the sugar moieties on the acidic hydrolysis of the glycosidic bond of 2'-deoxyadenosine: a kinetic and ^{15}N NMR spectroscopic study. *Tetrahedron* **43,** 4453–4461.
113. Blank, H.-U., Frahne, D., Myles, A., and Pfleiderer, W. (1970) Über die Tritylierung und Benzylierung von Adenosin-Derivaten. *Liebigs Ann. Chem.* **742,** 34–42.
114. Barton, J. K. and Lippard, S. J. (1980) Heavy metal interactions with nucleic acids. *Nucleic Acid-Metal Ion Interactions* (Spiro, T. G., ed.), Wiley, New York, pp. 32–113.
115. Benoit, R. L. and Fréchette, M. (1984) Protonation de l'adénine, de la purine et de l'adénosine en milieu acide fort. *Can. J. Chem.* **62,** 995–1000.
116. Caruthers, M. H., McBride, L. J., Bracco, L. P., and Dubendorff, J. W. (1985) Studies on nucleotide chemistry 15. Synthesis of oligodeoxynucleotides using amidine protected nucleosides. *Nucleosides, Nucleotides* **4,** 95–105.
117a. Kume, A., Sekine, M., and Hata, T. (1982) Phthaloyl group: a new amino protecting group of deoxyadenosine in oligonucleotide synthesis. *Tetrahedron Lett.* **23,** 4365–4368.
117b. Kume, A., Sekine, M., and Hata, T. (1983) Further improvements of oligodeoxyribonucleotides synthesis: synthesis of tetradeoxyadenylate on a new silica gel support using N^6-phthaloyldeoxyadenosine. *Chem. Lett.* **1983,** 1597–1600.
118. Mc Bride, L. J. and Caruthers, M. H. (1983) N^6(*N*-methyl-2-pyrrolidine amidine) deoxyadenosine—a new deoxynucleoside protecting group. *Tetrahedron Lett.* **24,** 2953–2956.
119. Froehler, B. C. and Matteuci, M. D. (1983) Dialkylformamidines: depurination resistant N^6-protecting group for deoxyadenosine. *Nucleic Acids Res.* **11,** 8031–8036.
120. Lindhal, T. and Andersson, A. (1972) Rate of chain breakage at apurinic sites in double-stranded deoxyribonucleic acid. *Biochemistry* **11,** 3618–3623.
121. Vasseur, J.-J., Rayner, B., Imbach, J.-L., Verma, S., McCloskey, J. A., Lee, M., Chang, D.-K., and Lown, J. W. (1987) Structure of the adduct formed

between 3-aminocarbazole and the apurinic site oligonucleotide model d[Tp(Ap)pT]. *J. Org. Chem.* **52,** 4994–4998.

122. Horn, T. and Urdea, M. S. (1988) Solid supported hydrolysis of apurinic sites in synthetic oligonucleotides for rapid and efficient purification on reverse-phase cartridges. *Nucleic Acids Res.* **16,** 11,559–11,571.

123. *See also* the brief note by Efcavitch, J. W. and Heiner, C. (1985) Depurination as a yield decreasing mechanism in oligodeoxynucleotide synthesis. *Nucleosides, Nucleotides* **4,** 267.

124. Gait, M. J., Popov, S. G., Singh, M., and Titmas, R. C. (1980) Rapid synthesis of oligodeoxyribonucleotides V. Further studies in solid phase synthesis of oligodeoxyribonucleotides through phosphotriester intermediates. *Nucl. Acids Symp. Ser.* **7,** 243–257.

125. Patel, T. P., Millican, T. A., Bose, C. C., Titmas, R. C., Mock, G. A., and Eaton, M. A. W. (1982) Improvements to solid phase phosphotriester synthesis of deoxyoligonucleotides. *Nucleic Acids Res.* **10,** 5605–5620.

126a. Anzai, K., Hunt, J. B., Zon, G., and Egan, W. (1982) Reactions of ethyl and phenyl chloroformate with adenosine derivatives as an entry to N^6-ureido-linked spin-labeled adenosine and other modified adenosines. *J. Org. Chem.* **47,** 4281–4285.

126b. Anzai, K. and Uzawa, J. (1984) Cyclonucleoside formation and ring cleavage in the reaction of 2',3'-*O*-isopropylideneadenosine with benzoyl chloride and its substituted derivatives. *J. Org. Chem.* **49,** 5076–5080.

127. Lyon, P. A. and Reese, C. B. (1978) Reaction between 2',3',5'-tri-*O*-acetyl-adenosine and aryl chloroformates. 2',3',5'-tri-*O*-acetyl-N(6)-phenoxycarbonyl-adenosine as an intermediate in the synthesis of 6-ureidopurine ribosides. *J. Chem. Soc. Perkin Trans.* **1,** 131–137.

128. Himmelsbach, F. and Pfleiderer, W. (1983) The use of the *p*-nitrophenylethoxy-carbonyl group for amino protection in cytidine and adenosine chemistry. *Tetrahedron Lett.* **24,** 3583–3586.

129. Nishino, S., Takamura, H., and Ishido, Y. (1986) Regioselective protection of carbohydrate derivatives. Part 20. Simple, efficient 2'-*O*-deacylation of fully acylated purine and pyrimidine ribonucleosides through tert-butoxide. *Tetrahedron* **42,** 1995–2004.

130. Takaku, H., Morita, K., and Sumiuchi, T. (1983) Selective removal of terminal dimethoxytrityl groups. *Chem. Lett.* **1983,** 1661–1664.

131a. Sproat, B. S. and Gait, M. J. (1985) Chemical synthesis of a gene for somatomedin C. *Nucleic Acids Res.* **13,** 2959–2977.

131b. Sproat, B. S. and Brown, D. M. (1985) A new linkage for solid phase synthesis of oligodeoxyribonucleotides. *Nucleic Acids Res.* **13,** 2979–2987.

132. Kume, A., Iwase, R., Sekine, M., and Hata, T. (1984) Cyclic diacyl groups for protection of N^6-amino group of deoxyadenosine in oligodeoxynucleotide synthesis. *Nucleic Acids Res.* **12,** 8525–8538.

133. Anzai, K., and Uzawa, J. (1986) Acyl migration from N^6 to N^7 of a 2', 3'-*O*-isopropylideneadenosine derivative accompanied by cyclonucleoside formation. *Can. J. Chem. Soc.* **64,** 2109–2114.

134. Jones, A. S., Mian, A. M., and Walker, R. T. (1966) The action of alkali on some purines and their derivatives. *J. Chem. Soc. (C)* **1966,** 692–695.
135. Garrett, E. R. and Mehta, P. J. (1972) Solvolysis of adenine nucleosides. II. Effects of sugars and adenine substituents on alkaline solvolyses. *J. Am. Chem. Soc.* **94,** 8542–8547.
136. Lehikoinen, P., Mattinen, J., and Lönnberg, H. (1986) Reactions of adenine nucleosides with aqueous alkalies: kinetics and mechanism. *J. Org. Chem.* **51,** 3819–3823.
137. Gordon, M. P., Weliky, V. S., and Brown, G. B. (1957) A study of the action of acid and alkali on certain purines and purine nucleosides. *J. Chem. Soc.* **79,** 3245–3251.
138. Magrath, D. I. and Brown, G. B. (1957) The synthesis of 9-β-D-ribofuranosylpurine-5'-phosphate and its behavior toward aqueous alkali. *J. Am. Chem. Soc.* **79,** 3252–3255.
139. Montgomery, J. A. and Thomas, H. J. (1971) A new approach to the synthesis of nucleosides of 8-azapurines (3-glycosyl-v-triazolo[4,5-d]pyrimidines) *J. Org. Chem.* **36,** 1962–1967.
140. Lönnberg, H. and Lehikoinen, P. (1984) Reaction of 9-(β-D-ribofuranosyl)purine with alkalies: kinetics and mechanism. *J. Org. Chem.* **49,** 4964–4969.
141. Tener, G. M. (1961) 2-Cyanoethyl phosphate and its use in the synthesis of phosphate esters. *J. Am. Chem. Soc.* **83,** 159–168.
142. Charubala, R. and Pfleiderer, W. (1981) Nucleotides, XIII: phosphorylations of adenosine and 2'-deoxyadenosine by phosphorochloridates. *Heterocycles* **15,** 761–776.
143. Finnan, J. L., Varshney, A., and Letsinger, R. L. (1980) Developments in the phosphite-triester method of synthesis of oligonucleotides. *Nucl. Acids Symp. Ser.* **7,** 133–145.
144. Imai, J. and Torrence, P. F. (1981) Bis(2,2,2-trichloroethyl) phosphorochloridite as a reagent for the phosphorylation of oligonucleotides: preparation of 5'-phosphorylated 2',5'-oligoadenylates. *J. Org. Chem.* **46,** 4015–4021.
145. Fourrey, J.-L. and Varenne, J. (1985) Preparation and phosphorylation reactivity of *N*-nonacylated nucleoside phosphoramidites. *Tetrahedron Lett.* **26,** 2663–2666.
146. Narang, S. A., Itakura, K., and Wightman, R. H. (1972) A simplification in the synthesis of deoxyribooligonucleotides. *Can. J. Chem.* **50,** 769–770.
147. Adamiak, R. W., Biala, E., Grzekowiak, K., Kierzek, R., Kraszewski, A., Markiewicz, W. T., Okupniak, J., Stawinski, J., and Wiewiorowski, M. (1978) The chemical synthesis of the anticodon loop of an eukaryotic initiator tRNA containing the hypermodified nucleoside N^6-/*N*-threonylcarbonyl/-adenosine/t^6A/1. *Nucleic Acids Res.* **5,** 1889–1905.
148. Hayakawa, Y., Uchiyama, M., Nobori, T., and Noyori, R. (1984) A convenient synthesis of adenylyl-(2'–5')-adenylyl-(2'–5')-adenosine (2–5A core) *Nucl. Acids Symp. Ser.* **15,** 85–88.
149. Balgobin, N., Josephson, S., and Chattopadhyaya, J. B. (1981) A general approach to the chemical synthesis of oligodeoxyribonucleotides. *Acta Chem. Scand.* **B35,** 201–212.

150. Letsinger, R. L., Miller, P. S., and Grams, G. W. (1968) Selective *N*-debenzoylation of *N,O*-polybenzoylnucleosides. *Tetrahedron Lett.* **9,** 2621–2624.
151. Urdea, M. S. and Horn, T. (1986) Solid-supported synthesis, deprotection and enzymatic purification of oligodeoxyribonucleotides. *Tetrahedron Lett.* **27,** 2933–2936.
152. van Boom, J. H. and Burgers, P. M. J. (1976) Use of levulinic acid in the protection of oligonucleotides via the modified phosphotriester method: synthesis of decaribonucleotide UAUAUAUAUA. *Tetrahedron Lett.* **17,** 4875–4878.
153. As a large excess of benzoyl chloride is used to synthesize *N*-6-benzoyl-dA, in ref. *79a,* a considerable amount of benzamide is formed at the next step, when the mixture is treated with aqueous ammonia. Benzamide seriously contaminates the crude product. It is possible to remove it by taking advantage of its larger solubility in water, as compared to the protected nucleoside. Kierzek adapted Jones' method to multigram quantities, ref. *79b*.
154. Brown, E. L., Belagaje, R., Ryan, M. J., and Khorana, H. G. (1979) Chemical synthesis and cloning of a tyrosine tRNA gene. *Methods in Enzymology, vol. 68, Recombinant DNA* (Wu, R., ed.), Academic, New York, pp. 109–151.
155. Chattopadhyaya, J. B. and Reese, C. B. (1980) Chemical synthesis of a tridecanucleoside dodecaphosphate sequence of SV40 DNA. *Nucleic Acids Res.* **8,** 2039–2053.
156. Schneiderwind, R. G. K. and Ugi, I. (1981) Die 2,2,2-Trichlor-tert-butyloxycarbonyl-Gruppe als *N*-Schutzgruppe bei Oligonukleotidsynthesen. *Z. Naturforsch. B: Chem. Sci.* **36B,** 1173–1175.
157. Kamimura, T., Tsuchiya, M., Urakami, K., Koura, K., Sekine, M., Shinozaki, K., Miura, K., and Hata, T. (1984) Synthesis of a dodecaribonucleotide, GUAUCAAUAAUG, by use of "fully" protected ribonucleotide building blocks. *J. Am. Chem. Soc.* **106,** 4552–4557.
158. Kuijpers, W. H. A., Huskens, J., Koole, L. H., and van Boeckel, C. A. A. (1990) Synthesis of well-defined phosphate-methylated DNA fragments: the application of potassium carbonate in methanol as deprotecting reagent. *Nucleic Acids Res.* **18,** 5197–5205.
159. Bridson, P. K., Markiewicz, W. T., and Reese, C. B. (1977) Acylation of 2',3',5'-tri-*O*-acetylguanosine. *J. Chem. Soc. Chem. Commun.* **1977,** 791,792.
160. Robins, M. J. and Uznanski, B. (1981) Nucleic acid related compounds. 33. Conversions of adenosine and guanosine to 2,6-dichloro, 2-amino-6-chloro, and derived purine nucleosides. *Can. J. Chem.* **59,** 2601–2607.
161. Matsuda, A., Shinozaki, M., Suzuki, M., Watanabe, K., and Miyasaka, T. (1986) A convenient method for the selective acylation of guanine nucleosides. *Synthesis* **1986,** 385–386.
162. There are chemical evidences that the site of tritylation of the aglycone residue is indeed *N*-2. *See* refs. *169* and *170*.
163. Büchi, H. and Khorana, H. G. (1972) Total synthesis of the structural gene of an alanine transfer ribonucleic acid from yeast. Chemical synthesis of an icosadeoxyribonucleotide corresponding to the nucleotide sequence 31 to 50. *J. Mol. Biol.* **72,** 251–288.

164. Benseler, F. and McLaughlin, L. W. (1986) An improved procedure for the protection of 2'-deoxyguanosine. *Synthesis* **1986,** 45,46.
165. Marugg, J. E., Tromp, M., Jhurani, P., Hoyng, C. F., Van der Marel, G. A., and van Boom, J. H. (1984) Synthesis of DNA fragments by the hydroxybenzotriazole phosphotriester approach. *Tetrahedron* **40,** 73–78.
166. Takaku, H., Ito, T., and Imai, K. (1986) Use of 3,4-dimethoxybenzyl group as a protecting group for the 2'-hydroxyl group in the synthesis of oligoribonucleotides. *Chem. Lett.* **1986,** 1005–1008.
167. Borowy-Borowski, H. and Chambers, R. W. (1987) A study of side reactions occuring during synthesis of oligodeoxynucleotides containing O^6-alkyldeoxyguanosine residues at preselected sites. *Biochemistry* **26,** 2465–2471.
168. Reese, C. B. (1978) The chemical synthesis of oligo- and poly-nucleotides by the phosphotriester approach. *Tetrahedron* **34,** 3143–3179.
169. Daskalov, H. P., Sekine, M., and Hata, T. (1980) New guanosine derivatives: facile O^6-phosphorylation, thiophosphinylation, sulfonylation and silylation of guanosine derivatives by 4-dimethylaminopyridine catalized reaction. *Tetrahedron Lett.* **21,** 3899–3902.
170. Daskalov, H. P., Sekine, M., and Hata, T. (1981) Synthesis and properties of O^6-substituted guanosine derivatives. *Bull. Chem. Soc. Jpn.* **54,** 3076–3083.
171. Takaku, H., Kamaike, K., and Kasuga, K. (1982) 4-Chlorophenyl 5-chloro-8-quinolyl hydrogen phosphate: a useful phosphorylating agent for guanosine 3'-phosphotriester. *Chem. Lett.* **1982,** 197–200.
172. Sekine, M., Matsuzaki, J., Satoh, M., and Hata, T. (1982) Improved 3'-*O*-phosphorylation of guanosine derivatives by O^6-oxygen protection. *J. Org. Chem.* **47,** 571–573.
173. Gaffney, B. L. and Jones, R. A. (1982) Synthesis of 0-6-alkylated deoxyguanosine nucleosides. *Tetrahedron Lett.* **23,** 2253–2256.
174. Gaffney, B. L., Marky, L. A., and Jones, R. A. (1984) The influence of the purine 2-amino group on DNA conformation and stability, II. Synthesis and physical characterization of d[CGT(2-NH_2)ACG], d[CGU(2-NH_2)ACG], and d[CGT(2-NH_2)AT(2-NH_2)ACG]. *Tetrahedron* **40,** 3–13.
175. Sekine, M., Matsuzaki, J., and Hata, T. (1985) Oligodeoxyribonucleotide synthesis by use of S,S-diphenyl deoxyribonucleoside 3'-phosphorodithioates and bifunctional condensing reagents in the phosphotriester approach. *Tetrahedron* **41,** 5279–5288.
176. Adamiak, R. W., Biala, E., and Skalski, B. (1985) Synthesis of 6-substituted purines and ribonucleosides with *N*-(6-purinyl)pyridinium salts. *Angew Chem. Internat. Edit.* **24,** 1054,1055.
177. Gdaniec, Z., Mielewczyk, S., and Adamiak, R. W. (1988) Synthesis and carbon-13 magnetic resonance spectra of pyridinium salts derived from nucleosides and nucleobases. *Heterocycles* **27,** 2807–2814.
178. François, P., Hamoir, G., Sonveaux, E;, Vermeersch, H., and Ma, Y. (1985) On the phosphorylation of deoxyribonucleosides and the protection of deoxyguanosine. *Bull. Soc. Chim. Belg.* **94,** 821–823.

179. Pon, R. T., Damha, M. J., and Ogilvie, K. K. (1985) Necessary protection of the O^6-position of guanine during the solid phase synthesis of oligonucleotides by the phosphoramidite approach. *Tetrahedron Lett.* **26,** 2525–2528.
180. Damha, M. J. and Ogilvie, K. K. (1986) Modification of guanine bases: reaction of N^2-acylated guanine nucleosides with dichloro-(*N,N*-diisopropylamino) phosphine. *J. Org. Chem.* **51,** 3559,3560.
181. Nielsen, J., Taagaard, M., Marugg, J. E., van Boom, J. H., and Dahl, O. (1986) Application of 2-cyanoethyl N,N,N',N'-tetraisopropylphosphorodiamidite for *in situ* preparation of deoxyribonucleoside phosphoramidites and their use in polymer-supported synthesis of oligodeoxyribonucleotides. *Nucleic Acids Res.* **14,** 7391–7403.
182. Nielsen, J., Dahl, O., Remaud, G., and Chattopadhyaya, J. (1987) Phosphitylation of guanine or inosine bases during the preparation of nucleoside phosphoramidites. Isolation of model products as thiophosphoric amide derivatives and structure elucidation by ^{15}N NMR spectroscopy. *Acta Chem. Scand.* **B41,** 633–639.
183. Pon, R. T., Damha, M. J., and Ogilvie, K. K. (1985) Modification of guanine bases by nucleoside phosphoramidite reagents during the solid phase synthesis of oligonucleotides. *Nucleic Acids Res.* **13,** 6447–6465.
184. Pon, R. T., Usman, N., Damha, M. J., and Ogilvie, K. K. (1986) Prevention of guanine modification and chain cleavage during the solid phase synthesis of oligonucleotides using phosphoramidite derivatives. *Nucleic Acids Res.* **14,** 6453–6470.
185. Eadie, J. S. and Davidson, D. S. (1987) Guanine modification during chemical DNA synthesis. *Nucleic Acids Res.* **15,** 8333–8349.
186. Yeung, A. T., Dinehart, W. J., and Jones, B. K. (1988) Modifications of guanine bases during oligonucleotide synthesis. *Nucleic Acids Res.* **16,** 4539–4554.
187. Zhou, X.-X., Sandström, A., and Chattopadhyaya, J. (1986) A convenient preparation of 2-*N*-(4-*t*-butylbenzoyl)-6-*O*-(2-nitrophenyl)guanosine and its application in the synthesis of 5'(GpGpGpU)3' constituting the 3'-anticodon stem of *E. coli* tRNAIle. *Chem. Scr.* **26,** 241–249.
188. Ogilvie, K. K., Usman, N., Nicoghosian, K., and Cedergren, R. J. (1988) Total chemical synthesis of a 77-nucleotide-long RNA sequence having methionine-acceptance activity. *Proc. Natl. Acad. Sci. USA* **85,** 5764–5768.
189. Hagen, M. D. and Chladek, S. (1989) General synthesis of 2'(3')-*O*-aminoacyl oligoribonucleotides. The protection of the guanine moiety. *J. Org. Chem.* **54,** 3189–3195.
190. Kamimura, T., Tsuchiya, M., Koura, K., Sekine, M., and Hata, T. (1983) Diphenylcarbamoyl and propionyl groups: a new combination of protecting groups for the guanine residue. *Tetrahedron Lett.* **24,** 2775–2778.
191. Kwiatkowski, M., Heikkilä, J., Björkman, S., and Chattopadhyaya, J. (1983) Chemical synthesis of an undecaribonucleoside decaphosphate constituting the 3'-terminal acceptor stem sequence of yeast tRNAPhe. *Chem. Scr.* **22,** 30–48.
192. den Hartog, J. A. J., Wille, G., and van Boom, J. H. (1981) Synthesis of oligoribonucleotides with sequences identical to the nucleation region of Tobacco

Mosaic Virus RNA: preparation of AAG, AAGAAG and AAGAAGUUG via phosphotriester methods. *Recl. Trav. Chim. Pays-Bas* **100,** 320–330.

193. Fujii, M., Yamakage, S., Takaku, H., and Hata, T. (1987) (Butylthio)carbonyl group: a new protecting group for the guanine residue in oligoribonucleotide synthesis. *Tetrahedron Lett.* **28,** 5713–5716.
194. Hata, T., Sekine, M., Honda, S., and Kamimura, T. (1980) A new approach to the synthesis of oligoribonucleotides bearing 'cap' structure. *Nucl. Acids Symp. Ser.* **7,** 151–155.
195. Shimidzu, T. and Letsinger, R. L. (1968) Synthesis of deoxyguanylyl-deoxyguanosine on an insoluble polymer support. *J. Org. Chem.* **33,** 708–711.
196. Tanimura, H., Sekine, M., and Hata, T. (1986) An effective method for the preparation of O^6-substituted guanosine and N^3-substituted uridine derivatives via the corresponding stannylated intermediates. *Tetrahedron Lett.* **27,** 4047–4050.
197. Gaffney, B. L. and Jones, R. A. (1982) A new strategy for the protection of deoxyguanosine during oligonucleotide synthesis. *Tetrahedron Lett.* **23,** 2257–2260.
198. Gao, X., Gaffney, B. L., Hadden, S., and Jones, R. A. (1986) Transient protection. 2. One-flask synthesis of 6-*O*-[(4-nitrophenyl)ethyl]-2'-deoxyguanosine nucleosides. *J. Org. Chem.* **51,** 755–758.
199. Chollet, A., Ayala, E., and Kawashima, E. H. (1984) Improved synthesis of oligodeoxyribonucleotides by solid-phase phosphotriester method utilizing O^6-[2-(*p*-nitrophenyl)ethyl]-2'-deoxyguanosine derivatives. *Helv. Chim. Acta* **67,** 1356–1364.
200. Trichtinger, T., Charubala, R., and Pfleiderer, W. (1983) Synthesis of O^6-*p*-nitrophenylethyl guanosine and 2'-deoxyguanosine derivatives. *Tetrahedron Lett.* **24,** 711–714.
201. Kuzmich, S., Marky, L. A., and Jones, R. A. (1983) Specifically alkylated DNA fragments. Synthesis and physical characterization of d[CGC(O^6Me)GCG] and d[CGT(0^6Me)GCG]. *Nucleic Acids Res.* **11,** 3393–3404.
202. Kuzmich, S., Marky, L. A., and Jones, R. A. (1983) Synthesis and physical characterization of the self-complementary, alternating pyrimidine/purine hexanucleotide d[CGTACG]. *Nucleic Acids Res.* **10,** 6265–6271.
203. Sekine, M., Matsuzaki, J., and Hata, T. (1982) A new type of protection mode for the guanine residue by using 1,2-diisobutyryloxyethylene group. *Tetrahedron Lett.* **23,** 5287–5290.
204. Matsuzaki, J., Kohno, K., Tahara, S., Sekine, M., and Hata, T. (1987) Solid phase synthesis of oligodeoxyribonucleotides utilizing the phenylthio group as a phosphate protecting group. *Bull. Chem. Soc. Jpn.* **60,** 1407–1413.
205. Reese, C. B. and Skone, P. A. (1985) Action of acid on oligoribonucleotide phosphotriester intermediates. Effect of released vicinal hydroxy functions. *Nucleic Acids Res.* **13,** 5215–5231.
206. Griffin, B. E., Jarman, M., and Reese, C. B. (1968) The synthesis of oligoribonucleotides-IV. Preparation of dinucleoside phosphates from 2',5'-protected ribonucleoside derivatives. *Tetrahedron* **24,** 639–662.

207. de Rooij, J. F. M., Wille-Hazeleger, G., Burgers, P. M. J., and van Boom, J. H. (1979) Neighbouring group participation in the unblocking of phosphotriesters of nucleic acids. *Nucleic Acids Res.* **6,** 2237–2258.
208. Brown, D. M., Magrath, D. I., Neilson, A. H., and Todd, A. R. (1956) Hydrolysis of esters of monoribonucleotides. *Nature* **177,** 1124,1125.
209. Pathak, T. and Chattopadhyaya, J. (1985) The 2'-hydroxyl function assisted cleavage of the internucleotide phosphotriester bond of a ribonucleotide under acidic conditions. *Acta Chem. Scand.* **B39,** 799–806.
210. Taira, K. (1987) Stereoelectronic control in the hydrolysis of RNA by imidazole. *Bull. Chem. Soc. Jpn.* **60,** 1903–1909.
211. Reese, C. B. and Trentham, D. R. (1965) Acyl migration in ribonucleoside derivatives. *Tetrahedron Lett.* **6,** 2467–2472.
212. Rammler, D. H. and Khorana, H. G. (1962) Studies on polynucleotides. XVI. Specific synthesis of the C_3'-C_5' interribonucleotidic linkage. Examination of routes involving protected ribonucleosides and ribonucleoside-3' phosphates. Syntheses of uridylyl-(3'-5')-adenosine, uridylyl-(3'-5')cytidine, adenylyl-(3'-5')-adenosine and related compounds. *J. Am. Chem. Soc.* **84,** 3112–3122.
213. Rammler, D. H. and Khorana, H. G. (1962) Studies on polynucleotides. XX. Amino acid acceptor ribonucleic acids (1) The synthesis and properties of 2'(or 3')-*O*-(DL-phenylalanyl)-adenosine, 2'(or 3')-*O*-(DL-phenylalanyl)-uridine and related compounds. *J. Am. Chem. Soc.* **85,** 1997–2002.
214. Kempe, T., Chow, F., Sundquist, W. I., Nardi, T. J., Paulson, B., and Peterson, S. M. (1982) Selective 2'-benzoylation at the cis 2',3'-diols of protected ribonucleosides. New solid phase synthesis of RNA and DNA-RNA mixtures. *Nucleic Acids Res.* **10,** 6695–6714.
215. Norman, D. G., Reese, C. B., and Serafinowska, H. T. (1984) The protection of 2'-hydroxy functions in oligoribonucleotide synthesis. *Tetrahedron Lett.* **25,** 3015–3018.
216. Reese, C. B., Saffhill, R., and Sulston, J. E. (1967) A symmetrical alternative to the tetrahydropyranyl protecting group. *J. Am. Chem. Soc.* **89,** 3366–3368.
217. Reese, C. B., Saffhill, R., and Sulston, J. E. (1970) 4-Methoxytetrahydropyran-4-yl. A symmetrical alternative to the tetrahydropyranyl protecting group. *Tetrahedron* **26,** 1023–1030.
218. Griffin, B. E. and Reese, C. B. (1964) Oligoribonucleotide synthesis via 2',5'-protected ribonucleoside derivatives. *Tetrahedron Lett.* **5,** 2925–2931.
219. Sandström, A., Kwiatkowski, M., and Chattopadhyaya, J. (1985) Chemical synthesis of a pentaribonucleoside tetraphosphate constituting the 3'-acceptor stem sequence of *E. coli* $tRNA^{Ile}$ using 2'-*O*-(3-methoxy-1,5-dicarbomethoxypentan-3-yl)-ribonucleoside building blocks. Application of a new achiral and acid-labile 2'-hydroxyl protecting group in tRNA synthesis. *Acta Chem. Scand.* **B39,** 273–290.
220. Ohtsuka, E., Yamane, A., Doi, T., and Ikehara, M. (1984) Chemical synthesis of the 5'-half molecule of *E. coli* $tRNA_2^{Gly}$. *Tetrahedron* **40,** 47–57.

221. Ohtsuka, E., Yamane, A., and Ikehara, M. (1983) Studies on transfer ribonucleic acids and related compounds. XLV. Block condensation of ribooligonucleotides containing 2'-*O*-tetrahydrofuranyl–5'-*O*-dimethoxytritylnucleosides. *Nucleic Acids Res.* **11,** 1325–1335.
222. Tanaka, T. Fujino, K., Tamatsukuri, S., and Ikehara, M. (1986) Synthesis of oligoribonucleotides via the phosphite-triester approach on a polymer support. *Chem. Pharm. Bull.* **34,** 4126–4132.
223. Smith, M., Rammler, D. H., Goldberg, I. H., and Khorana, H. G. (1962) Studies on polynucleotides. XIV. Specific synthesis of the C_3'-C_5' interribonucleotidic linkage. Synthesis of uridylyl-(3'–5')-uridine and uridylyl-(3'–5')-adenosine. *J. Am. Chem. Soc.* **84,** 430–440.
224. Smrt, J. (1973) Oligonucleotidic compounds. XLIV. Protection of the internucleotidic bond after its synthesis as approach to the synthesis of an oligonucleotidic chain. *Collect. Czech. Chem. Commun.* **38,** 3189–3197.
225. Seliger, H., Zeh, D., Azuru, G., and Chattopadhyaya, J. (1983) Two new and efficient routes to the preparation of oligoribonucleotides of defined sequence. *Chem. Scr.* **22,** 95–101.
226. Takaku, H., Yoshida, M., and Nomoto, T. (1983) A convenient synthesis of the 3'-terminal nonaoligoribonucleotide of Rous Sarcoma Virus 35S RNA via the modified phosphotriester approach. *J. Org. Chem.* **48,** 1399–1403.
227a. Hirao, I., Ishikawa, M., and Miura, K. (1986) Partial synthesis of leader sequence of phage f1 coat protein mRNA. *Chem. Lett.* **1986,** 1929–1932.
227b. Hirao, I. and Miura, K. (1989) Synthesis of oligoribonucleotides by the hydroxybenzotriazole-activated phosphotriester/dicyclohexylcarbodiimide system. *Chem. Lett.* **1989,** 1799–1802.
227c. Hirao, I., Ishikawa, M., Hori, H., Watanabe, K., and Miura, K. (1989) Synthesis of nonadeca- and octadecaribonucleotides using the solid-phase phosphotriester with tetrahydropyranyl groups as the 2'-hydroxyl-protecting group. *Bull. Chem. Soc. Jpn.* **62,** 1995–2001.
228. Kierzek, R., Caruthers, M. H., Longfellow, C. E., Swinton, D., Turner, D. H., and Freier, S. M. (1986) Polymer-supported RNA synthesis and its application to test the nearest-neighbor model for duplex stability. *Biochemistry* **25,** 7840–7846.
229. Tanimura, H., Maeda, M., Fukazawa, T., Sekine, M., and Hata, T. (1989) Chemical synthesis of the 24 RNA fragments corresponding to hop stunt viroid. *Nucleic Acids Res.* **17,** 8135–8147.
230. Christodoulou, C., Agrawal, S., and Gait, M. J. (1986) Incompatibility of acid-labile 2' and 5' protecting groups for solid-phase synthesis of oligoribonucleotides. *Tetrahedron Lett.* **27,** 1521,1522.
231. Chou, S.-H., Flynn, P., and Reid, B. (1989) Solid-phase synthesis and high-resolution NMR studies of two synthetic double-helical RNA dodecamers: r(CGCGAAUUCGCG) and r(CGCGUAUACGCG) *Biochemistry* **28,** 2422–2435.
232. Tanimura, H., Sekine, M., and Hata, T. (1987) Feature of internucleotidic 2', 3'-cyclicphosphate intermediates during liquid and solid phase synthesis of RNA

fragments and practical synthesis of octaadenylate in the phosphoramidite approach. *Chem. Lett.* 1057–1060.

233. At variance. *See* ref. *214.*

234. Reese, C. B., Rao, M. V., Serafinowska, H. T., Thompson, E. A., and Yu, P. S. (1991) Studies in the solid phase synthesis of oligo- and poly-ribonucleotides. *Nucleosides, Nucleotides* **10,** 81–97.

235. Ma, Y. and Sonveaux, E. (1989) The 9-fluorenylmethyloxycarbonyl group as a 5'-OH protection in oligonucleotide synthesis. *Biopolymers* **28,** 965–973.

236. Lehmann, C., Xu, Y.-Z., Christodoulou, C., Tan, Z.-K., and Gait, M. J. (1989) Solid-phase synthesis of oligoribonucleotides using 9-fluorenylmethoxycarbonyl (Fmoc) for 5'-hydroxyl protection. *Nucleic Acids Res.* **17,** 2379–2390.

237. Fukuda, T., Hamana, T., and Marmoto, R. (1988) Synthesis of RNA oligomer using 9-fluorenylmethoxycarbonyl (Fmoc) group for 5'-hydroxyl protection. *Nucl. Acids Symp. Ser.* **19,** 13–16.

238a. Iwai, S. and Ohtsuka, E. (1988) Synthesis of oligoribonucleotides by the phosphoramidite approach using 5'-levulinyl and 2'-tetrahydrofuranyl protection. *Tetrahedron Lett.* **29,** 5383–5386.

238b. Iwai, S. and Ohtsuka, E. (1988) 5'-Levulinyl and 2'-tetrahydrofuranyl protection for the synthesis of oligoribonucleotides by the phosphoramidite approach. *Nucleic Acids Res.* **16,** 9443–9456.

239. van Boom, J. H. (1977) Synthesis of oligoribonucleotides via phosphotriester intermediates. *Heterocycles* **7,** 1197–1226.

240. van Boom, J. H. and Burgers, P. M. J. (1978) The application of levulinic acid as protective group to the synthesis of tetradecaribonucleotide UAUAUAUAUAUAUA via the modified phosphotriester method. *Recl. Trav. Chim. Pays-Bas* **97,** 73–80.

241. van Boom, J. H., Burgers, P. M. J., Verdegaal, C. H. M., and Wille, G. (1978) Synthesis of oligonucleotides with sequences identical with or analogous to the 3'-end of 16S ribosomal RNA of *Escherichia coli*: preparation of UCCUUA and ACCUCCUUA via the modified phosphotriester method. *Tetrahedron* **34,** 1999–2007.

242. den Hartog, J. A. J. and van Boom, J. H. (1981) Chemical synthesis of a 5' and 3'/2'-phosphorylated heptamer sequence of *E. coli* $tRNA_f^{met}$: pAGCCUGG(34/2')p via phosphotriester methods. *Recl. Trav. Chim. Pays-Bas* **100,** 275–284.

243. van der Marel, G. A., Wille, G., and van Boom, J. H. (1982) Solid-phase synthesis of the RNA fragment: rAAGAAGAAGA. *Recl. Trav. Chim. Pays-Bas* **101,** 241–246.

244. Neilson, T. and Werstiuk, E. S. (1974) Synthesis of the anticodon loop of *Escherichia coli* Methionine transfer ribonucleic acid. *J. Am. Chem. Soc.* **96,** 2295–2297.

245. Robins, M. J., Wilson, J. S., and Hansske, F. (1983) Nucleic acid related compounds. 42. A general procedure for the efficient deoxygenation of secondary alcohols. Regiospecific and stereoselective conversion of ribonucleosides to 2'-deoxynucleosides. *J. Am. Chem. Soc.*,**105,** 4059–4065.

246. van Boom, J. H. and Wreesmann, C. T. J. (1984) Chemical synthesis of small oligoribonucleotides in solution. *Oligonucleotides Synthesis: A Practical Approach* (Gait, M. J., ed.), IRL, Oxford and Washington DC, pp. 153–183.
247. Verdegaal, C. H. M., Jansse, P. L., de Rooij, J. F. M., Veeneman, G., and van Boom, J. H. (1981) A convenient synthesis of 2'-*O*-acetal-N^2-acyl derivatives of riboguanosine. *Recl. Trav. Chim. Pays-Bas* **100,** 200–204.
248. Ohtsuka, E., Ohkubo, M., Yamane, A., and Ikehara, M. (1983) Studies on transfer ribonucleic acids and related compounds. XLIV. A large-scale synthesis of the anticodon heptanucleotide of formyl-methionine transfer ribonucleic acid by using 2'-*O*-tetrahydrofuranylnucleosides. *Chem. Pharm. Bull.* **31,** 1910–1916.
249. Markiewicz, W. T., Nowakowska, B., and Adrych, K. (1988) Tetra-*t*-butoxydisiloxane-1,3-diyl, a new type of bifunctional silyl protective group. *Tetrahedron Lett.* **29,** 1561–1564.
250a. Markiewicz, W. T. (1979) Tetraisopropyldisiloxane-1,3-diyl, a group for simultaneous protection of 3'- and 5'-hydroxy functions of nucleosides. *J. Chem. Res. Synop.* **1979,** 24,25; *J. Chem. Res. (M)* **1979,** 181–197.
250b. Markiewicz, W. T., Biala, E., and Kierzek, R. (1984) Application of the tetraisopropyldisiloxane-1,3-diyl group in the chemical synthesis of oligoribonucleotides. *Bull. Pol. Acad. Sci. Chem.* **32,** 433–451.
251. Tanaka, T., Tamatsukuri, S., and Ikehara, M. (1987) Solid phase synthesis of oligoribonucleotides using the *o*-nitrobenzyl group for 2'-hydroxyl protection and H-phosphonate chemistry. *Nucleic Acids Res.* **15,** 7235–7248.
252a. Hayes, J. A., Brunden, M. J., Gilham, P. T., and Gough, G. R. (1985) High-yield synthesis of oligoribonucleotides using *o*-nitrobenzyl protection of 2'-hydroxyls. *Tetrahedron Lett.* **26,** 2407–2410.
252b. Hayes, J. A., Gough, G. R., and Gilham, P. T. (1989) Photochemistry of the *o*-nitrobenzyl protecting group in RNA synthesis. *Nucleosides, Nucleotides* **8,** 1071,1072.
253. Tanaka, T., Orita, M., Uesugi, S., and Ikehara, M. (1988) Solid phase synthesis of oligoribonucleotides using *o*-nitrobenzyl protection of 2'-hydroxyl via a phosphotriester approach. *Tetrahedron* **44,** 4331–4338.
254. Bartholomew, D. G. and Broom, A. D. (1975) One-step chemical synthesis of ribonucleosides bearing a photolabile ether protecting group. *J. Chem. Soc. Chem. Commun.* **1975,** 38.
255. Ohtsuka, E., Fujiyama, K., and Ikehara, M. (1981) Studies on transfer ribonucleic acids and related compounds. XL. Synthesis of an eicosaribonucleotide corresponding to residues 35-54 of $tRNA_f^{Met}$ from *E. coli*. *Nucleic Acids Res.* **9,** 3503–3522.
256. Ohtsuka, E., Tanaka, S., and Ikehara, M. (1974) Studies on transfer ribonucleic acids and related compounds. IX. Ribooligonucleotide synthesis using a photosensitive *o*-nitrobenzyl protection at the 2'-hydroxyl group. *Nucleic Acids Res.* **1,** 1351–1357.
257. Ohtsuka, E., Tanaka, S., and Ikehara, M. (1978) Studies on transfer ribonucleic acids and related compounds. 23. Synthesis of a heptanucleotide corresponding to a eukaryotic initiator tRNA loop sequence. *J. Am. Chem. Soc.* **100,** 8210–8213.

258. Ohtsuka, E., Tanaka, S., and Ikehara, M. (1979) Studies on transfer ribonucleic acids and related compounds. XXXII. Synthesis of ribonucleotides corresponding to residues 1-5 and 6-10 of $tRNA_f^{Met}$ from *E. coli* and their base conversion analogs. *Nucleic Acids Res.* **7,** 1283–1296.
259. Ohtsuka, E., Tanaka, S., and Ikehara, M. (1979) Studies on transfer ribonucleic acids and related compounds. 29. Synthesis of a decaribonucleotide of *Escherichia coli* $tRNA_f^{Met}$ (bases 11-20) using a new phosphorylating reagent. *J. Am. Chem. Soc.* **101,** 6409–6414.
260. Ohtsuka, E., Takashima, H., and Ikehara, M. (1981) Solid phase synthesis of ribo-oligonucleotides on a polyacrylmorpholide support. *Tetrahedron Lett.* **22,** 765–768.
261. Tanaka, T., Tamatsukuri, S., and Ikehara, M. (1986) Solid phase synthesis of oligoribonucleotides using *o*-nitrobenzyl protection of 2'-hydroxyl via a phosphite triester approach. *Nucleic Acids Res.* **14,** 6265–6279.
262. Ohtsuka, E., Tanaka, S., and Ikehara, M. (1977) Studies on transfer ribonucleic acids and related compounds. XVIII. A photolabile 2'-ether of guanosine as an intermediate for oligonucleotide synthesis. *Synthesis* **1977,** 453–454.
263. Ohtsuka, E., Tanaka, T., Tanaka, S., and Ikehara, M. (1978) Studies on transfer ribonucleic acids and related compounds. 20. A new versatile ribooligonucleotide block with 2'-(*o*-nitrobenzyl) and 3'-phosphorodianilidate groups suitable for elongation of chains in the 3' and 5' directions. *J. Am. Chem. Soc.* **100,** 4580–4584.
264. Ohtsuka, E., Tanaka, S., and Ikehara, M. (1977) Studies on transfer ribonucleic acids and related compounds. XVI. Synthesis of ribooligonucleotides using a photosensitive *O*-nitrobenzyl protection for the 2'-hydroxyl group. *Chem. Pharm. Bull.* **25,** 949–959.
265. Ohtsuka, E., Wakabayashi, T., Tanaka, S.,Tanaka, T., Oshie, K., Hasegawa, M., and Ikehara, M. (1981) Studies on tRNA and related compounds. XXXVII. Synthesis and physical properties of 2'- or 3'-*O*-(*o*-nitrobenzyl)nucleosides: the use of *o*-nitrophenyldiazomethane as a synthetic reagent. *Chem. Pharm. Bull.* **29,** 318–324.
266. Flockerzi, D., Silber, G., Charubala, R., Schlosser, W., Varma, R. S., Creegan, F., and Pfleiderer, W. (1981) Nucleoside, XXXVII. Synthese und Eigenschaften von 2'-*O*- und 3'-*O*-(tert-Butyldimethylsilyl)–5'-*O*-(4-methoxytrityl)- sowie 2',3'-bis-*O*-(tert-Butyldimethylsilyl)ribonucleosiden, Ausgangssubstanzen für Oligoribonucleotid-Synthesen. *Liebigs Ann. Chem.* **1981,** 1568–1585.
267. Ogilvie, K. K., Theriault, N., and Sadana, K. L. (1977) Synthesis of oligoribonucleotides. *J. Am. Chem. Soc.* **99,** 7741–7743.
268. Ogilvie, K. K., Beaucage, S. L., Schifman, A. L., Theriault, N. Y., and Sadana, K. L. (1978) The synthesis of oligoribonucleotides. II. The use of silyl protecting groups in nucleoside and nucleotide chemistry. VII. *Can. J. Chem.* **56,** 2768–2780.
269. Ogilvie, K. K., Sadana, K. L., Thompson, E. A., Quilliam, M. A., and Westmore, J. B. (1974) The use of silyl groups in protecting the hydroxyl functions of ribonucleosides. *Tetrahedron Lett.* **15,** 2861–2863.
270. Usman, N., Ogilvie, K. K., Jiang, M. Y., and Cedergren, R. J. (1987) Automated chemical synthesis of long oligoribonucleotides using 2'-*O*-silylated

ribonucleoside 3'-*O*-phosphoramidites on a controlled-pore glass support: synthesis of a 43-nucleotide sequence similar to the 3'-half molecule of an *Escherichia coli* formylmethionone tRNA. *J. Am. Chem. Soc.* **109,** 7845–7854.
271. Hakimelahi, G. H., Proba, Z. A., and Ogilvie, K. K. (1982) New catalysts and procedures for the dimethoxytritylation and selective silylation of ribonucleosides. *Can. J. Chem.* **60,** 1106–1113.
272. Sadana, K. L., and Loewen, P. C. (1978) A rapid and efficient synthesis of ribonucleotides. *Tetrahedron Lett.* **19,** 5095–5098.
273. Köhler, W. and Pfleiderer, W. (1979) Nucleoside, XXX. Synthese und Eigenschaften von 4-*N*-Benzoyl-*O*-(tert-butyldimethylsilyl)cytidinen. *Liebigs Ann. Chem.* **1979,** 1855–1871.
274. Jones, S. S. and Reese, C. B. (1979) Migration of *t*-butyldimethylsilyl protecting groups. *J. Chem. Soc. Perkin Trans.* **1,** 2762–2764.
275. Silber v. G., Flockerzi, D., Varma, R. S., Charubala, R., Uhlmann, E., and Pfleiderer, W. (1981) Nucleotide, XV. Synthese und Eigenschaften von 2'-*O*-*t*-Butyldimethylsilyl-5'-*O*-monomethoxytritylribonucleosid-3'-phosphotriestern, Ausgangssubstanzen für Oligonucleotid-Synthesen. *Helv. Chim. Acta* **64,** 1704–1716.
276. Flockerzi, D., Schlosser, W., and Pfleiderer, W. (1983) Nucleotide, XVIII. Synthese und Eigenschaften von (tert-Butyldimethylsilyl)guanosinen, Guanosin-3'-phosphotriestern und Guanosin-haltigen Oligonucleotiden. *Helv. Chim. Acta* **66,** 2069–2085.
277. Scaringe, S. A., Francklyn, C., and Usman, N. (1990) Chemical synthesis of biologically active oligoribonucleotides using β-cyanoethyl protected ribonucleoside phosphoramidites. *Nucleic Acids Res.* **18,** 5433–5441.
278. Wu, T. and Ogilvie, K. K. (1990) A study on the alkylsilyl groups in oligoribonucleotide synthesis. *J. Org. Chem.* **55,** 4717–4724.
279. Milecki, J., Dembek, P., and Antkowiak, W. Z. (1989) On the application of *t*-butyldimethylsilyl group in chemical RNA synthesis. Part I. 31P NMR study of 2'-*O*-*t*-BDMSi group migration during nucleoside 3'-OH phosphorylation and phosphitylation reactions. *Nucleosides Nucleotides* **8,** 463–474.
280. Flockerzi, D., Uhlmann, E., and Pfleiderer, W. (1983) Nucleotide, XVII. Synthese von homogenen Adenosyl-3',5'-Oligomeren nach dem Phosphotriester-Verfahren. *Helv. Chim. Acta* **66,** 2018–2030.
281. Ichiba, M., Charubala, R., Varma, R. S., and Pfleiderer, W. (1986) Nucleotides, part XXV. Chemical synthesis of the hexaribonucleotide CAACCA. *Helv. Chim. Acta* **69,** 1768–1777.
282. Ogilvie, K. K., Theriault, N. Y., Seifert, J.-M., Pon, R. T., and Nemer, M. J. (1980) The chemical synthesis of oligoribonucleotides. IX. A comparison of protecting groups in the dichloridite procedure. *Can. J. Chem.* **58,** 2686–2693.
283. Usman, N., Pon, R. T., and Ogilvie, K. K. (1985) Preparation of ribonucleoside 3'-*O*-phosphoramidites and their application to the automated solid phase synthesis of oligonucleotides. *Tetrahedron Lett.* **26,** 4567–4570.
284. Iwai, S., Sasaki, T., and Ohtsuka, E. (1990) Large scale synthesis of oligoribonucleotides on a solid support: synthesis of a catalytic RNA duplex. *Tetrahedron* **46,** 6673–6688.

285. Wu, T., Ogilvie, K. K., Perreault, J.-P., and Cedergren, R. J. (1989) Convenient procedure for the preparation of specific mixed DNA-RNA polymers. *J. Am. Chem. Soc.* **111,** 8531–8533.
286a. Takaku, H. and Kamaike, K. (1982) Synthesis of oligoribonucleotides using 4-methoxybenzyl group as a new protecting group of the 2'-hydroxyl group of adenosine. *Chemistry Lett.* **1982,** 189–192.
286b. Takaku, H., Kamaike, K., and Tsuchiya, H. (1984) Synthesis of ribonucleotides using the 4-methoxybenzyl group as a new protecting group for the 2'-hydroxyl group. *J. Org. Chem.* **49,** 51–56.
287. Garegg, P. J., Lindh, I., Regberg, T., Stawinski, J., Strömberg, R., and Henrichson, C. (1986) Nucleoside H-phosphonate. IV. Automated solid phase synthesis of oligoribonucleotides by the hydrogenphosphonate approach. *Tetrahedron Lett.* **27,** 4055–4058.
288. Agrawal, S. and Tang, J.-Y. (1990) Efficient synthesis of oligoribonucleotide and its phosphorothioate analog using H-phosphonate approach. *Tetrahedron Lett.* **31,** 7541–7544.
289. *See*, however, ref. *14*.
290. Ogilvie, K. K. and Theriault, N. Y. (1979) The synthesis of oligoribonucleotides V. The stepwise synthesis of the 3'-terminal heptanucleotide sequence of $tRNA^{fMet}$ from *E. coli*. *Can. J. Chem.* **57,** 3140–3144.
291. Ogilvie, K. K. and Nemer, M. J. (1980) The synthesis of oligoribonucleotides VI. The synthesis of a hexadecamer by a block condensation approach. *Can. J. Chem.* **58,** 1389–1397.
292. Ogilvie, K. K. and Nemer, M. J. (1980) Silica gel as solid support in the synthesis of oligoribonucleotides. *Tetrahedron Lett.* **21,** 4159–4162.
293. Ogilvie, K. K., Nemer, M. J., and Gillen, M. F. (1984) Large scale bench-top synthesis of a nineteen unit ribonucleotide on silica gel. *Tetrahedron Lett.* **25,** 1669–1672.
294. Stawinski, J. Strömberg, R., Thelin, M., and Westman, E. (1988) Studies on the *t*-butyldimethylsilyl group as 2'-*O*-protection in oligoribonucleotide synthesis via H-phosphonate approach. *Nucleic Acids Res.* **16,** 9285–9298.
295. Wang, Y., Lyttle, M. H., and Borer, P. N. (1990) Enzymatic and NMR analysis of oligoribonucleotides synthesized with 2'-tert-butyldimethylsilyl protected cyanoethylphosphoramidite monomers. *Nucleic Acids Res.* **18,** 3347–3352.
296. Milligen/Biosearch, a division of Millipore, and Isogen Bioscience (Amsterdam, the Netherlands)
297. van Boom, J. H., Burgers, P. M. J., and Haasnoot, C. A. G. (1977) General method for the synthesis of 3', 5'-diesters and 2'-acetals of the four common nucleosides. *Recl. Trav. Chim. Pays-Bas* **96,** 91–120.
298a. Neilson, T. and Werstiuk, E. S. (1971) Oligoribonucleotide synthesis II. Preparation of 2'-tetrahydropyranyl derivatives of adenosine and cytidine necessary for insertion in stepwise synthesis. *Can. J. Chem.* **49,** 493–499.
298b. Neilson, T. and Werstiuk, E. S. (1971) Oligoribonucleotide synthesis III. Synthesis of trinucleotides using a stepwise phosphotriester method. *Can. J. Chem.* **49,** 3004–3011.

298c. Neilson, T., Wastrodowski, E. V., and Werstiuk, E. S. (1973) Oligoribonucleotide synthesis V. Preparation of 2'-tetrahydropyranyl derivatives of guanosine and their insertion into a general stepwise synthesis. *Can. J. Chem.* **51,** 1068–1074.

298d. Gregoire, R. J. and Neilson, T. (1978) Oligoribonucleotide synthesis. XI. Improved preparation of 2'-tetrahydropyranyl derivatives of guanosine and adenosine necessary for insertion in phosphotriester synthesis. *Can. J. Chem.* **56,** 487–490.

299a. England, T. E. and Neilson, T. (1976) Oligoribonucleotide synthesis. IX. Synthesis of sequences corresponding to the dihydrouridine loop neck region common in several transfer RNA molecules. *Can. J. Chem.* **54,** 1714–1721.

299b. Werstiuk, E. S. and Neilson, T. (1976) Oligoribonucleotide synthesis. X. An improved synthesis of the anticodon loop region of methionine transfer ribonucleic acid from *E. coli. Can. J. Chem.* **54,** 2689–2696.

300. van Boom, J. H., Owen, G. R., Preston, J., Ravindranathan, T., and Reese, C. B. (1971) The synthesis of oligoribonucleotides. Part IX. Preparation of ribonucleoside 2'-acetal 5'-esters. *J. Chem. Soc. (C),* **1971,** 3230–3237.

301. Jones, S. S., Rayner, B., Reese, C. B., Ubasawa, A., and Ubasawa, M. (1980) Synthesis of the 3'-terminal decaribonucleoside nonaphosphate of yeast alanine transfer ribonucleic acid. *Tetrahedron* **36,** 3075–3085.

302. Reese, C. B., Serafinowska, H. T., and Zappia, G. (1986) An acetal group suitable for the protection of 2'-hydroxy functions in rapid oligoribonucleotide synthesis. *Tetrahedron Lett.* **27,** 2291–2294.

303. Rao, T. S., Reese, C. B., Serafinowska, H. T., Takaku, H., and Zappia, G. (1987) Solid phase synthesis of the 3'-terminal nonadecaribonucleoside octadecaphosphate sequence of yeast alanine transfer ribonucleic acid. *Tetrahedron Lett.* **28,** 4897–4900.

304. Sekine, M. and Hata, T. (1983) Cyclic orthoester functions as new protecting groups in nucleosides. *J. Am. Chem. Soc.* **105,** 2044–2049.

305. Sekine, M. and Nakanishi, T. (1991) Oligoribonucleotide synthesis by use of the [[2-(methylthio)phenyl]thio]methyl (MPTM) group as the 2'-hydroxyl protecting group. *Chem. Lett.* **1991,** 121–124.

306. Iwai, S., Yamada, E., Asaka, M., Hayase, Y., Inoue, H., and Ohtsuka, E. (1987) A new solid-phase synthesis of oligoribonucleotides by the phosphoro-*p*-anisidate method using tetrahydrofuranyl protection of 2'-hydroxyl groups. *Nucleic Acids Res.* **15,** 3761–3772.

307. Ohtsuka, E., Yamane, A., and Ikehara, M. (1983) Studies on transfer ribonucleic acids and related compounds. XLIII. Synthesis of oligoribonucleotides by using 5'-selective phosphorylation of 2'-*O*-tetrahydrofuranyl nucleosides. *Chem. Pharm. Bull.* **31,** 1534–1543.

308. Yamakage, S., Sakatsume, O., Furuyama, E., and Takaku, H. (1989) 1-(2-Chloroethoxy)ethyl group for the protection of 2'-hydroxyl group in the synthesis of oligoribonucleotides. *Tetrahedron Lett.* **30,** 6361–6364.

309. Sakatsume, O., Ogawa, T., Hosaka, H., Kawashima, M., Takaki, M., and Takaku, H. (1991) Synthesis and properties of non-hammerhead RNA using 1-

(2-chloroethoxy)-ethyl group for the protection of 2'-hydroxyl function. *Nucleosides, Nucleotides* **10,** 141–153.
310. Kwiatkowski, M. and Chattopadhyaya, J. (1982) An efficient synthesis of adenylyl-(3'–5')-adenosine through the phosphotriester approach. *Chem. Scr.* **20,** 139–141.
311. Balgobin, N., Kwiatkowski, M., and Chattopadhyaya, J. (1982) A novel strategy for the chemical synthesis of DNA and RNA fragments using 2-oxymethyleneanthraquinone (MAQ) as a 3'-terminal phosphate protecting group. *Chem. Scr.* **20,** 198–200.
312. Griffin, B. E., Reese, C. B., Stephenson, G. F., and Trentham, D. R. (1966) Oligoribonucleotide synthesis from nucleoside 2'-*O*-benzyl ethers. *Tetrahedron Lett.* **7,** 4349–4354.
313. Charubala, R. and Pfleiderer, W. (1987) 2'-OH protection by the *p*-nitrophenylethylsulfonyl (NPES) group in oligoribonucleotide synthesis. *Nucleosides, Nucleotides* **6,** 517–520.
314. Chattopadhyaya, J. B. and Reese, C. B. (1978) The 9-phenylxanthen-9-yl protecting group. *J. Chem. Soc. Chem. Commun.* **1978,** 639,640.
315. Kwiatkowski, M. and Chattopadhyaya, J. (1984) The 9-(4-octadecyloxyphenylxanthen)-9-yl- group. A new acid-labile hydroxyl protective group and its application in the preparative reverse-phase chromatographic separation of oligoribonucleotides. *Acta Chem. Scand.* **B38,** 657–671.
316. Kohli, V., Blöcker, H., and Köster, H. (1980) The triphenylmethyl (trityl) group and its uses in nucleotide chemistry. *Tetrahedron Lett.* **21,** 2683–2686.
317. Chaudhary, S. K. and Hernandez, O. (1979) A simplified procedure for the preparation of triphenylmethyl ethers. *Tetrahedron Lett.* **20,** 95–98.
318. Reddy, M. P., Rampal, J. B., and Beaucage, S. L. (1987) An efficient procedure for the solid phase tritylation of nucleosides and nucleotides. *Tetrahedron Lett.* **28,** 23–26.
319. Gough, G. R., Collier, K. J., Weith, H. L., and Gilham, P. T. (1979) The use of barium salts of protected deoxyribonucleoside-3' *p*-chlorophenyl phosphates for construction of oligonucleotides by the phosphotriester method: high-yield synthesis of dinucleotide blocks. *Nucleic Acids Res.* **7,** 1955–1964.
320. Kierzek, R., Ito, H., Bhatt, R., and Itakura, K. (1981) Selective *N*-deacylation of *N,O*-protected nucleosides by zinc bromide. *Tetrahedron Lett.* **22,** 3761–3764.
321. Matteuci, M. D., and Caruthers, M. H. (1980) The use of zinc bromide for removal of dimethoxytrityl ethers from deoxynucleosides. *Tetrahedron Lett.* **21,** 3243–3246.
322. Ito, H., Ike, Y., Ikuta, S., and Itakura, K. (1982) Solid phase synthesis of polynucleotides. VI. Further studies on polystyrene copolymers for the solid support. *Nucleic Acids Res.* **10,** 1755–1769.
323. Sproat, B. S. and Bannwarth, W. (1983) Improved synthesis of oligodeoxynucleotides on controlled pore glass using phosphotriester chemistry and a flow system. *Tetrahedron Lett.* **24,** 5771–5774.

324. Adams, S. P., Kavka, K. S., Wykes, E. J., Holder, S. B., and Galluppi, G. R. (1983) Hindered dialkylamino nucleoside phosphite reagents in the synthesis of two DNA 51-mers. *J. Am. Chem. Soc.* **105,** 661–663.
325. Letsinger, R. L., Groody, E. P., Lander, N., and Tanaka, T. (1984) Some developments in the phosphite-triester method for synthesis of oligonucleotides. *Tetrahedron* **40,** 137–143.
326. Mitchell, M. J., Hirschowitz, W., Rastinejad, F., and Lu, P. (1990) Boron trifluoride-methanol complex as a non-depurinating detritylating agent in DNA synthesis. *Nucleic Acids Res.* **18,** 5321.
327. de Rooij, J. F. M., Burgers, P. M. J., Wille-Hazeleger, G., and van Boom, J. H. (1978) Formation of 5'-amino-deoxyribonucleoside monophosphates in the de-blocking of oligonucleotides via intermediate aryl phosphotriesters. *Nucleic Acids Res.* Spec. Publ. **4,** s37–s40.
328a. Gioeli, C. and Chattopadhyaya, J. B. (1982) The fluoren-9-yl-methoxycarbonyl group for the protection of hydroxy-groups; its application in the synthesis of an octathymidylic acid fragment. *J. Chem. Soc. Chem. Commun.* **1982,** 672–674.
328b. Balgobin, N. and Chattopadhyaya, J. B. (1987) Solid phase synthesis of DNA under a non-depurinating condition with a base labile 5'-protecting group (Fmoc) using phosphiteamidite approach. *Nucleosides, Nucleotides* **6,** 461–463.
329. Reese, C. B., Titmas, R. C., and Yau, L. (1978) Oximate ion promoted unblocking of oligonucleotide phosphotriester intermediates. *Tetrahedron Lett.* **19,** 2727–2730.
330. Ogilvie, K. K., Beaucage, S. L., and Entwistle, D. W. (1976) A facile method for the removal of phosphate protecting groups in nucleotide synthesis. *Tetrahedron Lett.* **17,** 1255,1256.
331. Reese, C. B. and Stewart, J. C. M. (1968) Methoxyacetyl as a protecting group in ribonucleotide chemistry. *Tetrahedron Lett.* **9,** 4273–4276.
332. Arentzen, R. and Reese, C. B. (1977) The phosphotriester approach to oligonucleotide synthesis: preparation of oligo- and poly-thymidylic acids. *J. Chem. Soc. Perkin Trans.* **1,** 445–460.
333. Letsinger, R. L. and Lunsford, W. B. (1976) Synthesis of thymidine oligonucleotides by phosphite triester intermediates. *J. Am. Chem. Soc.* **98,** 3655–3661.
334. Sproat, B. S. and Gait, M. J. (1984) Solid-phase synthesis of oligodeoxyribonucleotides by the phosphotriester method. *Oligonucleotides Synthesis: A Practical Approach* (M. J. Gait, ed.), IRL, Oxford and Washington DC, pp. 83–115.
335. Werstiuk, E. S. and Neilson, T. (1972) Oligoribonucleotide synthesis. IV. Approach to block synthesis. *Can. J. Chem.* **50,** 1283–1291.
336. Werstiuk, E. S. and Neilson, T. (1973) Oligoribonucleotide synthesis. VI. Selective deblocking of the 5'-*O*-triphenylmethoxyacetyl grouping in protected dinucleotides. *Can. J. Chem.* **51,** 1889–1892.
337. Neilson, T., Deugau, K. V., England, T. E., and Werstiuk, E. S. (1975) Oligoribonucleotide synthesis. VIII. Insertion of terminal 5'-phosphate groupings. *Can. J. Chem.* **53,** 1093–1098.
338. van der Marel, G. A., Marugg, J. E., de Vroom, E., Wille, G., Tromp, M., van Boeckel, C. A. A., and van Boom, J. H. (1982) Phosphotriester synthesis of

DNA fragments on cellulose and polystyrene solid supports. *Recl. Trav. Chim. Pays-Bas* **101,** 234–241.
339. Sekine, M. and Hata, T. (1985) 4,4',4"-tris(levulinoyloxy)trityl as a new type of primary hydroxyl protecting group. *Bull. Chem. Soc. Jpn.* **58,** 336–339.
340. Sekine, M. and Hata, T. (1984) 4,4',4"-tris(4, 5-dichlorophtalimido)trityl: a new type of hydrazine-labile group as a protecting group of primary alcohols. *J. Am. Chem. Soc.* **106,** 5763,5764.
341. Chattopadhyaya, J. B., Reese, C. B., and Todd, A. H. (1979) 2-Dibromomethylbenzoyl: an acyl protecting group removable under exceptionally mild conditions. *J. Chem. Soc. Chem. Commun.* **1979,** 987,988.
342. Brown, J. M., Christodoulou, C., Reese, C. B., and Sindona, G. (1984) Two new protected acyl protecting groups for alcoholic hydroxy functions. *J. Chem. Soc. Perkin Trans.* **1,** 1785–1790.
343. Brown, J. M., Christodoulou, C., Modak, A. S., Reese, C. B., and Serafinowska, H. T. (1989) Synthesis of the 3'-terminal half of yeast alanine transfer ribonucleic acid ($tRNA^{Ala}$) by the phosphotriester approach in solution. Part 2. *J. Chem. Soc. Perkin Trans.* **1,** 1751–1767.
344. Balgobin, N., Welch, C., and Chattopadhyaya, J. (1982) The complementarity of two β-eliminating protecting groups in the synthesis of octathymidylic acid through the phosphotriester approach. *Chem. Scr.* **20,** 196,197.
345a. Balgobin, N., Josephson, S., and Chattopadhyaya, J. B. (1981) The 2-phenylsulfonylethylcarbonyl (PSEC) group for the protection of the hydroxyl function. *Tetrahedron Lett.* **22,** 3667–3670.
345b. Josephson, S., Balgobin, N., and Chattopadhyaya, J. (1981) The application of 2-(4-chlorophenyl)-sulfonylethoxycarbonyl (CPSEC) group in the synthesis of a DNA segment using the phosphotriester approach. *Tetrahedron Lett.* **22,** 4537–4540.
346. Ma, Y. and Sonveaux, E. (1987) The 9-fluorenylmethyloxycarbonyl (Fmoc) group as a 5'-O base labile protecting group in solid supported oligonucleotide synthesis. *Nucleosides Nucleotides* **6,** 491–493.
347. Seliger, H., Gupta, K. C., Kotschi, U., Spaney, T., and Zeh, D. (1986) New preparative methods in oligonucleotide chemistry and their application to gene synthesis. *Chem. Scr.* **26,** 561–567.
348. Biernat, J., Wolter, A., and Köster, H. (1983) Purification orientated synthesis of oligodeoxynucleotides in solution. *Tetrahedron Lett.* **24,** 751–754.
349a. Görtz, H.-H. and Seliger, H. (1981) New hydrophobic protecting groups for the chemical synthesis of oligonucleotides. *Angew Chem. Int. Edit.* **20,** 681,682.
349b. Seliger, H. and Görtz, H.-H. (1981) Specific separation of products in supported oligonucleotide syntheses using the triester method. *Angew Chem. Int. Edit.* **20,** 683,684.
350. Fourrey, J. L., Varenne, J., Blonski, C., Dousset, P., and Shire, D. (1987) 1,1-bis-(4-methoxyphenyl)-1'-pyrenyl methyl (bmpm): a new fluorescent 5' protecting group for the purification of unmodified and modified oligonucleotides. *Tetrahedron Lett.* **28,** 5157–5160.
351. Fisher, E. F. and Caruthers, M. H. (1983) Color coded triarylmethyl protecting groups for deoxypolynucleotide synthesis. *Nucleic Acids Res.* **11,** 1589–1599.

352a. Schott, H. and Ruess, H. (1986) Synthesis of fragments of the terminal repeating units of macronuclear DNA from hypotrichous ciliates. *Makromol. Chem.* **187,** 81–104.
352b. Schott, H., Semmler, R., Closs, K., and Eckstein, H. (1987) Preparative synthesis of guanylate-rich fragments of the terminal sequence of macronuclear DNA from hypotrichous ciliates using the phosphotriester method in solution. *Makromol. Chem.* **188,** 1313–1346.
353. Denny, W. A., Leupin, W., and Kearns, D. R. (1982) Simplified liquid-phase preparation of four decadeoxyribonucleotides and their preliminary spectroscopic characterization. *Helv. Chim. Acta* **65,** 2372–2393.
354a. Takaku, H., Watanabe, T., and Hamamoto, S. (1988) Use of 1,1,1,3,3,3-hexafluoro-2-propyl protecting group in the synthesis of DNA fragments via phosphoramidite intermediates. *Tetrahedron Lett.* **29,** 81–84.
354b. Yamakage, S., Fujii, M., Takaku, H., and Uemura, M. (1989) 1,1,1,3,3,3-hexafluoro-2-propyl group as a new phosphate protecting group for oligoribonucleotide synthesis in the phosphotriester approach. *Tetrahedron* **45,** 5459–5468.
355. Katagiri, N., Itakura, K., and Narang, S. A. (1975) The use of arylsulfonyltriazoles for the synthesis of oligonucleotides by the triester approach. *J. Am. Chem. Soc.* **97,** 7332–7337.
356. Reese, C. B. and Zard, L. (1981) Some observations relating to the oximate ion promoted unblocking of oligonucleotide aryl esters. *Nucleic Acids Res.* **9,** 4611–4626.
357. Reese, C. B. and Yau, L. (1978) Reaction between 4-nitrobenzaldoximate ion and phosphotriesters. *Tetrahedron Lett.* **19,** 4443–4446.
358. Patel, T. P., Chauncey, M. A., Millican, T. A., Bose, C. C., and Eaton, M. A. W. (1984) A rapid deprotection procedure for phosphotriester DNA synthesis. *Nucleic Acids Res.* **12,** 6853–6859.
359. Takaku, H., Yamaguchi, R., and Nomoto, T. (1979) 5-chloro-8-quinolyl group as high efficient phosphate protecting group for the synthesis of oligoribonucleotides. *Tetrahedron Lett.* **20,** 3857–3860.
360. Takaku, H., Kato, M., Yoshida, M., and Yamaguchi, R. (1980) A convenient method for insertion of the 5'-terminal phosphate group in the triester approach to oligoribonucleotide synthesis. *J. Org. Chem.* **45,** 3347–3350.
361. Takaku, H., Kamaike, K., and Kasuga, K. (1982) Synthesis of bis(5-chloro-8-quinolyl) nucleoside 5'-phosphates in oligoribonucleotide synthesis by the phosphotriester approach. *J. Org. Chem.* **47,** 4937–4940.
362. Asseline, U. and Thuong, N. T. (1985) L'ion benzohydroxamate: nouveau réactif de désarylation en synthèse d'oligonucléotide. *Tetrahedron Lett.* **26,** 1005–1008.
363. Hotoda, H., Wada, T., Sekine, M., and Hata, T. (1989) Pre-activation strategy for oligodeoxyribonucleotide synthesis using triaryloxydichlorophosphoranes in the phosphotriester method. *Nucleic Acids Res.* **17,** 5291–5305.
364. Matsuzaki, J., Hotoda, H., Sekine, M., and Hata, T. (1989) Bis(2,4,6-tribromophenyl) phosphorochloridate: a new type of condensing reagent in oligonucleotide synthesis. *Nucleosides, Nucleotides* **8,** 367–382.

365. Tanimura, H., Sekine, M., and Hata, T. (1986) Further development of oligoribonucleotide: bis(tributyltin)oxide as a reagent for removal of the internucleotidic phenylthio group via the phosphotriester approach. *Tetrahedron* **42,** 4179–4186.
366. Sekine, M., Tanimura, H., and Hata, T. (1985) An effective method for removal of the internucleotidic phenylthio group from fully protected oligonucleotides by the use of bis(tributyltin) oxide. *Tetrahedron Lett.* **26,** 4621–4624.
367. Sood, A. K. and Narang, S. A. (1977) A rapid and convenient synthesis of poly-thymidylic acid by the modified triester approach. *Nucleic Acids Res.* **4,** 2757–2765.
368. Adamiak, R. W., Barciszewska, M. Z., Biala, E., Grzeskowiak, K., Kierzek, R., Kraszewski, A., Markiewicz, W. T., and Wiewiorowski, M. (1976) Nucleoside 3'-phosphotriesters as key intermediates for the oligoribonucleotide synthesis. III. An improved preparation of nucleoside 3'-phosphotriesters, their 1H NMR characterization and new conditions for removal of 2-cyanoethyl group. *Nucleic Acids Res.* **3,** 3397–3408.
369. Hsiung, H. M. (1982) Improvements in the phosphotriester synthesis of deoxyribooligonucleotides, the use of hindered primary amines and a new isolation procedure. *Tetrahedron Lett.* **23,** 5119–5122.
370. Hsiung, H. M., Inouye, S., West, J. Sturm, B., and Inouye, M. (1983) Further improvements on the phosphotriester synthesis of deoxyribooligonucleotides and the oligonucleotide directed site-specific mutagenesis of *E. coli* lipoprotein gene. *Nucleic Acids Res.* **11,** 3227–3239.
371a. Sinha, N. D., Biernat, J., and Köster, H. (1983) β-Cyanoethyl *N,N*-dialkylamino/*N*-morpholinomonochloro phosphoamidites, new phosphitylating agents facilitating ease of deprotection and work-up of synthesized oligonucleotides. *Tetrahedron Lett.* **24,** 5843–5846.
371b. Sinha, N. D., Biernat, J., McManus, J., and Köster, H. (1984) Polymer support oligonucleotide synthesis XVIII: use of β-cyanoethyl-*N,N*-dialkylamino-/*N*-morpholino phosphoramidite of deoxynucleosides for the synthesis of DNA fragments simplifying deprotection and isolation of the final product. *Nucleic Acids Res.* **12,** 4539–4557.
372. Letsinger, R. L. and Ogilvie, K. K. (1969) Synthesis of oligothymidylates via phosphotriester intermediates. *J. Am. Chem. Soc.* **91,** 3350–3355.
373. Letsinger, R. L., Ogilvie, K. K., and Miller, P. S. (1969) Developments in syntheses of oligodeoxyribonucleotides and their organic derivatives. *J. Am. Chem. Soc.* **91,** 3360–3365.
374. Broka, C., Hozumi, T., Arentzen, R., and Itakura, K. (1980) Simplifications in the synthesis of short oligonucleotide blocks. *Nucleic Acids Res.* **8,** 5461–5471.
375. Efimov, V. A., Reverdatto, S. V., and Chakhmakhcheva, O. G. (1982) New effective method for the synthesis of oligonucleotides via phosphotriester intermediates. *Nucleic Acids Res.* **10,** 6675–6694.
376. Uhlmann, E. and Pfleiderer, W. (1980) New improvements in oligonucleotide synthesis by use of the *p*-nitrophenylethyl phosphate blocking group and its deprotection by DBU or DBN. *Tetrahedron Lett.* **21,** 1181–1184.

377. Beiter, A. H. and Pfleiderer, W. (1984) Solution synthesis of protected di-2'-deoxynucleoside phosphotriesters via the phosphoramidite approach. *Tetrahedron Lett.* **25,** 1975–1978.
378. Claesen, C. A. A., Segers, R. P. A. M., and Tesser, G. I. (1985) A comparison of β-functionalized ethyl groups for the protection of the phospho function in decathymidylate synthesis using a phosphite triester approach. *Recl. Trav. Chim. Pays-Bas* **104,** 209–214.
379. Claesen, C. A. A., Segers, R. P. A. M., and Tesser, G. I. (1985) Ar(alk)ylsulfonyl ethyl groups as phosphorus-protecting functions. *Recl. Trav. Chim. Pays-Bas* **104,** 119–122.
380. Takaku, H., Hamamoto, S., and Watanabe, T. (1986) Use of 2-(2-pyridyl)ethyl group as a new protecting group of internucleotidic phosphates in oligonucleotide synthesis. *Chem. Lett.* **1986,** 699–702.
381. Hamamoto, S., Shishido, Y., Furuta, M., Takaku, H., Kawashima, M., and Takaki, M. (1989) Use of the 2-(4-pyridyl)ethyl protecting group in the synthesis of DNA fragments via phosphoramidite intermediates. *Nucleosides, Nucleotides* **8,** 317–326.
382. van Boom, J. H., Burgers, P. M. J., Crea, R., van der Marel, G., and Wille, G. (1977) Synthesis of oligonucleotides with sequences identical with or analogous to the 3'-end of 16S ribosomal RNA of *Escherichia coli*: preparation of $m^6{}_2$ACCUCC and ACCUC$m^4{}_2$C via phosphotriester intermediates. *Nucleic Acids Res.* **4,** 747–759.
383. Daub, G. W. and van Tamelen, E. E. (1977) Synthesis of oligoribonucleotides based on the facile cleavage of methyl phosphotriester intermediates. *J. Am. Chem. Soc.* **99,** 3526–3528.
384. Smith, D. J. H., Ogilvie, K. K., and Gillen, M. F. (1980) The methyl group as phosphate protecting group in nucleotide synthesis. *Tetrahedron Lett.* **21,** 861–864.
385. Dahl, B. H., Bjergarde, K., Henriksen, L., and Dahl, O. (1990) A highly reactive, odourless substitute for thiophenol/triethylamine as a deprotection reagent in the synthesis of oligonucleotides and their analogs. *Acta Chem. Scand.* **44,** 639–641.
386. Caruthers, M. H. (1982) Chemical synthesis of oligodeoxynucleotides using the phosphite triester intermediates. *Chemical and Enzymatic Synthesis of Gene Fragments, a Laboratory Manual* (Gassen, H. G. and Lang, A., eds.), Verlag Chemie, Weinheim, pp. 71–79.
387. Efimov, V. A., Buryakova, A. A., Dubey, I. Y., Polushin, N. N., Chakhmakhcheva, O. G., and Ovchinnikov, Y. A. (1986) Application of new catalytic phosphate protecting groups for the highly efficient phosphotriester oligonucleotide synthesis. *Nucleic Acids Res.* **14,** 6525–6540.
388. Froehler, B. C. and Matteuci, M. D. (1985) 1–Methyl-2-(2-hydroxyphenyl)imidazole: a catalytic phosphate protecting group in deoxyoligonucleotide synthesis. *J. Am. Chem. Soc.* **107,** 278,279.
389. Sproat, B. S., Rider, P., and Beijer, B. (1986) Highly efficient oligodeoxyribonucleotide synthesis using fully base protected phosphodiester building blocks

carrying 2-(1-methylimidazol-2-yl)phenyl protection of the phosphate. *Nucleic Acids Res.* **14,** 1811–1824.
390. Catlin, J. C. and Cramer, F. (1973) Deoxyoligonucleotide synthesis via the triester method. *J. Org. Chem.* **38,** 245–250.
391. Li, B. F. L., Reese, C. B., and Swann, P. F. (1987) Synthesis and characterization of oligodeoxynucleotides containing 4-*O*-methylthymine. *Biochemistry* **26,** 1086–1093.
392. Itakura, K., Katagiri, N., Narang, S. A., Bahl, C. P., Marians, K. J., and Wu, R. (1975) Chemical synthesis and sequence studies of deoxyribooligonucleotides which constitute the duplex sequence of the lactose operator of *Escherichia coli.* J. Biol. *Chem.* **250,** 4592–4600.
393. Bahl, C. P., Wu, R., Itakura, K., Katagiri, N., and Narang, S. A. (1976) Chemical and enzymatic synthesis of lactose operator of *Escherichia coli* and its binding to lactose repressor. *Proc. Natl. Acad. Sci. USA* **73,** 91–94.
394. Crea, R., Kraszewski, A., Hirose, T., and Itakura, K. (1978) Chemical synthesis of genes for human insulin. *Proc. Natl. Acad. Sci. USA* **75,** 5765–5769.
395. Seliger, H., Bach, T.-C., Siewert, G., Boidol, W., Töpert, M., Schulten, H.-R., and Schiebel, H. M. (1984) Synthesis of deoxyoligonucleotide linker fragments for genetic engineering using improved preparative and analytical techniques. *Liebigs Ann. Chem.* **1984,** 835–853.
396. Gough, G. R., Singleton, C. K., Weith, H. L., and Gilham, P. T. (1979) Protected deoxyribonucleoside-3' aryl phosphodiesters as key intermediates in polynucleotide synthesis. Construction of an icosanucleotide analogous to the sequence at the ends of Rous sarcoma virus 35S RNA. *Nucleic Acids Res.* **6,** 1557–1570.
397. Gioeli, C. and Chattopadhyaya, J. (1982) Fluorene-9-methyl-, a phosphate protecting group: its application in the phosphotriester approach through the synthesis of tetracosathymidilic acid. *Chem. Scr.* **19,** 235–237.
398. Balgobin, N. and Chattopadhyaya, J. (1982) An efficient chemical synthesis of a biologically functional DNA molecule, 5'd(ATGGGTTTCTTCGC-)3', through the phosphotriester approach. *Chem. Scr.* **20,** 133–138.
399. Itakura, K., Katagiri, N., and Narang, S. A. (1974) Synthesis of lactose-operator gene fragments by the improved triester method. *Can. J. Chem.* **52,** 3689–3693.
400. Marugg, J. E., Nielsen, J., Dahl, O., Burik, A., van der Marel, G. A., and van Boom, J. H. (1987) (2-Cyano-1,1-dimethylethoxy)bis(diethylamino)phosphine: a convenient reagent for the synthesis of DNA fragments. *Recl. Trav. Chim. Pays-Bas.* **106,** 72–76.
401. De Bernardini, S., Waldmeier, F., and Tamm, C. (1981) Nucleosides and Nucleotides. Part 17. A simple preparation of protected deoxynucleoside-3'-phosphates. *Helv. Chim. Acta* **64,** 2142–2147.
402. Chattopadhyaya, J. B. and Reese, C. B. (1979) Some observations relating to phosphorylation methods in oligonucleotide synthesis. *Tetrahedron Lett.* **20,** 5059–5062.

403. Sekine, M., Hamaoki, K., and Hata, T. (1979) Synthesis and properties of S,S-diaryl nucleoside phosphorodithioates in oligoribonucleotide synthesis. *J. Org. Chem.* **44,** 2325,2326.
404. Christodoulou, C. and Reese, C. B. (1983) Dealkylation of nucleoside arylmethyl 2-chlorophenyl phosphates: the 2,4-dinitrobenzyl protecting group. *Tetrahedron Lett.* **24,** 951–954.
405. Reese, C. B., Titmas, R. C., and Valente, L. (1981) Action of toluene-*p*-thiol and triethylamine on fully protected thymidylyl-(3'-5')-thymidine. Possible occurrence of thiolate ion-promoted internucleotide cleavage in the synthesis of oligonucleotides by the phosphotriester approach. *J. Chem. Soc. Perkin Trans.* **1,** 2451–2455.
406. Takaku, H., Yoshida, M., Kamaike, K., and Hata, T. (1981) 4-chlorophenyl 5-chloro-8-quinolyl phosphorochloridate: a practically useful phosphorylating agent for oligoribonucleotide synthesis via phosphotriester approach. *Chem. Lett.* **1981,** 197–200.
407. Takaku, H., Nomoto, T., and Kamaike, K. (1981) 4-chlorophenyl 5-chloro-8-quinolyl phosphorotetrazolide: a highly efficient phosphorylating agent for oligoribonucleotide synthesis. *Chem. Lett.* **1981,** 543–546.
408. Takaku, H., Kamaike, K., and Suetake, M. (1983) A simple synthetic method of deoxyribodinucleotide blocks. *Chem. Lett.* **1983,** 111–114.
409. Ohtsuka, E., Tanaka, T., Wakabayashi, T., Taniyama, Y., and Ikehara, M. (1978) A new ribo-oligonucleotide block synthesized by phosphorylation with *p*-chlorophenyl *N*-phenylchlorophosphoramidate. *J. Chem. Soc. Chem. Commun.* **1978,** 824,825.
410. Ohtsuka, E., Murao, K., Ubasawa, M., and Ikehara, M. (1970) Studies on transfer ribonucleic acids and related compounds. I. Synthesis of ribooligonucleotides using aromatic phosphoramidates as protecting group. *J. Am. Chem. Soc.* **92,** 3441–3445.
411. Ohtsuka, E., Shibahara, S., Ono, T., Fukui, T., and Ikehara, M. (1981) Synthesis of deoxyribooligonucleotides by using aromatic phosphoramidates as the protecting group for the 3'-phospho ends. *Heterocycles* **15,** 395–398.
412. van Boom, J. H., Burgers, P. M. J., van der Marel, G., Verdegaal, C. H. M., and Wille, G. (1977) Synthesis of oligonucleotides with sequences identical with or analogous to the 3'-end of 16S ribosomal RNA of *Escherichia coli*: preparation of ACCUCC via the modified phosphotriester method. *Nucleic Acids Res.* **4,** 1047–1063.
413. de Rooij, J. F. M., Wille-Hazeleger, G., van Deursen, P. H., Serdijn, J., and van Boom, J. H. (1979) Synthesis of complementary DNA fragments via phosphotriester intermediates. *Recl. Trav. Chim. Pays-Bas* **98,** 537–548.
414. de Rooij, J. F. M., Wille-Hazeleger, G., Vink, A. B. J., and van Boom, J. H. (1979) Synthesis of functionalized DNA fragments suitable for reversible attachment to activated cellulose. *Tetrahedron* **35,** 2913–2926.
415. Arentzen, R., van Boeckel, C. A. A., van der Marel, G., and van Boom, J. H. (1979) 2,2,2-tribromomethyl 2-chloro-4-*t*-butylphenyl phosphorochloridate: a convenient phosphorylating agent for the synthesis of DNA-fragments by the phosphotriester approach. *Synthesis* **1979,** 137–139.

416. Balgobin, N. and Chattopadhyaya, J. (1982) 2-(4-nitrophenyl)thioethyl, A phosphate protecting group and its application in conjunction with 5'-*O*-2,2-dibromomethylbenzoyl group in the synthesis of dodecathymidilic acid through the phosphotriester approach. *Chem. Scr.* **20,** 144–146.
417. Balgobin, N. and Chattopadhyaya, J. (1982) 5–benzisoxazolylmethylene (BIM) A new phosphate protecting group; its application in DNA synthesis through the phosphotriester approach. *Chem. Scr.* **20,** 142,143.
418. Josephson, S. and Chattopadhyaya, J. (1981) The application of the 2-phenylsulfonylethyl-, a novel phosphate protecting group, in the synthesis of DNA fragments of defined sequences. *Chem. Scr.* **18,** 184–188.
419. Balgobin, N., Josephson, S., and Chattopadhyaya, J. (1981) 2-Phenylsulfonylethyl, a new phosphate protecting group: its application in the synthesis of dodecathymidilic acid. *Tetrahedron Lett.* **22,** 1915–1918.
420. Hayakawa, Y., Uchiyama, M., Kato, H., and Noyori, R. (1985) Allyl protection of internucleotide linkage. *Tetrahedron Lett.* **26,** 6505–6508.
421. Farrance, I. K., Eadie, J. S., and Ivarie, R. (1989) Improved chemistry for oligodeoxyribonucleotide synthesis substantially improves restriction enzyme cleavage of a synthetic 35mer. *Nucleic Acids Res.* **17,** 1231–1245.
422. Gryaznov, S. M. and Letsinger, R. L. (1991) Synthesis of oligonucleotides via monomers with unprotected bases. *J. Am. Chem. Soc.* **113,** 5876–5877.

CHAPTER 2

Incorporation of Modified Bases into Oligonucleotides

Rich B. Meyer, Jr.

1. Introduction

Nucleobases with attached side chains have been incorporated into oligodeoxynucleotides (ODNs) to achieve a number of goals. The functionalized side chains have been used to attach reporter groups (protein ligands as well as fluorescent moieties) and metal-chelating groups. Electrophilic and photochemical alkylating functions have been attached to react with the other strand(s) in a duplex (or triplex).

The incorporation of nucleosides with modified bases into oligonucleotides presents several major issues:

1. The protection of reactive groups in the modified moiety;
2. The ability of the functional groups to withstand the conditions of the automated DNA synthesis cycles;
3. The relative reactivity of the unblocked side arm functions, if appropriate; and
4. The effect of the modification on the stability of hybrids formed by the ODN.

In general, nucleosides bearing side arms terminating in nucleophilic groups such as $-NH_2$ have often been used as linkers for postsynthesis modification. Several types of reporter groups have been added to ODNs in this manner; the preparation of addition of a sensitive electrophilic crosslinking group *(1)* will be described here.

Figure 1 shows the positions that have been modified with greatest success for introduction of side arms onto the nucleic acid bases. The 5-substituted 2'-deoxyuridines **1** have been most often used for incor-

From: *Methods in Molecular Biology, Vol. 26: Protocols for Oligonucleotide Conjugates*
Edited by: S. Agrawal Copyright ©1994 Humana Press Inc., Totowa, NJ

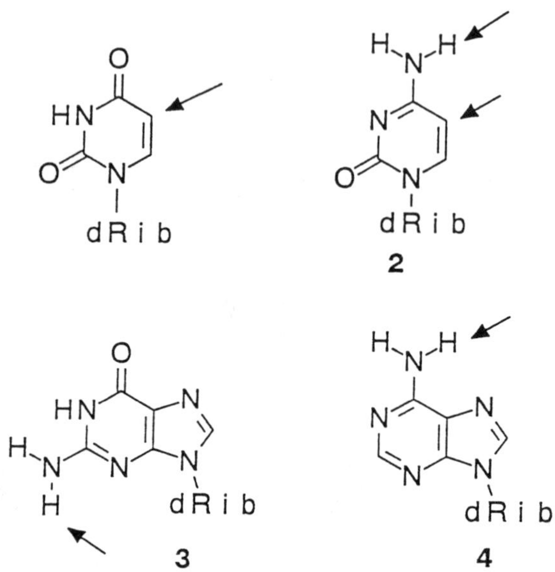

Fig. 1. Sites of possible modification of the DNA bases.

poration into ODNs because of the ease of introduction of a side arm in the 5-position. In particular, the 5-(3-aminopropynyl) *(2)* **5** and (3-aminopropyl)-2'-deoxyuridines *(3)* **6** and their *N*-substituted derivatives have seen considerable use, as have *N*-substituted derivatives of 5-(carbamoylethyl)-2'-deoxyuridine **7** *(4; see* Fig. 2). Although it is possible to modify the 5-position on deoxycytidine **2**, the predominant modifications reported are on the *N*-4 *(5)*. The 6-position of the pyrimidine cannot be functionalized without compromising the energetic preference for the *anti*-conformation in the hybrid. Molecular model studies indicate that, in both A- and B-form DNA, and in certain triple helices, all nucleosides are in the *anti*-conformation. Incorporation of 6-methylthymidine into an ODN significantly diminished the stability of its hybrids *(6)*.

Few derivitized purine nucleosides (deoxyguanosines, **3** or deoxyadenosines, **4**) have seen use in ODN preparation until recently. If a substituent is placed on the most readily modified position in the purine nucleosides, the 8-position, the nucleoside is forced into the *syn*-conformation, whereas the *anti*-conformation is required for hybridization. Certain oligonucleotides that contain deoxyguanosines

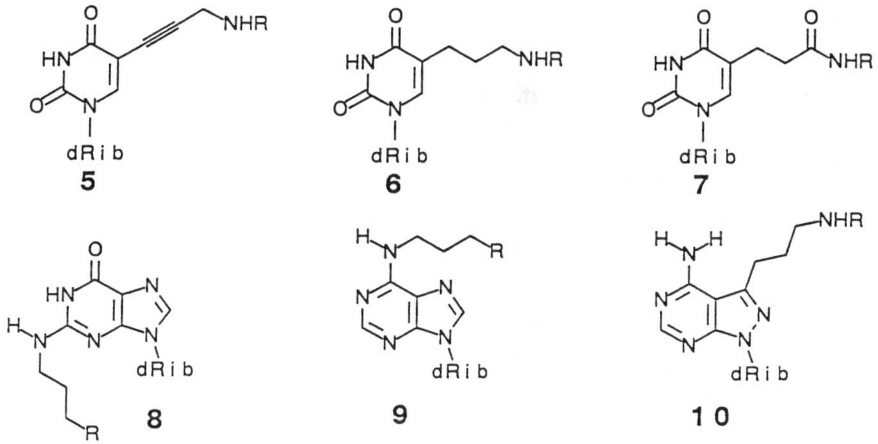

Fig. 2. Modified bases that have been used in modified oligonucleotide synthesis.

with *N*-2 substituents **8** show improved hybridization strength, however (7). Several 3-deaza-3-substituted-2'-deoxyguanosines have been incorporated into oligonucleotides, but lower hybrid stability of these oligomers was observed.

A deoxyadenosine reagent with biotin connected by a short chain to the *N*-6 position **9** is commercially available for incorporation into polynucleotides by nick translation, and this type of modification could be used for synthetic ODNs. A pyrazolo[3,4-*d*]pyrimidine derivative **10** of deoxyadenosine with a biotin on a chain attached to the 3-position (equivalent to the 7-position of adenine) has been prepared, and the hybridization properties of oligomers containing it have been compared to those containing **9** for hybridization properties *(8)*. Analog **9** was found to destabilize the hybrid (compared to oligomers with no substituted base), whereas oligomers with **10** were about as stable as the unsubstituted ones. This analog has not yet been incorporated into synthetic oligonucleotides, although that should not be problematic.

In this chapter the preparation of some representative base-modified nucleosides has been discussed. The ones presented were chosen because of first-hand experience with the synthetic methods, versatility of application, and favorable (or, at least, nondetrimental) effect of the substitution on hybridization properties of the oligonucleotides containing them. Two 5-substituted deoxyuridines and an N^2 substituted deoxyguanosine are shown. Our standard ODN synthesis and purification methods are given, although they do not significantly differ from standard procedures, and a method is given for postsynthesis modification of one of the amine side chains with a sensitive electrophilic group.

2. Materials

1. Melting points were determined on a Mel-Temp melting point apparatus in open capillary tubes and are uncorrected.
2. Nuclear magnetic resonance (NMR) spectra were obtained at 300 MHz on a Varian VXR-300 spectrometer or at 200 MHz on a Gemini 200 spectrometer. The chemical shift values are expressed in δ values (parts per million) relative to tetramethylsilane as an internal standard.
3. Thin-layer chromatography was run on silica gel 60 F-254 (EM Reagents) aluminum backed plates. EM Reagents silica gel (230–400 mesh) was used for flash column chromatography *(9)*. Components were detected on TLC by UV light and by charring by exposure to H_2SO_4 in MeOH spray followed by heating.
4. Oligodeoxynucleotide synthesis was done on a 1 μM scale on an Applied Biosystems Model 390B Synthesizer (Foster City, CA) controlled by Software System version 1.34 or on a 1 μM scale on a Milligen 7500 synthesizer (Millipore, Bedford, MA) controlled by DNA Express version 1.9. DNA synthesis reagents were purchased from Glenn Research (Herndon, VA) or from Milligen.
5. HPLC analyses and semipreparative ODN purifications were performed with a Ranin Gradient HPLC system (Emeryville, CA) equipped with 10 mL/min pump heads and a Gilson (Worthington, OH) Model 116 Dual UV wavelength detector unless otherwise indicated. Analytical HPLC used a Ranin Dynamax-300A C18, 4.6 × 250 mm column eluted with a linear gradient of 9–13.5% acetonitrile in 0.1*M* triethylammonium acetate (pH 7.5) over 30 min monitored at 260 nm. Semipreparative HPLC was performed with identical equipment and buffers using a linear gradient of 9–12% over 30 min.
6. Correct elemental analyses were obtained on all new compounds. Centricon microconcentrators were purchased from Amicon (Beverly, MA).
7. ODNs are labeled with [γ-^{32}P]ATP from DuPont (NEN Research Products; Boston, MA) and T4 polynucleotide kinase from United States Biochemical (Cleveland, OH) using the procedure of Maxam and Gilbert *(10)*. The ^{32}P-labeled product is purified using DuPont Nensorb™ 20 columns (Wilmington, DE).
8. Cerenkov counting is done on a Beckman LS 5000TD from Beckman Instruments, Inc. (Fullerton, CA).

9. ODN concentrations are calculated from A_{260} values using a concentration constant of 8.8 mL/µmol for the modified 2'-deoxyuridines.
10. T_Ms are calculated with the computer program *OLIGO 3.3* from National Biosciences (Hamel, MN) using the concentration of the strand in excess.
11. The 20% polyacrylamide-7M urea gels are prepared as described by Maniatas *(11)* and had dimensions of 0.4 × 200 × 400 mm with 4 mm square slots. The gels are run at 50–55°C with 0.045*M* Tris-HCl and 0.001*M* EDTA buffer (TBE) at 55 watts until the xylene cyanol marker dye had traveled 7.5 cm.

3. Methods

3.1. Synthesis of 5-(4-Phthalimidobutyl)-2'-Deoxyuridine (13b)

3.1.1. 4-Phthalimido-1-Butyne

To a solution of triphenylphosphine (26 g, 0.1 mol) in dioxane (200 mL) in an ice bath is added diisopropylazidodicarboxylate (20 mL, 90 mmol). After addition is complete a solid formed. A mixture of phthalimide (14.7 g, 0.1 mol) and 3-butyn-1-ol (7 g, 100 mmol) in dioxane (50 mL) are poured in and stirred overnight. The butyne precipitated from CH_2Cl_2 as long prismatic needles (8.25 g, 42%): mp 134–138°C.

3.1.2. 5-(4-Phthalimido-1-Butynyl)-2'-Deoxyuridine ***(12b)***

A mixture of 5-iodo-2'-deoxyuridine (**11**, 354 mg, 1 mmol) and CuI (76 mg, 0.4 mmol) is flushed with argon (3×) and suspended in DMF (10 mL). Et_3N (200 mg, 2 mmol) and 4-phthalimido-1-butyne (300 mg, 1.5 mmol) are added and formed a blue solution. The reaction is flushed with argon and tetrakis(triphenylphosphine)palladium(0) (230 mg, 0.2 mmol) is added. The reaction is warmed to 50°C for 2 h and stirred at room temperature overnight. After reaction is complete, the solvent is evaporated and the yellow gum dissolved in CH_2Cl_2. Scratching of the flask induced crystallization. The material is recrystallized from 95% EtOH to give 335 mg (78%) of **4b** as needles: mp 175–178°C; ^1H NMR (300 MHz, d_6-DMSO) 11.56 (br s, 1, NH), 8.07 (s, 1, H6), 7.91–7.83 (m, 4, phenyl), 6.091 (t, 1, J = 6.6 Hz, 1, H-4'), 2.733 (t, J = 7.2 Hz, 2, propargyl CH_2).

3.1.3. 5-(4-Phthalimidobutyl)-2'-Deoxyuridine ***(13b)***

A solution of **12b** (1 g, 2.4 mmol) in 95% EtOH is refluxed with Raney Nickel (3 g). After 48 h the UV absorption peak had shifted from 290 to 263 nm. Catalyst is carefully removed by filtration. The filtrate is evaporated to dryness and the solid is recrystallized from MeOH:H_2O to give 960 mg (97%) of **5b**: mp 180–181°C; ^{1}H NMR (300 MHz, d_6-DMSO) 11.35 (br s, 1, NH), 7.89–7.82 (m, 4, Ph), 7.69 (s, 1, H-6), 6.16 (t, J = 6.6 Hz, 1, H-1'), 2.22 (t, J = 6.3 Hz, 2, UCH_2).

3.2. Synthesis of 5'-O-Dimethoxytrityl-5-(3-Trifluoroacetamidopropyl)-2'-Deoxyuridine (13a) and its 3'-O-(2-Cyanoethoxy)-N,N'-Diisopropylaminophosphoramidite

3.2.1. N-Propargyltrifluoroacetamide

Propargylamine (8.81 g, 0.16 mol) is stirred and chilled in an *i*-PrOH-dry ice bath while excess of trifluoroacetic anhydride (30 mL, 0.21 mol) is added dropwise. The mixture is allowed to room temperature and distilled. 3-Trifluoroacetamidopropyne is distilled at 73–74°C/13 torr (lit b.p. 68.5–69.5°C/11 torr) *(12)* as an oil and solidified on refrigeration; yield 22.5 g (93%), mp 17°C, n_D^{20} 1.3915, IR ($CHCl_3$) 2130 (C ≡ CH) and 1730 cm^{-1} (C = O).

3.2.2. 5-(3-Trifluoroacetamidopropynyl)-2'-Deoxyuridine (12a)

A mixture of 5-iodo-2'-deoxyuridine (**11**, 3.54 g, 10 mmol), copper(I) iodide (0.19 g, 1 mmol) and tetrakis-(triphenylphosphine)palladium(0) (0.58 g, 0.5 mmol) is dried *in vacuo* at 60°C for 3 h and placed under argon. A suspension of the mixture in dry DMF (20 mL) is stirred under argon and treated with dry triethylamine (1.7 mL, 12 mmol) followed by *N*-propargyltrifluoroacetamide (2.9 g, 19 mmol). The mixture is stirred at room temperature for 17 h (TLC on silica with 12:1:1:1 ethyl acetate-acetone-water-methanol indicates that the reaction is then complete). The mixture is treated with 2% acetic acid (100 mL), and the catalyst is removed by filtration and flushed with two 30 mL portions of 50% methanol. The filtrates are combined and flushed through LiChroprep RP-18 column (5 × 25 cm, 25–40 μm, EM Science). The column is washed with 100 mL portions of 1% HOAc/H_2O, followed by 10, 20, 30, and 40% (v/v) aqueous methanol, and then eluted with 1% acetic acid in 50% (v/v) methanol (700 mL). The fractions with the main product (R_f 0.5) are combined and evaporated to 30 mL to give pure 5-(3-trifluoroacetamidopropynyl)-2'-deoxyuridine as a precipitate which is collected on filter, flushed with ice water and dried *in vacuo;* yield 2.95 g (78%); mp 173–174°C; λ_{max} (nm, pH 7.5) 232, 288.

3.2.3. 5-(3-Trifluoroacetamidopropyl)-2'-Deoxyuridine ***(13a)***

A solution of 5-(3-trifluoroacetamidopropynyl)-2'-deoxyuridine (2.5 g, 6.6 mmol) in methanol (20 mL) is stirred with ammonium formate, pH 5 (prepared by addition of 2 mL, 52 mmol, of cold 98% formic acid into 1.3 mL, 33 mmol, of dry ice frozen 25% ammonia), and 10% palladium on activated carbon for 15 min at 50°C. The reaction is monitored by HPLC (Microsorb C18 column, Rainin Instrument Co., Inc.) and a Waters 994 Programmable Photodiode Array detector to make sure that all starting material with λ_{max} 288 nm transferred into reduced compound with λ_{max} 268 nm. The catalyst is removed by filtration, the filtrate evaporated, and product is purified on LiChroprep RP-18 column by the above procedure for 5-(3-trifluoroacetamido-propynyl)-2'-deoxyuridine. Fractions containing the desired product are combined and evaporated to dryness *in vacuo*, and the resulted solid is triturated with dry ether, collected on filter and flushed with ether, and dried *in vacuo;* yield 2.13 g (84%); mp 160–162°C; λ_{max} (nm, pH 7.5) 220, 268.

3.2.4. 5'-O-Dimethoxytrityl-5-(3-Trifluoroacetamidopropyl)-2'-Deoxyuridine

5-(3-Trifluoroacetamidopropyl)-2'-deoxyuridine (**13a**, 1.91 g, 5 mmol) is thoroughly evaporated twice from dry pyridine, then stirred in 30 mL of dry pyridine. 4,4'-Dimethoxytrityl chloride (2.37 g, 7 mmol) is added, and the mixture stirred at room temperature for 4 h. Water (30 mL) is added and the desired product is extracted with ether (2 × 300 mL). The ether layers are combined and dried by stirring with sodium sulfate for 1 h. The mixture is filtered, the solid is washed with ether, and filtrate and washings are combined and evaporated. The residue is dissolved in methanol (10 mL) and this solution is applied to LiChroprep RP-18 column (5 × 25 cm, 25–40 μm, EM Science), the column is washed with 400 mL of 70% (v/v) methanol and then eluted with 90% (v/v) methanol (700 mL). The residue, after evaporation of appropriate fractions with the product (R_f 0.5 in 500:20:2.5 ethyl acetate-methanol-triethylamine) is dissolved in methanol (30 mL) and dropped into a stirred 50% (v/v) methanol (100 mL). After beginning of crystallization more water (100 mL) is added. The mixture is stirred on an ice bath, precipitate is collected on filter, washed with water, and dried *in vacuo*; yield 3.0 g (88%); mp 80–90°C.

*3.2.5. 3'-O-(2-Cyanoethoxy)-*N,N'-*Diisopropylaminophosphoramidite of 5'-O-Dimethoxytrityl-5-(3-Trifluoroacetamidopropyl)-2'-Deoxyuridine*

5'-*O*-Dimethoxytrityl-5-(3-trifluoroacetamidopropyl)-2'-deoxyuridine (2.05 g, 3 mmol) is evaporated with dry pyridine (2 × 10 mL) and dried *in vacuo* overnight. The resulting solid is transferred to an argon atmosphere and dissolved in a mixture of anhydrous *N,N*-diisopropylethylamine (2.6 mL, 15 mmol) and dichloromethane (10 mL). 2-Cyanoethoxy-*N,N'*-diisopropylaminochlorophosphine (1.0 mL, 5.1 mmol) is added to the mixture dropwise for 1 min by syringe with swirling. The resulting solution is stirred for 1.5 h (TLC in 5:5:1 ethylacetate-dichloromethane-triethylamine showed two spots, R_f 0.5 and 0.56) and treated with methanol (0.2 mL). The mixture is diluted with ethylacetate (300 mL), and the organic layer is washed with 10% sodium hydrogen carbonate (2 × 150 mL) and saturated sodium chloride (2 × 150 mL). The organic layer is dried with sodium sulfate and filtered, and the solvent is completely removed *in vacuo*. The product is purified by pouring a xylene (10 mL) solution into hexanes (300 mL). A solid precipitate is filtered, flushed with hexanes, and dried *in vacuo*; yield 2.2 g (83%); mp 55–65°C.

3.3. Synthesis of 6-O-[2-(4-Nitro-Phenyl)Ethyl]-N²-Isobutyryl-N²-[Imidazol-1-yl(Propyl)]-2'-Deoxyguanosine (19) and its 5'-O-(4,4'-Dimethoxytrityl)-3'-O-(2-Cyanoethoxy)-N,N'-Diisopropylaminophosphoramidite

3.3.1. 2-Chloro-2'-Deoxyinosine **(15)**

To a stirred suspension of 2,6-dichloro-9-(2-deoxy-3,5-di-O-*p*-toluoyl-β-**D**-erythropentofuranosyl)purine *(13)* (**14**,10.3 **2**,19.04 mmol) in allyl alcohol (150 mL) is added sodium hydride (60%, 0.8 g, 20.0 mmol) in small portions over a 10-min period at room temperature. The reaction mixture is stirred at 55°C for 20 min with exclusion of moisture, cooled, filtered, and the filtrate washed with allyl alcohol

(50 mL). To the filtrate IRC-50 (weakly acidic) H^+ resin is added until the pH of the solution reaches 4–5. The resin is filtered, washed with methanol (100 mL) and the filtrate is evaporated to dryness. The residue is adsorbed on silica gel (10 g, 60–100 mesh) and evaporated to dryness. The dried silica gel is placed on top of silica column (5 × 25 cm, 100–250 mesh) packed in dichloromethane. The column is then eluted with CH_2Cl_2/acetone (1:1). The fractions having the product are pooled together and evaporated to dryness to give 6 g (96%) of the 6-allyloxy derivative as a foam: ^{1}H NMR (Me_2SO-d_6) 6.35 (t, ^{1}H, J1',2' = 6.20 Hz, C1'H), 8.64 (s, ^{1}H, C8H).

This compound (6 g, 18.4 mmol), Pd/C (10%, 1 g) and triethylamine (1.92 g, 19.0 mmol) in ethyl alcohol (200 mL) is hydrogenated at atmospheric pressure for 30 min period at room temperature, at which time hydrogen uptake is complete. The reaction mixture is filtered, washed with methanol (50 mL), and the filtrate evaporated to dryness. The product 5.26 g (100%) is found to be moisture sensitive. The oil is used as such for further reaction without purification. A small portion on crystallization from MeOH/CH_2Cl_2 gave colorless crystalline solid. ^{1}H NMR (Me_2SO-d_6) 6.23 (t, ^{1}H, J1',2' = 6.20 Hz, C1'H), 8.32 (s, ^{1}H, C8H).

3.3.2. N^2*-[Imidazol-1-yl(Propyl)]-2'-Deoxyguanosine* ***(16)***

A solution of the nucleoside 3 (10.3 g, 36.0 mmol) and 1-(3-aminopropyl)imidazole (9.0 g, 72.0 mmol) in 2-methoxyethanol (60 mL) is heated in a steel bomb at 100°C (oil bath) for 12 h. The bomb is cooled to 0°C, and the precipitated solid is filtered. The solid is washed with methanol (50 mL), acetone (50 mL), and dried over sodium hydroxide to give g (67%) of pure **16**. A small amount is recrystallized from DMF for analytical purposes: mp 245–47°C: ^{1}H NMR (Me_2SO-d_6) 6.12 (t, ^{1}H, J1',2' = 6.20 Hz, C1'H), 6.46 (b s, ^{1}H, NH), 6.91 (s, ^{1}H, ImH), 7.18 (s, ^{1}H, ImH), 7.66 (s, ^{1}H, ImH), 7.91 (s, ^{1}H, C8H), 10.60 (b s, ^{1}H, NH).

3.3.3. N^2*,3',5'-Triisobutyryl-*N^2*-[Imidazol-1-yl (Propyl)]-2'-Deoxyguanosine* ***(17)***

To a well dried solution of **16** (1.5 g, 4.0 mmol) and triethylamine (1.62 g, 16.0 mmol) in dry pyridine (30 mL) and dry DMF (30 mL) is added isobutyryl chloride (1.69 g, 16.0 mmol) at room temperature. The reaction mixture is allowed to stir at room temperature for 12 h and evaporated to dryness. The residue is partitioned between dichloromethane (100 mL) and water (50 mL) and extracted with CH_2Cl_2 (2 × 100 mL). The organic extract is washed with brine (100 mL) and dried over anhydrous $MgSO_4$. The dried organic extract is evaporated to dryness and the residue is purified by flash chromatography using CH_2Cl_2/MeOH as eluent. The pure fractions are pooled, evaporated to dryness and on crystallization from CH_2Cl_2/MeOH give 1.8 g (77%) of **17** as colorless crystalline solid: mp 210–12°C: ^{1}H NMR (Me_2SO-d_6) β 1.04 (2 m, 18 H), 6.30 (t, ^{1}H, J1',2' = 6.20 Hz, C1'H), 6.84 (s, ^{1}H, ImH), 7.18 (s, ^{1}H, ImH), 7.34 (s, ^{1}H, ImH), 8.34 (s, ^{1}H, C8H), 10.60 (br s, ^{1}H, NH).

3.3.4. 6-O-[2-(4-Nitrophenyl)Ethyl]-N^2,3'-O-,5'-O-Triisobutyryl-N^2-[Imidazol-1-yl(Propyl)]-2'-Deoxyguanosine **(18)**

To a stirred solution of **17** (2.0 g, 3.42 mmol), triphenylphosphine (2.68 g, 10.26 mmol) and *p*-nitrophenyl ethanol (1.72 g, 10.26 mmol) in dry dioxane (50 mL) is added diethyl azodicarboxylate (1.78 g, 10.26 mmol) at room temperature. The reaction mixture is stirred at room temperature for 12 h and evaporated to dryness. The residue is purified by flash chromatography on silica gel using CH_2Cl_2/acetone as the eluent. The pure fractions are pooled together and evaporated to dryness to give 2.4 g (96%) of **18** as amorphous solid: ^{1}H NMR (Me_2SO-d_6) 1.04 (2 m, 18H), 6.34 (t, ^{1}H, 1',2' = 6.20 Hz, C1'H), 6.82 (s, ^{1}H, ImH), 7.08 (s, ^{1}H, ImH), 7.56 (s, ^{1}H, ImH), 7.62 (d, ^{2}H, ArH), 8.1 (d, ^{2}H, ArH), 8.52 (s, ^{1}H, C8H).

3.3.5. 6-O-[2-(4-Nitrophenyl)Ethyl]-N^2-Isobutyryl-N^2-[Imidazol-1-yl(Propyl)-2'-Deoxyguanosine **(19)**

To a stirred solution of **18** (9.0 g, 12.26 mmol) in methanol (250 mL) is treated with conc. ammonium hydroxide (30%, 150 mL) at room temperature. The reaction mixture is stirred at room temperature for 4 h and evaporated to dryness under reduced pressure. The residue is purified by flash chromatography over silica gel using CH_2Cl_2/MeOH as the eluent. The pure fractions are pooled together and evaporated to dryness to give 5.92 g (81%) of the title compound: ^{1}H NMR (Me_2SO-d_6) δ1.04 (m, ^{6}H), 6.34 (t, ^{1}H, J1',2' = 6.20 Hz, C1'H), 6.82 (s, ^{1}H, ImH), 7.12 (s, ^{1}H, ImH), 7.54 (s, ^{1}H, ImH), 7.62 (d, ^{2}H, ArH), 8.16 (d, ^{2}H, ArH), 8.56 (s, ^{1}H, C8H).

3.3.6. 5'-O-(4,4'-Dimethoxytrityl)-6-O-[2-(4-Nitrophenyl)Ethyl]- N^2*-Isobutyryl-*N^2*-[Imidazol-1-yl(Propyl)]-2'-Deoxyguanosine*

The substrate **19** (5.94 g, 10 mmol) is dissolved in dry pyridine (75 mL) and evaporated to dryness. This is repeated three times to remove traces of moisture. To a solution of this residue in dry pyridine (100 mL) is added dry triethylamine (4.04 g, 40 mmol), 4-(dimethyl-amino)pyridine (1.2 g) and 4,4'-dimethoxytrityl chloride (10.14 g, 30 mmol) at room temperature. The reaction mixture is stirred at room temperature for 12 h under argon atmosphere. Methanol (50 mL) is added and the stirring is continued for additional 15 min and evaporated to dryness. The residue is purified by flash chromatography on a silica gel column using CH_2Cl_2/Me_2CO containing 1% triethylamine as eluent. The pure fractions are pooled together and evaporated to dryness to give 7.2 g (80%) as a colorless foam: ^{1}H NMR (Me_2SO-d_6) δ1.04 (m, ^{6}H), 3.62 (s, ^{3}H, OCH 3), 3.66 (s, ^{3}H, OCH 3), 6.32 (t, ^{1}H, J1',2' = 6.20 Hz, C1'H), 6.64–7.32 (m, ^{15}H, ImH and ArH), 7.52 (s, ^{1}H, ImH), 7.62 (d, ^{2}H, ArH), 8.16 (d, ^{2}H, ArH), 8.42 (s, ^{1}H, C8H).

*3.3.7. 3'-O-(2-Cyanoethoxy)-*N,N'*-Diisopropylaminophosphoramidite of 5'-*O*-(4,4'-Dimethoxytrityl)-6-O-[2-(4-Nitro-Phenyl)Ethyl]-*N^2*-Isobutyryl-*N^2*-[Imidazol-1-yl(Propyl)]-2'-Deoxyguanosine* ***(9)***

The tritylated nucleoside (2.5 g, 2.79 mmol) is dissolved in dry pyridine (30 mL) and evaporated to dryness. This is repeated three times and the residue is dried over solid sodium hydroxide overnight. The dried **8** is dissolved in dry CH_2Cl_2 (30 mL) and cooled to 0°C under argon atmosphere. To this cold stirred solution is added *N,N*-diisopropylethylamine (0.72 g, 5.6 mmol) followed by 2-cyanoethoxy-*N,N'*-diisopropylaminochlorophosphine (1.32 g, 5.6 mmol) dropwise over a period of 15 min. The reaction mixture is stirred at 0°C for 1 h and at room temperature for 2 h. The reaction mixture is diluted with CH_2Cl_2 (100 mL) and washed with brine (50 mL). The organic extract is dried over anhydrous $MgSO_4$ and the solvent is removed under reduced pressure. The residue is purified by flash chromatography over silica gel using hexane/acetone containing 1% triethylamine as eluent. The main fractions are collected and evaporated to dryness. The residue is dissolved in dry CH_2Cl_2 (10 mL) and added dropwise into a stirred solution of hexane (1500 mL) during 30 min. After the addition, the stirring is continued for additional 1 h at room temperature under argon. The precipitated solid is filtered, washed with hexane and dried over solid NaOH under vacuum overnight to give 2.0 g (65%) of the title compound as colorless powder.

3.4. ODN Synthesis and Modification

3.4.1. Synthesis

Oligonucleotides are synthesized on a 1 µmol scale on either an ABI 380B (ODNs with the modified dGuo analogs) or a MilliGen 7500 (ODNs with modified dUrd analogs) automated DNA synthesizer using the standard cyanoethyl-*N,N*-diisopropylamino-phosphoramidite (CED-phosphoramidite) chemistry. With the modified dGuo analog, a 0.1*M* solution of reagent is used and the time for that cycle is increased to 360 s.

3.4.2. Removal from the Column

ODNs are purified by adaptations of standard methods *(14)*. Removal of the protecting groups and detachment from the column is accomplished in the usual manner, with concentrated ammonia at 25°C for 3 d, or 40°C for 24 h. This treatment efficiently cleaves the trifluoroacetyl and phthalimido groups on the side arm modifications. ODNs with 5'-dimethoxytrityl groups and modified dUrd analogs are chromatographed by HPLC using a Hamilton PRP-1 (7.0 × 305 mm) reversed-phase column employing a gradient of 5–45% CH_3CN in 0.1*M* $Et_3NH^+OAc^-$, pH 7.5, over 20 min.

3.4.3. Detritylation

After detritylation with 80% acetic acid for 1 h at 80°C, the ODNs are precipitated by addition of 3*M* sodium acetate and 1-butanol *(15)*.

3.4.4. Deprotection of ODNs Containing p-Nitrophenylethyl-Protected dGuo

For the ODNs with modified dGuo analogs with and 5'-dimethoxytrityl groups, the CPG column is removed from the synthesizer and flushed with pyridine, and a solution of 1*M* DBU in pyridine is added directly to the column and allowed to stand overnight. The column is flushed with ether, dried with argon, and treated with conc ammonia as usual to remove the ODN and deprotect. These ODNs are chromatographed and detritylated in a manner similar to above.

3.4.5. Final Purification

The oligonucleotides are purified using semipreparative HPLC conditions: 4.5 mm × 50 cm C-18 reverse phase column, eluted with a gradient of 9–13.5% MeCN in 0.1*M* $Et_3NH^+OAc^-$. A 15-mer ODN **20** containing one 5-(3-aminopropyl)-dUrd **6**, for example, gives a product peak eluting near 14 min, which is collected and evaporated to dryness. Figure 3a shows a chromatograph of this 15-mer **20**.

3.4.6. Derivatization of a Modified ODN: Preparation of Bromoacetamidopropyl Oligodeoxynucleotides

One hundred µg (22 nmol) of this purified material **20** is dissolved in 80 µL of H_2O, 10 µL of 1*M* sodium borate (pH 8.3) and mixed with 10 µL of 0.22*M* of *N*-hydroxysuccinimidyl bromoacetate **21** in acetonitrile. The reaction is followed by analytical HPLC (described above) and starting oligonucleotide is consumed within 1 h. Figure 3B shows an HPLC of the reaction mixture. For purification, 400 µL of H_2O is added. The solution is transferred to a Centricon 3 microconcentrator and centrifuged at 5500 rpm in a fixed angle rotor with a radius of 7 cm for 1 h. The Centricon 3 is washed with 2 additional aliquots of H_2O (500 µL). The final retentate had a volume between 35–45 µL. The yield of electrophilic ODN **22**, based on A_{260} of starting material, ranged from 40–75% and is greater than 95% pure based on HPLC analysis. Figure 3C shows the HPLC of the final product. This material is stored frozen and should not be concentrated further.

3.4.7. Formation of Sodium Thiophosphate Adduct

Adduct formation between selected ODNs with electrophilic modifications and sodium thiophosphate is used to indicate the integrity of electrophile. A 100 µL aliquot of 120 µ*M* **14b** is combined with 10 µL of 1*M* sodium borate (pH 8.3) and 10 µL of 12 m*M* sodium thiophosphate and heated at 37°C. After 2 h, analytical HPLC showed that greater than 80% of the starting electrophilic oligonucleotide is converted to the thiophosphate adduct.

4. Discussion

4.1. Synthesis of Modified Nucleosides

The addition of substituted alkynes to the 5-position of dUrd (Fig. 4) has proven to be a popular method for functionalization of a preformed deoxynucleoside. This has been accomplished by palladium-mediated

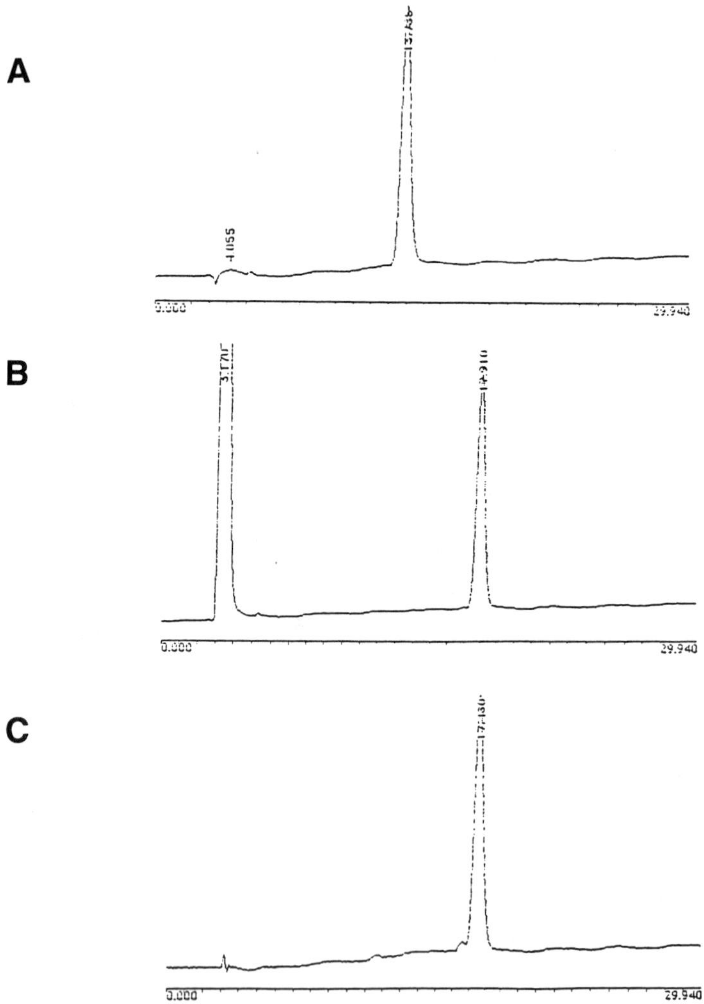

Fig. 3. HPLC elution profiles of ODN **20** (panel **A**), ODN **20** + reagent **21** (panel **B**), and of the purified product **22**. Elution of the 4.5 mm × 50 cm C-18 reverse phase column was with a gradient of 9–13.5% MeCN in 0.1*M* $Et_3NH^+OAc^-$.

coupling of the alkyne with 5-I-dUrd in the presence of CuI, using either Pd(2)[3] or Pd(0) *(16)*. Both methods are effective; the latter may be less sensitive to variation in experimental conditions. Both the phthalimide and trifluoroacetyl blocking groups on the terminal amine seem to be effective; the CF_3CO derivative **12a, 13a** is preferred

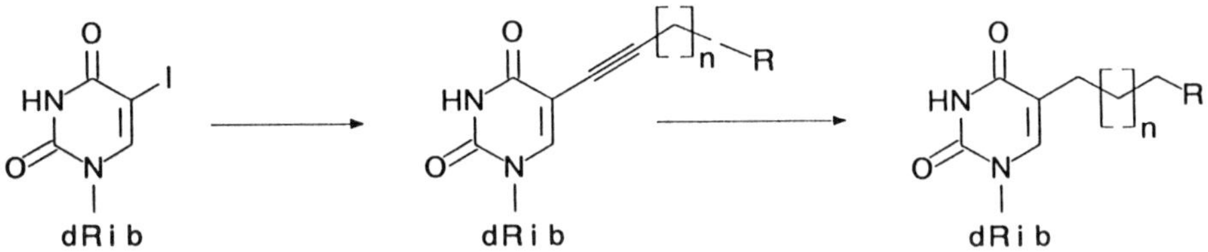

Fig. 4. Synthesis of 5-substituted-dUrd.

in the series with the 3-carbon side arm because the amidite and other intermediates are crystalline. Preparation of 4-phthalimido-1-butyne was accomplished by direct conversion from an alkynol using the DEAD methodology *(17)*. The 5-alkynyl-dUrd **12** can be used directly in ODN synthesis *(2)*, or reduced to a 5-(substituted alkyl)-dUrd **13**. In our hands, this reduction can lead to side products unless care is taken. Several methods for reduction can be used; the ones given were found to be generally applicable to compounds in this class. In the ammonium formate reduction (formate is the hydrogen source), care must be taken that the solution remains slightly acidic to avoid amine deprotection.

The 5-(ω-(blocked amino)alkyl)-2'-deoxyuridines **13a** and **13b** were converted to the corresponding 5'-dimethoxytrityl-3'-*O*-cyanoethyl-diisopropylphosphoramidites by standard methods *(18)*.

For the preparation of the N^2-substituted deoxyguanosines (Fig. 5), 2-chlorodeoxyinosine (**15**, prepared from blocked 2,6-dichloropurine 2'-deoxyriboside **14**) was treated with the appropriate substituted amine to give the *N*-substituted dGuo **16**. This was protected as the *N*-*i*-butyryl-O^6-*p*-nitrophenylethyl derivative **19**. The nitrophenethyl group was used to protect the O^6 position to avoid potential problems in the automated synthesis cycles, although the commercially available dGuo amidites were used for unmodified dGuo introduction. Dimethoxy-tritylation and phosphitylation by standard methods gave the reagent for the automated synthesizer.

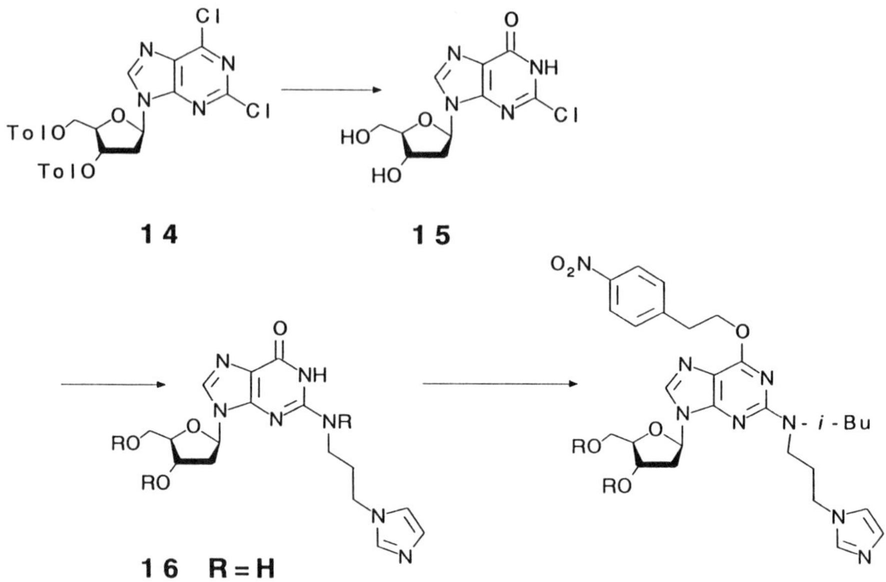

16 R = H
17 R = *i*-Bu

Fig. 5. Synthesis of protected N^2-substituted dGuo.

4.2. Synthesis of Modified Oligodeoxynucleotides

Amidites of the modified dUrd derivatives were dissolved in acetonitrile used on an automated Milligen 7500 synthesizer using standard 1 µmol scale protocols. The coupling yields based on trityl release for all the modified nucleosides were equivalent to the commercial thymidine amidite (>99%). The trifluoroacetyl and phthalimide groups were removed during routine ammonia treatment. The ammonia solutions containing the dimethoxytrityl ODN were applied directly to a polystryene (PRP-1) column *(19)*, accomplishing initial purification and removal of ammonia in one step. After detritylation, the ODNs were purified a second time by reverse phase chromatography.

The modified dGuo derivatives were incorporated into oligonucleotides using the standard protocol on the ABI 380B, except that the treatment time with modified nucleoside was increased. The *p*-nitrophenethyl group was removed with DBU.

Reaction of the amine-modified ODNs **20** at a concentration of 1–1.5 mg/mL with a 100-fold molar excess of *N*-hydroxysuccinimidyl bromoacetate **21** (Fig. 6) in 0.1*M* sodium borate (pH 8.3) at room temperature

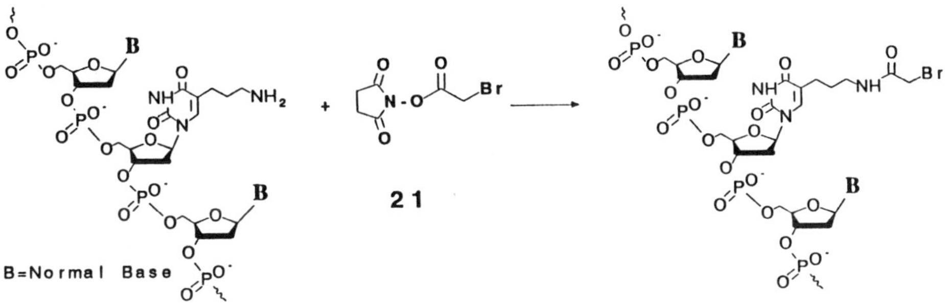

Fig. 6. Bromoacetylation of an oligonucleotide containing a 5-(3-amnopropyl)-dUrd residue.

caused complete disappearance of starting material within 1 h. Electrophilic ODN formation in the reaction was followed by HPLC and indicated by consumption of the amine-modified ODN starting material and appearance of the product **22** peak. The reactivity of the electrophilic moiety on the ODN was indicated by its quantitative conversion to a new product peak after treatment with sodium thiophosphate. The initial method used to isolate the electrophilic ODNs was to purify the product by chromatography. Both gel filtration and reverse phase chromatography resolved the reaction mixture. If the electrophilic ODNs are allowed to stand after concentration, degradation occurs (*see* Note 1). The electrophilic ODNs were isolated >95% pure by centrifugal dialysis directly from the reaction with a Centricon-3 microconcentrator. The centrifugally dialyzed ODNs were diluted with deionized water to 0.5 to 1.5 mg/mL and were >90% pure, based on HPLC, for up to 6 mo at –20°C (*see* Note 2).

5. Notes

1. The reactive electrophilic oligonucleotide 22 is stable frozen in dilute solution for several months. It should not be lyophilized, however, because the concentrated solids or solutions are quite unstable.
2. The oligonucleotides with aminoalkyl side chains have been derivatized by many functional groups. The placement of fluorescent labels, biotin, and large hydrophobic groups onto these side chains using the appropriate activated esters has been accomplished in our and other laboratories.

Acknowledgments

The author is indebted to K. Ramasamy and P. Dan Cook of Isis Pharmaceuticals for the details of the deoxyguanosine derivatives, and to Alexander Gall and John Tabone for helpful discussions. This work was supported in part by Grant CA45905 from the NIH, USPHS, HHS, and from a contract from the US Army, DAMD17-88-C-8201.

References

1. Meyer, R. B., Jr., Tabone, J. C., Hurst, G., Smith, T. M., and Gamper H. (1989) Efficient, specific crosslinking and cleavage of DNA by stable, synthetic complementary oligodeoxynucleotides. *J. Am. Chem. Soc.* **111,** 8517–8519.
2. Haralambidis, J., Chai, M., and Tregear, G. W. (1987) Preparation of base-modified nucleosides suitable for non-radioactive label attachment and their incorporation into synthetic oligodeoxyribonucleotides. *Nucleic Acids Res.* **15,** 4856–4876.
3. Gibson, K. J. and Benkovic, S. J. (1987) Synthesis and application of derivatizable oligonucleotides. *Nucleic Acids Res.* **15,** 6455-6467.
4. Dreyer, G. B. and Dervan, P. B. (1985) Sequence-specific cleavage of single-stranded DNA: oligodeoxynucleotide-EDTA-Fe(II). *Proc. Natl. Acad. Sci. USA* **82,** 968–972.
5. Horn, T. and Urdea, M. S. (1989) Forks and combs and DNA: the synthesis of branched oligodeoxyribonucleotides. *Nucleic Acids Res.* **17,** 6959–6967.
6. Sanghvi, Y. S., Hoke, G. D., Zounes, M., Chan, H., Acevedo, O., Ecker, D. J., Mirabelli, C. K., Crooke, S. T., and Cook, P. D. (1991) Synthesis and biological evaluation of antisense oligonucleotides containing modified pyrimidines. *Nucleosides Nucleotides* **10,** 345.
7. Ramasamy, K., Springer, R. S., Martin, J. F., Freier, S. M., Hoke, G. D., Bruice, T. W., and Cook, P. D. (1991) Synthesis and biophysical evaluation of N^2-substituted guanine and adenine modified oligonucleotides as catalytic cleavers of RNA. International Union of Biochemistry Conference on Nucleic Acid Therapeutics, Clearwater Beach, FL, No. 82.
8. Petrie, C. R., Adams, A. D., Stamm, M., Van Ness, J., Watanabe, S. M., and Meyer, R. B., Jr. (1991) A novel biotinylated adenylate analog derived from pyrazolo[3,4-*d*]pyrimidine for labeling DNA probes. *Bioconj. Chem.* **2,** 441–446.
9. Still, W. C., Kahn, M., and Mitra, A. (1978) Rapid chromatographic technique for preparative separations with moderate resolution. *Org. Chem.* **43,** 2923–2925.
10. Maxam, A. M. and Gilbert, W. (1980) Sequencing end-labeled DNA with base-specific chemical cleavages. *Meth. Enzymol.* **65,** 499–560.
11. Sambrook, J., Fritsch, E. F. II, and Maniatas, T. (1989) *Molecular Cloning: A Laboratory Manual,* 2nd Edition, Cold Spring Harbor Laboratory, Cold Spring Harbor, NY.

12. Eur. pat. appl. 87305848-1/01.07.87.
13. Kazimierczuk, Z., Cottam, H. B., Revankar, G. R., and Robins, R. K. (1984) Synthesis of 2'-deoxytubercidin, 2'-deoxyadenosine, and related 2'-deoxynucleosides via a novel direct stereospecific sodium salt glycosylation procedure. *J. Am. Chem. Soc.* **106,** 6379–6382.
14. McLaughlin, L. W. (1984) in Gait, M. J. (ed.), *Oligonucleotide Synthesis—A Practical Approach*, IRL, Oxford, UK.
15. Van Ness, J., Kalbfleisch, S., Petrie, C. R., Reed, M. W., Tabone, J. C., and Vermeulen, N. M. J. (1991) A versatile solid support system for oligonucleotide probe-based hybridization assays. *Nucleic Acids Res.* **19,** 3345–3350.
16. Hobbs, F. W., Jr. (1989) Palladium-catalyzed synthesis of alkynylamino nucleosides. A universal linker for nucleic acids. *J. Org. Chem.* **54,** 3420–3422.
17. Mitsunobu, O. (1981) The use of diethyl azodicarboxylate and triphenylphosphine in synthesis and transformation of natural products. *Synthesis* 1–23.
18. Jones, R. A. (1984) In *Oligonucleotide Synthesis* (Gait, M. J., ed.), IRL, Washington, DC, Chapter 2.
19. Germann, M. W., Pon, R. T., and van de Sande, J. H. (1987) A general method for the purification of synthetic oligodeoxyribonucleotides containing strong secondary structure by reversed-phase high-performance liquid chromatography on PRP-1 resin. *Anal. Biochem.* **165,** 399–405.

CHAPTER 3

Functionalization of Oligonucleotides with Amino Groups and Attachment of Amino Specific Reporter Groups

Sudhir Agrawal

1. Introduction

The most common technique to label oligodeoxynucleotides is the enzymatic incorporation of the radioisotope ^{32}P at the 5'-end. Although this method affords high sensitivity of detection, the use of the radiolabeled probe is precluded in many clinical and diagnostic applications. In the last few years, several labeling procedures for the detection of specific nucleic acid sequences using nonradioactive labels such as biotin, digoxigenin, dinitrophenol, acetylaminofluorene, and certain enzymes or fluorophores have been developed. The labels are incorporated into synthetic oligonucleotides either enzymatically or chemically before hybridization of the probe.

The enzymatic incorporation of biotin into DNA was accomplished by the reaction of DNA polymerase using d-TTP and d-UTP analogs, which contain a biotin or masked amino group at the C-5 position of the pyrimidine rings *(1–9)*. Analogs of dATP and dCTP functionalized at position 6 and at position 4, respectively, have also been incorporated *(10–12)*. This technique suffers from some disadvantages, such as the preparation of the large amount of probe is expensive in terms of enzymes and substrates.

Chemical methods for labeling of DNA with nonradioactive labels are more suitable. To date, the functionalization of oligonucleotides for attaching these radioactive labels has largely relied on:

From: *Methods in Molecular Biology, Vol. 26: Protocols for Oligonucleotide Conjugates*
Edited by: S. Agrawal Copyright ©1994 Humana Press Inc., Totowa, NJ

1. Functionalization of 5'- or 3'-termini by numerous chemical reactions using deprotected oligonucleotides in aqueous or largely aqueous medium;
2. Synthesizing the modified nucleoside containing the masked primary amino group on the base heterocycle and its incorporation into oligonucleotides during synthesis;
3. Use of suitably protected chemical moities, which can be coupled at the 5'- or 3'-termini of protected oligonucleotide during synthesis.

Functionalization of deprotected oligonucleotide has been carried out to introduce a primary amino group *(13,14),* a sulfydryl group *(15,16),* aldehyde and carboxyl groups *(17)* and also a hydrazide group *(18)* at the 5'-terminus. A sulfhydryl group at the 3'-terminus has also been introduced *(19).* The functionalization of deprotected oligonucleotide has been discussed in detail in Chapter 5.

Another approach is to incorporate suitably protected modified nucleoside phosphoramidites, which carry a masked primary amino group at the base heterocycle *(20–32).* The advantage of this approach is that multiple amino groups can be added through repetitive coupling cycles. However, the preparation of such reagents requires many synthetic steps, and modified nucleoside bases are added that might change the hybridization properties of oligonucleotides.

Suitably protected nucleosides modified at the 5'-position of a sugar such as 5'-amino-5'-deoxynucleoside phosphoramidites *(33–35)* and 5'-mercapto-5'-deoxynucleoside phosphoramidites *(36)* have been used as the last coupling to incorporate an amino or sulfhydryl group at the 5'-terminus of synthetic oligonucleotides. The use of these nucleosides, however is limited because they allow incorporation of a single primary amine or sulfhydryl group, only at the 5'-terminus. Incorporation of modified nucleosides in oligonucleotides has been discussed in detail in Chapters 2 and 7.

The procedure for introducing a primary aliphatic amino group at the 5'-end of oligonucleotides was simplified by the introduction of a conveniently prepared phosphoramidite amino modifying reagent constructed from amino ethanol *(37)* or amino hexanol *(38,39)* rather than a deoxyribonucleoside. Subsequently, various analogs of aminoalkyl alcohols have been prepared for the same purpose *(40–43).* A cyclic phosphoramidite reagent has also been reported *(44).* Similarly, linker molecules have been developed to incorporate a sulfhydryl group at the 5'-terminus *(45)*. These linker molecules have limited utility because only a single functional group can be incorporated and only at the 5'-terminus.

Other possible routes to attach a nonradioactive label at specific sites in oligonucleotides are at suitably functionalized phosphate backbones. Phosphoramidate analogs of dinucleotides have been synthesized, and reporter groups have been attached to oligonucleotides through a phosphoramidate linkage *(46,47)*. Recently, phosphorothioate diester internucleoside linkages have been incorporated at specific sites in oligonucleotides *(48,49)* and thiol specific reporter groups have been attached.

In the present chapter, various approaches for incorporating amino group onto oligonucleotides have been discussed (Fig. 1). These approaches are divided into three sections:

1. Incorporation of the amino group at the 5'-end of an oligonucleotide;
2. Incorporation of the amino group at the internucleotide phosphate of an oligonucleotide;
3. Incorporation of the amino group at the 3'-end of an oligonucleotide.

2. Materials

2.1. DNA Synthesis Reagents

1. The nucleoside phosphoramidites and related chemicals for DNA synthesis were obtained from Millipore.
2. The amino linker molecule was either synthesized (*see* Section 3) or obtained from commercial sources.
3. Biotin N-hydroxysuccinimide, fluorescein isothiocyanate, and tetramethylrhodamine isothiocyanate were obtained from Molecular Probes.

2.2. HPLC Buffers and Conditions

1. The HPLC system consisted of a 600E gradient programmer, a 481 variable wavelength UV detector, a 745 data module and Rheodyne a 7125 injector.
2. Ion exchange HPLC was carried out using Whatman partisphere SAX column.
3. Reversed phase HPLC was carried out using a Novapak C_{18} cartridge for use with an RCM 100 cartridge holder (Waters, Milford, MA).
4. Buffers for ion exchange HPLC were prepared from a stock solution of 1*M* KH_2 PO_4 adjusted to pH 6.3 with KOH to give (A) 1 m*M* and (B) 300 m*M*, both containing 60% formamide.

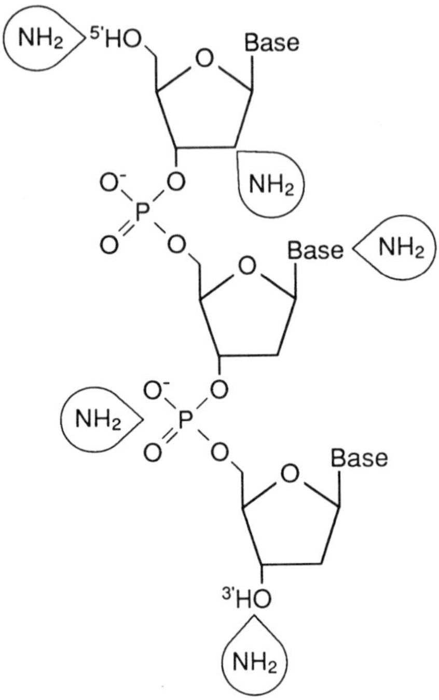

Fig. 1. Sites of amino group incorporation in an oligonucleotide. Amino groups can be introduced at 5'-terminus, at 3'-terminus, at internucleotide phosphate, at heterocyclic bases and also at 2'-position.

5. Buffers for reversed phase HPLC were 0.1*M* ammonium acetate (pH unadjusted) containing (A) 0% CH_3CN and (B) 80% CH_3CN.

3. Methods

3.1. Synthesis of the Linker Molecule for Incorporating an Amino Group at the 5'-End of Oligonucleotide (Fig. 2)

*3.1.1. 2-(9-Fluorenylmethoxycarbonyl) Aminoethanol **3***

1. Place ethanolamine, **1** (10% aqueous w/v, 22.5 mmoles) in a round bottomed flask, with aqueous sodium carbonate (30% solution w/v, 59 mmoles) and acetone (20 mL).
2. Add dropwise under stirring a solution of 9-fluorenylmethylchloroformate, **2** (23.65 mmoles) in acetone (15 mL) over 15 min. Continue stirring for another 30 min. A white solid will separate out.

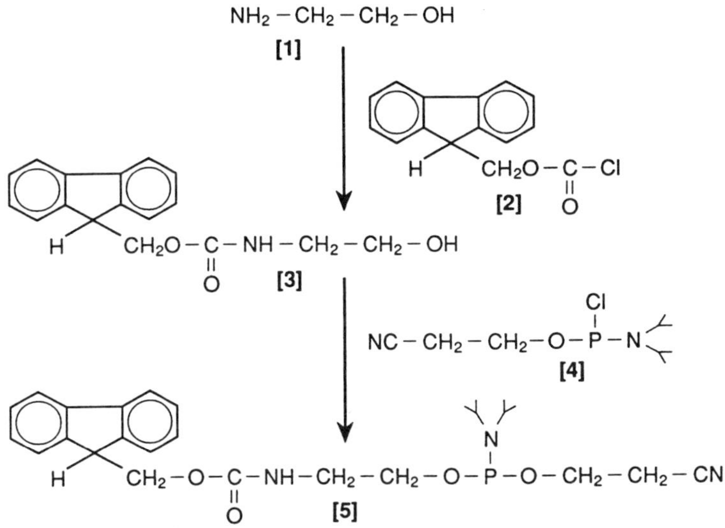

Fig. 2. Synthesis of linker molecule for introducing amino group at 5'-terminus of an oligonucleotide.

3. Filter the white solid using using sintered glass funnel. Wash the filtered solid with ice cold acetone-water (25 mL). Dry the solid mass in a vacuum dessicator to obtain approx 5.96 g of **3** (mpt 144°C). The yield is typically in the range of 85–95%. ^{1}H NMR (δ) (DMSO) 2.2 (S (b) OH, ^{1}H); 3.4 (m, CH_2O, ^{2}H); 3.6 (t, CH_2N, ^{2}H); 4.2–4.5 (m, CH_2OCO + H (C-9), ^{3}H); 5.2 (s (b), NH, ^{1}H); 7.2–7.9 (m,aromatic, ^{8}H).

3.1.2. 2-(9-Fluorenylmethoxycarbonyl) Aminoethyl (2-Cyanoethyl, N,N*-Diisopropylamino) Phosphite,* **5**

1. Take vacuum dried 2-(9-fluorenylmethoxycarbonyl) aminoethanol, **3** (0.239 g, 1 mmole), suspend in anhydrous dichloromethane (3 mL), add anhydrous diisopropylethylamine (0.78 mL, 4.6 mmole), and stir the mixture under an argon atmosphere at 0°C temperature (using an ice bath).
2. To the stirred mixture, add 2-cyanoethyl, *N,N*-diisopropylaminophosphochloridite, **4** (0.473 g, 2 mmole) dropwise through a syringe. After a few minutes of stirring , a white solid will precipitate out.

3. Remove the ice bath and continue stirring for 15 min at room temperature, and check the progress of the reaction by TLC (dichloromethane-ethylacetate and triethylamine; 9:9:1 v/v). When the reaction is complete, add ethylacetate (30 mL) to the reaction mixture. Wash the reaction mixture with cold saturated brine (2 × 25 mL) and dry the organic phase over anhydrous sodium sulfate.
4. Concentrate the reaction mixture using rotary evaporator to obtain the product, **5** a colorless oil, 470 mg (99% yield). ^{31}P NMR (δ) (DMSO) 146.4.

3.2. Synthesis of a Linker Molecule for Incorporation of an Amino Group at the Internucleotide Phosphate of an Oligonucleotide (Fig. 3)

3.2.1. N-1-Trifluoroacetylhexanediamine 7

1. Place hexanediamine, **6** (1.16 g; 10 mmole) in dry round bottomed flask, add triethyl amine (1 mL; 7 mmole) and methanol (20 mL). Stir the mixture at room temperature. To the stirred solution, add ethyltrifluoroacetate (1.2 mL, 10 mmole) dropwise over 1 h, using a dropping funnel. Continue stirring the reaction mixture overnight.
2. Concentrate the reaction mixture using a rotary evaporator to obtain a solid mass. The solid mass can be purified using flash chromatography using silica gel (solvent: 0–25% methanol in dichloromethane).
3. Pool the fractions after flash chromatography, containing the desired products, and concentrate to obtain a colorless powder of the product, **7,** 1.1 gm, yield 42.6%, melting point 52°C. ^{1}H NMR ($CDCl_3$, d, TMS = 0.00) 7.1–7.2 (m, ^{3}H, NH_2, NH) 3.2–3.3 (m, ^{2}H, CO-NH-CH_2) 2.8–2.9 (m, ^{2}H, CH_2-NH_2) 1.2–1.6 (m, ^{8}H, -CH_2-$(CH_2)_2$-CH_2-).

3.3. Synthesis of a Linker Molecule for Incorporation of an Amino Group at the 3'-End of an Oligonucleotide (Fig. 4)

3.3.1. 1-0-Dimethoxytrityl-1,3 Propanediol, 9 (Step a, Fig. 4)

1. Place a solution of 1,3 propanediol **8** (1.14 gm., 15 mmole) in pyridine (30 mL). To the stirred solution, add a solution of 4,4'-dimethoxytrityl chloride (1 g, 13 mmole) dropwise over 15 min. Continue stirring the reaction mixture for 3 h.
2. Add 5% aqueous $NaHCO_3$ solution (60 mL) to the reaction mixture. Extract the product with dicholoromethane (50 mL × 2). Wash the organic phase with water and dry over anhydrous Na_2SO_4.

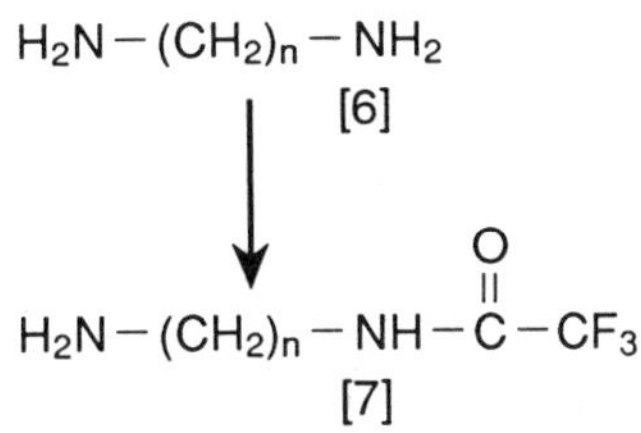

Fig. 3. Synthesis of linker molecule for introducing amino group at internucleotide linkage.

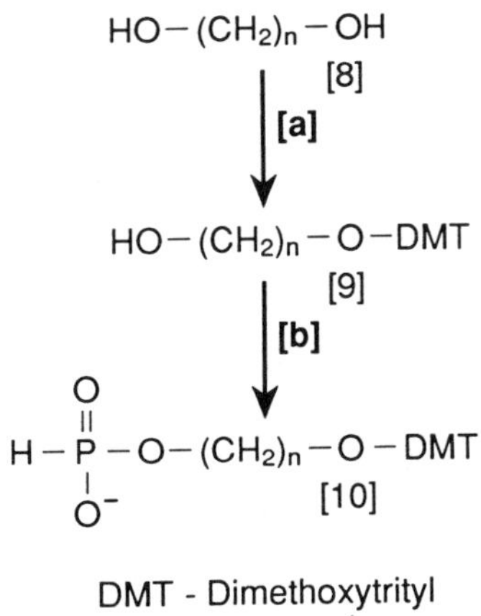

Fig. 4. Synthesis of linker molecule for introducing single or multiple amino groups at 3'-terminus of oligonucleotide.

3. Evaporate the solvent using a rotary evaporator to obtain the oily residue. Purify the residue by applying to a silica column and eluting with CH_2Cl_2. The final product is obtained as a yellow oil in 70% yield.

*3.3.2. 1-0-Dimethoxytrityl-1,3-Propanediol H-Phosphonate, **10** (Step b, Fig. 4)*

The H-phosphonate of 1-0-dimethoxytrityl-1,3-propanediol, **10** can be prepared by the same procedure as described in Chapter 4 in vol. 20 of this series.

3.3.3. 1-0-Dimethoxytrityl-1, 3-propanediol-CPG

1. Place 10 g of long chain alkyamine controlled pore glass support (LCAA-CPG) in pyridine (50 mL) in a round bottomed flask. Add succinic anhydride (2 g, 20 mmole), 4-dimethylaminopyridine (400 mg, 3.3 mmole).

Cap the round bottomed flask tightly and shake on a rotary shaker for 16–20 h. Filter off the CPG using a sintered funnel, wash with pyridine, followed by dichloromethane and ether. Dry the long chain alkylamidopropanoic acid CPG (LCAAP-CPG) under vacuum.

2. Place 1 gm of LCAAP-CPG in pyridine (15 mL) and add 1-0-dimethoxytrityl-1,3-propanediol (190 mg, 0.5 mmole), 4-dimethylaminopyridine (12 mg, 0.1 mmole), triethylamine (80 µL) and diethylcarbodiimide (380 mg, 2 mmole). Shake the mixture in a sealed round bottomed flask using a rotary shaker for 24 h. Add pentachlorophenol (130 mg, 0.5 mmole) and shake the mixture overnight. Filter off the CPG, and wash successively with pyridine, dichloromethane, and ether.
3. Place dry CPG from Step 2 in a round bottomed flask, add piperidine (10 mL) and shake for 6 h, to cap the unreacted sites. Wash the 1-0-dimethoxytrityl-1,3-propanediol-CPG with dichloromethane and dry in vacuum. The loading of the CPG will be in the range of 40–60 µmole/g.

3.4. Incorporation of an Amino Group at the 5'-End of an Oligonucleotide (Fig. 5)

3.4.1. Synthesis

1. Prepare a solution of linker molecule (vacuum dried) in anhydrous acetonitrile (25 mg/mL) and place into an extra reservoir (U-bottle) of automated synthesizer.
2. At the start of the synthesis of required oligonucleotide sequence, add a U base at the 5'-end. This will enable the linker molecule from the U reservoir to couple at the end of the oligonucleotide sequence.
3. Start the synthesis using the appropriate coupling cycle. The same coupling cycle will be used to carry out the linker molecule coupling.
4. At the end of the oligonucleotide synthesis, remove the column, wash the solid support with acetonitrile (5 mL) using a luer end syringe and dry the support.

3.4.2. Deprotection

1. Remove the solid support from the column and transfer into a screw capped vial (2 mL). Add 0.5 to 1 mL of concentrated ammonium hydroxide solution (30%). Seal the tube tightly to avoid ammonia leakage. Incubate the deprotection mixture at 55°C for 8 h.
2. Cool the deprotection mixture to 0°C, transfer the supernatant to either a small bottomed flask or an Eppendorf tube for removing the ammonia. After transferring the ammonia supernatant , rinse the support with 0.5–1 mL water and mix with the ammonia supernatant. Concentrate the ammonia solution to dryness either by rotary evaporation or SpeedVac.

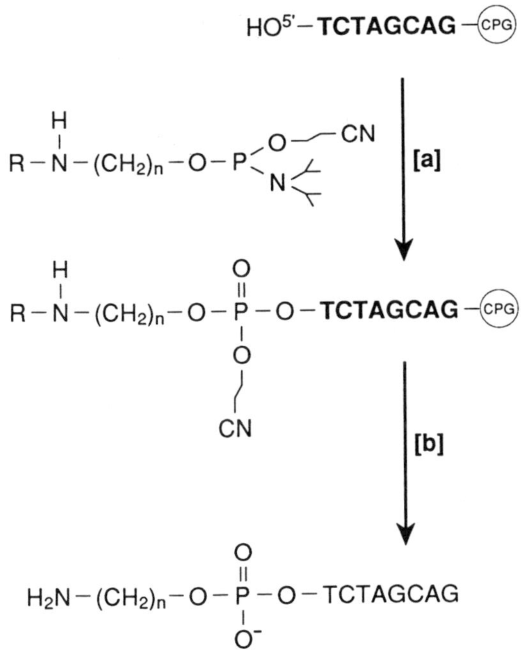

Fig. 5. Incorporation of amino group at 5'-terminus of oligonucleotide.

3. Dissolve the dried material in 1 mL of water. This crude mixture contains the full length oligonucleotide containing the 5'-amino group, along with failure oligonucleotide sequences.

3.4.3. Purification and Analysis

Purification of oligonucleotides carrying a 5'-amino group can be carried out by reversed phase HPLC, ethanol precipitation or polyacrylamide gel electrophoresis. The crude mixture can be desalted using a Sep pak C_{18} column. This desalted mixture can also be used for attaching a reporter group onto the oligonucleotide.

3.4.3.1. HPLC Analysis and Purification

1. Take 0.2 A_{260} U of crude, analyze on reversed phase HPLC using the condition described in the Materials Section.
2. Collect the slowest eluting peak. The change in retention time of derivatized oligonucleotide is caused by a change in hydrophobicity. Oligonucleotides derivatized with two carbon linkers (ethyl linkers) will elute earlier than oligonucleotides derivatized with six carbon linkers (hexyl linkers).

3. Remove the solvent by evaporation using a rotary evaporator. The purified derivatized oligonucleotide can be desalted using a Sep pak C_{18} cartridge or Sephadex G-25. Remove the solvent after desalting, resuspend the purified derivatized oligonucleotide in water and store at –20°C until further use.

3.5. Incorporation of an Amino Group at the Internucleotide Linkage of an Oligonucleotide

Incorporation of an amino group at a desired internucleotide linkage in an oligonucleotide can be carried out by using a combination of the phosphoramidite approach and the H-phosphonate approach for assembling the oligonucleotide. If the amino group is to be incorporated at the first internucleotide linkage at the 3'-end, only the H-phosphonate approach can be used. Protocols for the approaches are as follows:

3.5.1. Incorporation of an Amino Group at the Internucleotide Linkage at the 3'-End of an Oligonucleotide

3.5.1.1. Synthesis

1. Start the synthesis of the required oligonucleotide sequence using the H-phosphonate approach (*see* Chapter 4 of vol. 20 in this series). After the first coupling and washings (leave the DMT-group on), remove the column from the synthesizer. Purge helium or argon through the column to dry the CPG.
2. Prepare a solution of N-1-trifluoroacetylhexanediamine in carbon tetrachloride (10%, w/v) for a 1 µ*M* scale synthesis, take 2 mL of this solution in a dry glass or plastic syringe with a luer end fitting and attach it to a column containing CPG bound dinucleoside H-phosphonate. Attach an empty syringe at the other end of the column. Slowly push the carbon tetrachloride solution from one syringe through the column into an empty syringe. Leave the CPG in contact with the CCl_4 solution for 10 min. After 10 min, push the remaining CCl_4 solution into the empty syringe through the column and leave it for another 10 min. Remove one of the syringes, and wash the CPG with 1–2 mL CCl_4. Dry the CPG by flushing with dry helium or argon. Finally wash the column with 2–3 mL of acetonitrile.
3. Place the column back onto the DNA synthesizer, and continue synthesis of the required oligonucleotide by using either the phosphoramidite approach or the H-phosphonate approach. If the oligonucleotide is synthesized using the H-phosphonate approach, it needs to be oxidized with iodine to generate phosphodiester linkages.

Multiple amino groups can also be introduced at the 3'-end at contiguous internucleoside linkages. To carry this out, sequences of the oligonucleotide where amino groups are to be incorporated should be assembled using H-phosphonate chemistry, e.g., if two amino groups are to be incorporated, the first two couplings of oligonucleotide sequence should be carried out using H-phosphonate chemistry, followed by oxidation using N-1-trifluoroacetylhexanediamine in CCl_4.

3.5.2. Incorporation of an Amino Group at a Desired Internucleotide Linkage in an Oligonucleotide (Fig. 6)

3.5.2.1. Synthesis

For incorporation of an amino group at a desired internucleotide linkage in the middle of the oligonucleotide, a combination of the phosphoramidite approach and the H-phosphonate are used for assembling oligonucleotide. An example is given here for assembling an oligonucleotide carrying one amino group in the center (Fig. 6)

1. Start the synthesis of the oligonucleotide sequence using the phosphoramidite approach. At the site where the amine group is to be incorporated, remove the column to carry out one coupling using the H-phosphonate approach (step a, Fig. 6).
2. After the H-phosphonate coupling (leave DMT-on), dry the CPG bound oligonucleotide carrying one H-phosphonate linkage, and oxidize with N-1-trifluoroacetylhexanediamine, as described above in Section 3.5.1. (step b, Fig. 6).
3. After the oxidation and related washing steps, place the column back on the automated DNA synthesizer to continue the synthesis of oligonucleotide sequence using the phosphoramidite approach (step c, Fig. 6).

3.5.3. Deprotection

1. Remove the solid support from the column and transfer into a screw capped vial for treatment with ammonia. Follow all the steps as described in Section 3.4.2. (step d, Fig. 6).

3.5.4. Purification and Analysis

The crude mixture contains oligonucleotide carrying an amino group at a desired internucleotide linkage, in addition to failed sequences. This mixture should be desalted using a Sep pak C_{18} cartridge (reversed phase based) or Sephadex G-25 (gel permeation), before attaching reporter groups.

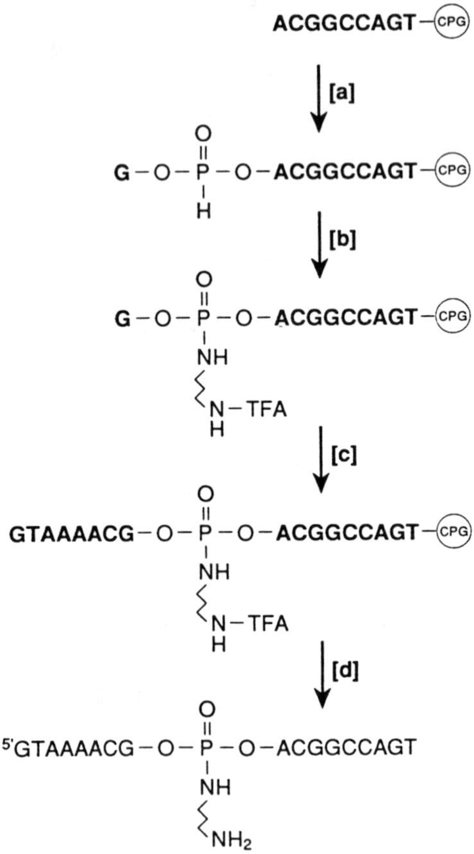

Fig. 6. Incorporation of amino group at internucleotide linkage in the center of the oligonucleotide.

An oligonucleotide carrying an amino group at the internucleotide linkage, (phosphoramidate linkage) contains diastereoisomers, which can be separated and purified before attaching reporter groups or after attaching reporter groups. If the oligonucleotide contains more than one phosphoramidate linkage, separation of diastereoisomers is not efficient.

3.5.4.1. HPLC Analysis and Purification

3.5.4.1.1. Ion Exchange HPLC

1. Take 0.2 A_{260} U of crude mixture, analyze using ion exchange HPLC using the condition described in Section 2.

2. An oligonucleotide carrying amino groups at an internucleotide linkage carries one less charge than the underivatized oligonucleotide, so it will elute earlier than the underivatized oligonucleotide. On ion exchange HPLC, diastereoisomers will not separate.

   Oligonucleotides carrying amino groups at internucleotide phosphates should be not be purified using ion exchange HPLC as the position of elution of an *N*-1 peak of underivatized oligonucleotide is the same as the derivatized oligonucleotide (both carry the same number of charges).

3.5.4.1.2. Reversed Phase HPLC

1. Take 0.2 A_{260} U of crude mixture in water, analyze on reversed phase column using the condition described in the Materials section.
2. An oligonucleotide carrying an amino group will elute later than an underivatized oligonucleotide. If the oligonucleotide is carrying one phosphoramidate linkage, the peak will elute as a singlet or doublet. For an oligonucleotide carrying more than one phosphoramidate linkage, the peak will elute as a multiplet.
3. Collect the peak, evaporate the solvent, and desalt using Sep pak C_{18} or Sephadex G-25. Evaporate the solvents of the desalted derivatized oligonucleotide. Suspend in water and store at –20°C until further use.

3.6. Incorporation of an Amino Group at the 3'-End of an Oligonucleotide

3.6.1. Dialdehyde Approach (Fig. 7)

3.6.1.1. Synthesis

1. Start the synthesis of the required oligonucleotide sequence using the phosphoramidite approach or the H-phosphonate approach on a ribonucleoside CPG. After the assembly, deprotect the oligonucleotide using concentrated ammonium hydroxide (as described above), to obtain the oligonucleotide carrying one ribonucleoside at the 3'-end. Purify the oligonucleotide either by HPLC (ion exchange or reversed phase) or PAGE and desalt the purified oligonucleotide (step a, Fig. 7).
2. Take 5–20 A_{260} U of oligonucleotide in 500 µL of 100 m*M* sodium acetate buffer (pH 5) and mix with 500 µL of a freshly prepared solution of sodium periodate solution (10 m*M*). Incubate the reaction mixture at 4°C for 1 h in dark (step b, Fig. 7).
3. Dialyze the oxidized oligonucleotide, at 4°C in dark, sequentially against two changes of 0.05*M* sodium acetate pH 5.1, 0.1*M* NaCl and two changes of 0.3*M* sodium borate pH 9, 0.1*M* NaCl (step c, Fig. 7).
4. To the dialyzed oligonucleotide, add 1,6-diaminohexane (stock solution of the diamine preadjusted to pH 9.3) to obtain an 0.4*M* concentra-

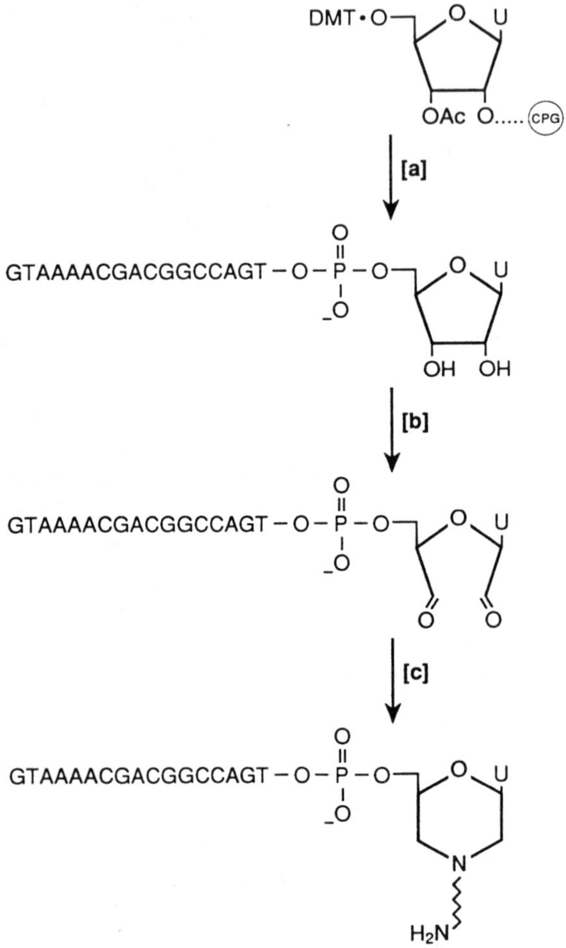

Fig. 7. Incorporation of amino group at 3'-terminus.

tion of amine. Incubate the reaction mixture for 45–60 min at 20°C in the dark.

5. Prepare a fresh stock solution of $NaBH_4$ in water. Add aliquots of $NaBH_4$ solution to the reaction mixture (step 4, Fig. 7) four times at 30-min intervals, resulting in increments of 0.025*M* to 0.1*M* $NaBH_4$ and continue incubation for a further 3 h at 20°C. Adjust the pH of the reaction mixture to pH 5–5.5 with 4*M* sodium acetate pH 5 to quench the residual $NaBH_4$.

6. The derivatized oligonucleotide can be purified by using HPLC or PAGE or can be used at this stage for attaching reporter groups. Instead of using the diamine in step 4, a variety of reporter groups carrying amino, hydrazide, thiosemicarbazide, semicarbazide can be attached directly to the oxidized oligonucleotides.

3.6.2 Linker Approach (Fig. 8)

Using this approach, one or more amino groups can be incorporated at the 3'-end of an oligonucleotide.

3.6.2.1. For a Single Amino Group Incorporation (Fig. 8, Steps a–e)

1. Start the synthesis using the H-phosphonate approach on Linker-CPG (*see* Sections 2 and 3). After the first coupling, (the first nucleoside of the oligonucleotide) and washings, remove the column from the synthesizer (leave DMTr-group on).
2. Oxidize the support with a solution of *N*-1-trifluoroacetylhexanediamine in CCl_4, as described above in Section 3.5.1. After oxidation and related washing steps, continue the synthesis using the phosphoramidite approach or the H-phosphonate approach.
3. Deprotect the oligonucleotide using concentrated ammonium hydroxide (*see* Section 3.4.2.). The deprotected oligonucleotide carries a single amino group at the 3'-end.

3.6.2.2. For Multiple Amino Group Incorporation (Fig. 8, Steps a, b, f–i)

1. Start the synthesis using the H-phosphonate approach on Linker-CPG (*see* Material and Methods section). Carry out couplings of Linker H-phosphonate molecules on Linker-CPG. The number of couplings to be carried out depends on the number of amino groups to be incorporated, e.g., if three amino groups are to be incorporated, carry out three couplings with Linker H-phosphonate.
2. After couplings with Linker-H-phosphonate, remove the column from the synthesizer. Flush dry helium or argon through the column to dry the CPG. Oxidize the CPG support with a solution of *N*-1-trifluoroacetylhexanediamine in CCl_4, as described above in Section 3.5.1. After the oxidation, continue the synthesis of the required oligonucleotide sequence on the above derivatized CPG, by using the phosphoramidite approach or the H-phosphonate approach.
3. Deprotect the assembled CPG bound oligonucleotide with concentrated ammonium hydroxide (*see* Section 3.4.2.). The deprotected crude mix-

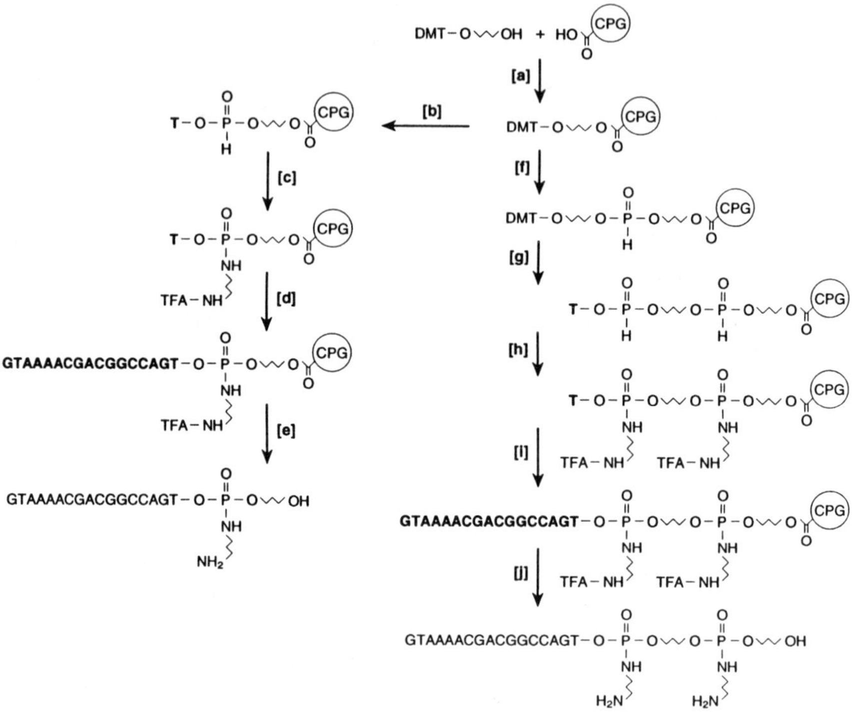

Fig. 8. Incorporation of single or multiple amino groups at 3'-terminus of oligonucleotide.

ture contains the oligonucleotide carrying the amino groups. The disadvantage of 3'-end amino incorporation is that all oligonucleotide lengths, failed or parent, carry the amino groups.

4. Purification of crude oligonucleotide obtained from step 5 should be carried out using ion exchange chromatography or PAGE.

4. Incorporation of Amino Specific Reporter Groups

4.1. Biotin Attachment

1. Take crude or purified amino functionalized oligonucleotide (2–5 A_{260} U) in 200 µL of 250 m*M* Tris-HCl buffer pH 7.6. (If the oligonucleotide is not properly desalted the pH will change, affecting the reaction adversely).
2. Add 150 µL of 15 m*M* *N*-hydroxysuccinimide ester of biotin in a mixture of water and dimethylformamide (1:1. v/v). Vortex the mixture, and incubate for 6–18 h at room temperature in the dark.

3. Excess unreacted biotin can be removed by passing the reaction mixture through Sephadex G-25. Pack a column 17 × 1.5 cm with preswollen Sephadex G-25 (in ethanol-water, 8:2 v/v) and equilibrate the column with water. Apply the reaction mixture onto the column.
4. Elute the column with water, collect the void volume. Concentrate the collected void volume and resuspend in water. The recovery of oligonucleotide at this stage is generally 80–90%. This material can be further purified by HPLC or gel PAGE.

4.2. Fluorophore Attachment

1. Take crude or purified amino functionalized oligonucleotides (2–5 A_{260} U) in 200 µL water.
2. Add 200 µL of a solution of the fluorophore derivative (1–2 mg) in a mixture of 1*M* sodium carbonate-bicarbonate buffer pH 9:water: dimethylformamide (5:3:3 v/v). The choice of organic solvent and volume depends on the nature of the fluorophore. Vortex the mixture, and incubate the reaction mixture at ambient temperature in the dark.
3. Excess unreacted fluorophore can be removed by applying the reaction mixture to a Sephadex G-25 column. Prepare a column of Sephadex G-25, as described above in Section 4.1. Equilibrate the column with 20% ethanol-water. (Use of ethanol helps in removing noncovalently associated dye from oligonucleotide).
4. Apply the reaction mixture onto the column, and elute with 20% ethanol-water. Collect the void volume (light colored band) and concentrate for further purification by HPLC or PAGE.

4.3. Purification of Reporter Group-Oligonucleotide Conjugates by HPLC (Figs. 9 and 10)

1. Take 1–2 A_{260} U of materials obtained from Section 4.1. or 4.2. and analyze using reversed phase HPLC using condition described in the Materials section.
2. The biotin oligonucleotide conjugates will elute later than the unreacted oligonucleotides. In the case of fluorophore-oligonucleotide conjugates, the elution time depends on the nature of the fluorophore. In the case of fluorophore-oligonucleotide conjugates the purification peak can be monitored at 260 nm (for oligonucleotide) as well as at the fluorophore absorbance wavelength (495 nm for fluorescein, 560 nm for rhodamine, and so on).
3. Collect the reporter group-oligonucleotide conjugate peak, concentrate, and desalt using Sep pak C_{18} cartridge or Sephadex G-25.

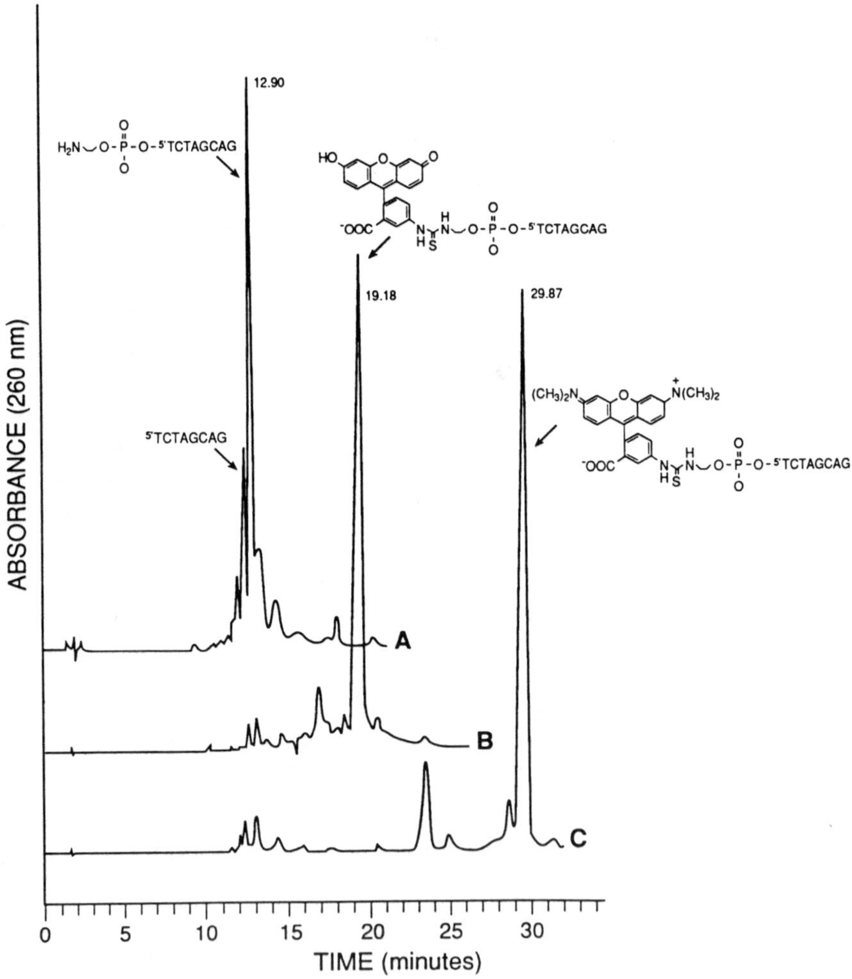

Fig. 9. Reversed phase HPLC analysis of (**A**) 5'-amino functionalized oligonucleotide, (**B**) fluorescein conjugated oligonucleotide, and (**C**) tetramethylrhodamine conjugated oligonucleotide. In HPLC profile (**A**), unfunctionalized oligonucleotide can be seen. The HPLC system and buffers used are described in the Materials section. The gradient used was 0%B for 2', 0–8%B for 5' and 8–18%B for 30', flowrate 1.5 mL min^{-1}.

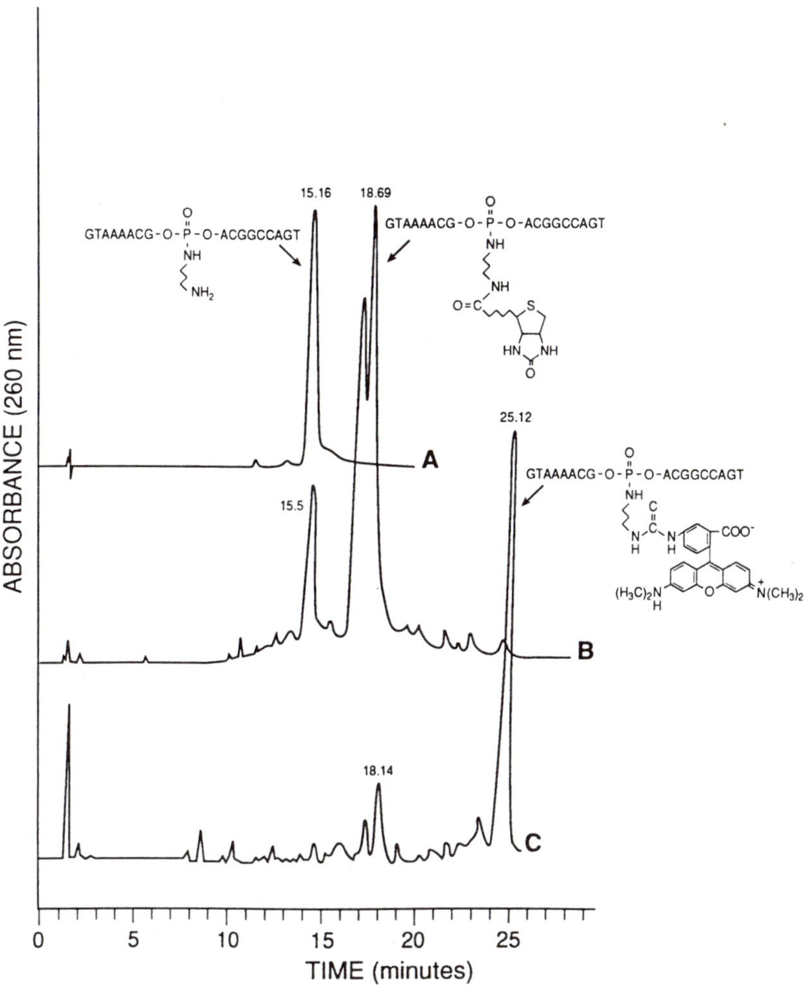

Fig. 10. Reversed phase HPLC analysis of (**A**) oligonucleotide carrying amino group at internucleotide linkage, (**B**) its conjugate with biotin, and (**C**) its conjugate with tetramethyl rhodamine. Sometimes product peak of the conjugate elutes as doublet because of the diastereomeric nature of the phosphoramidate linkage. HPLC conditions used were same as in Fig. 9.

4. Concentrate the desalted product, suspend in water and store at –20°C (protect the fluorophore-oligonucleotide conjugates from the light).

5. Discussion

Oligonucleotides carrying reporter groups have become important tools in basic as well as in applied science. In addition to their use as hybridization probes *(50)* they have been used for diagnostic purposes *(51),* automated sequencing *(52–54)* electron microscopy *(35),* fluorescence microscopy *(56–57),* X-ray crystallography, and affinity chromatography *(58–59).*

Incorporation of an amino group at the 5'-end of oligonucleotide has been described herein with an aminoethyl linker. The length of the linker molecule is very critical for a specific use, so various amino alkyl linkers containing 2–12 carbon chains have been used to incorporate amino groups *(60).* Synthesis of the linker molecule requires appropriate protection of the amino group. In the present chapter the fluorenylmethoxycarbonyl (Fmoc) group has been described. However, other base labile (e.g., trifloroacetyl) or acid labile (e.g., monomethoxytrityl) groups have also been used (Fig. 11). Aminoalkyl linkers have also been incorporated into oligonucleotides by using H-phosphonate *(61)* and methylphosphonamidite *(62)* derivatives. In most cases, reporter groups are attached to amino functionalized oligonucleotides after ammonia deprotection and purification, to avoid exposure of reporter groups to oligonucleotide deprotection conditions (in concentrated ammonia, 60°C, 8–16 h). Some reporter groups have been found to be stable under oligonucleotide deprotection conditions, so their suitable amidite derivatives have been synthesized (Fig. 12). These amidites can be used for the last coupling to incorporate reporter group directly onto oligonucleotides *(63).*

For incorporation of an amino group at 3'-end, amino functionalized CPG has also been introduced *(64).* Derivatization of the CPG is carried out so that the amino group is protected by a base labile group and the hydroxyl group is protected by an acid labile group (Fig. 13C). The oligonucleotide can be assembled onto the derivatized CPG through a hydroxyl group. After oligonucleotide assembly and deprotection, the linker molecule comes off from the CPG, leaving the amino group attached to the oligonucleotide. Multiple amino groups have also been incorporated by this approach *(66).* Reporter groups, which are stable

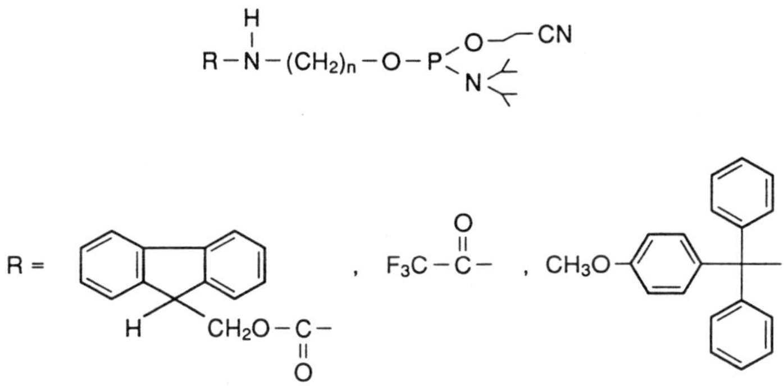

Fmoc TFA MMT

Fig. 11. Various protecting groups for amino groups that have been used in the synthesis of amino linker molecule. FMOC, Fluorenylmethoxycarbonyl; TFA, Trifluoroacetyl; and MMT, Monomethoxytrityl.

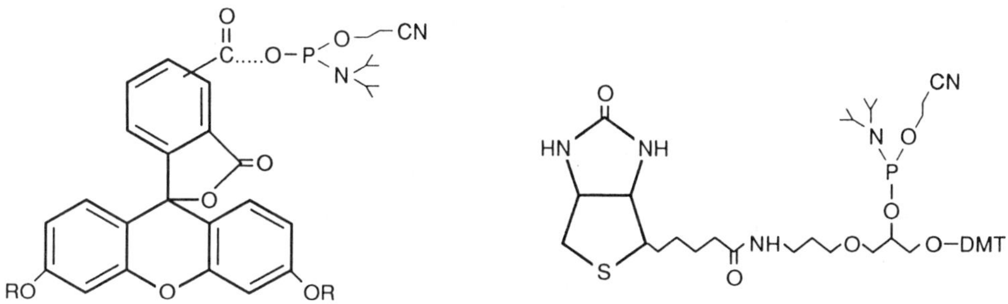

A **B**

Fig. 12. Suitably protected amidites of reporter groups, (**A**) fluorescein amidite (available from Pharmacia, Uppsala, Sweden) and (**B**) biotin amidite *(63)*.

under oligonucleotide assembly and deprotection conditions, have also been directly attached to CPG for incorporating them onto oligonucleotides. Some reporter groups attached to CPG have been listed in Fig. 13.

To amino functionalized oligonucleotides, a variety of reporter group derivatives can be attached (Fig. 14).

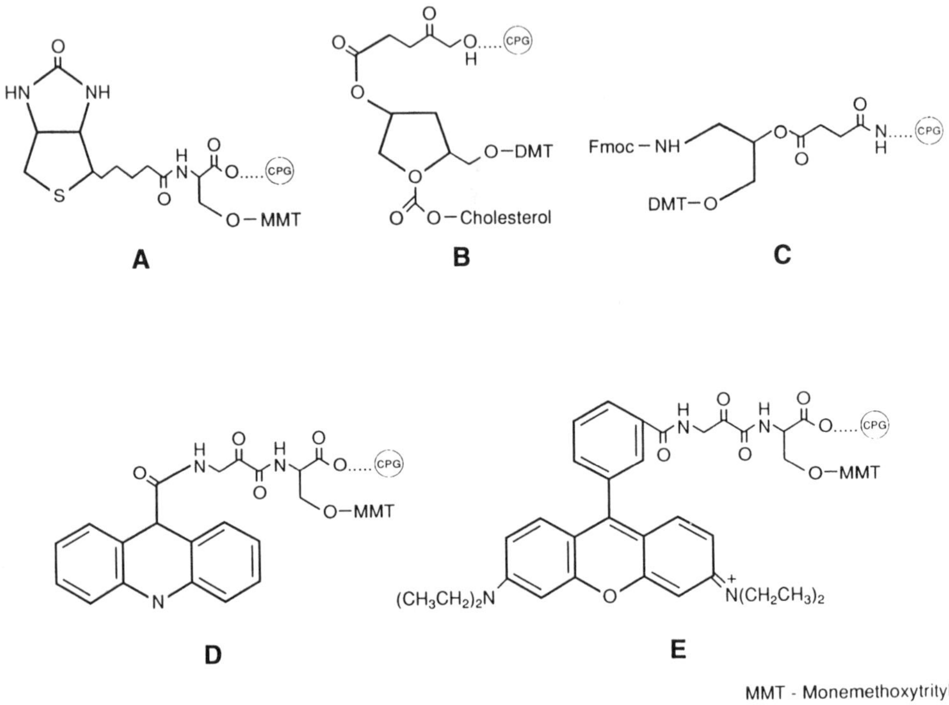

Fig. 13. Various reporter groups attached to controlled pore glass (CPG) support suitable for introducing reporter group at 3'-end of oligonucleotide. (**A**) biotin CPG, (**B**) cholesterol CPG *(65)*, (**C**) amino CPG *(66)*, (**D**) acridine CPG *(65)*, and (**E**) tetraethylrhodamine CPG (Peninsula Labs, Belmont, CA).

Acknowledgment

I am indebted to Paul Zamecnik for encouragement and support. The research work was supported by grants from NIAID cooperative drug discovery grant U01 A 124846 and by a grant from G. Harold and Leila Y. Mathers Foundation. I thank Carol Tuttle and Beth Hanninen for processing the manuscript and expert secretarial support.

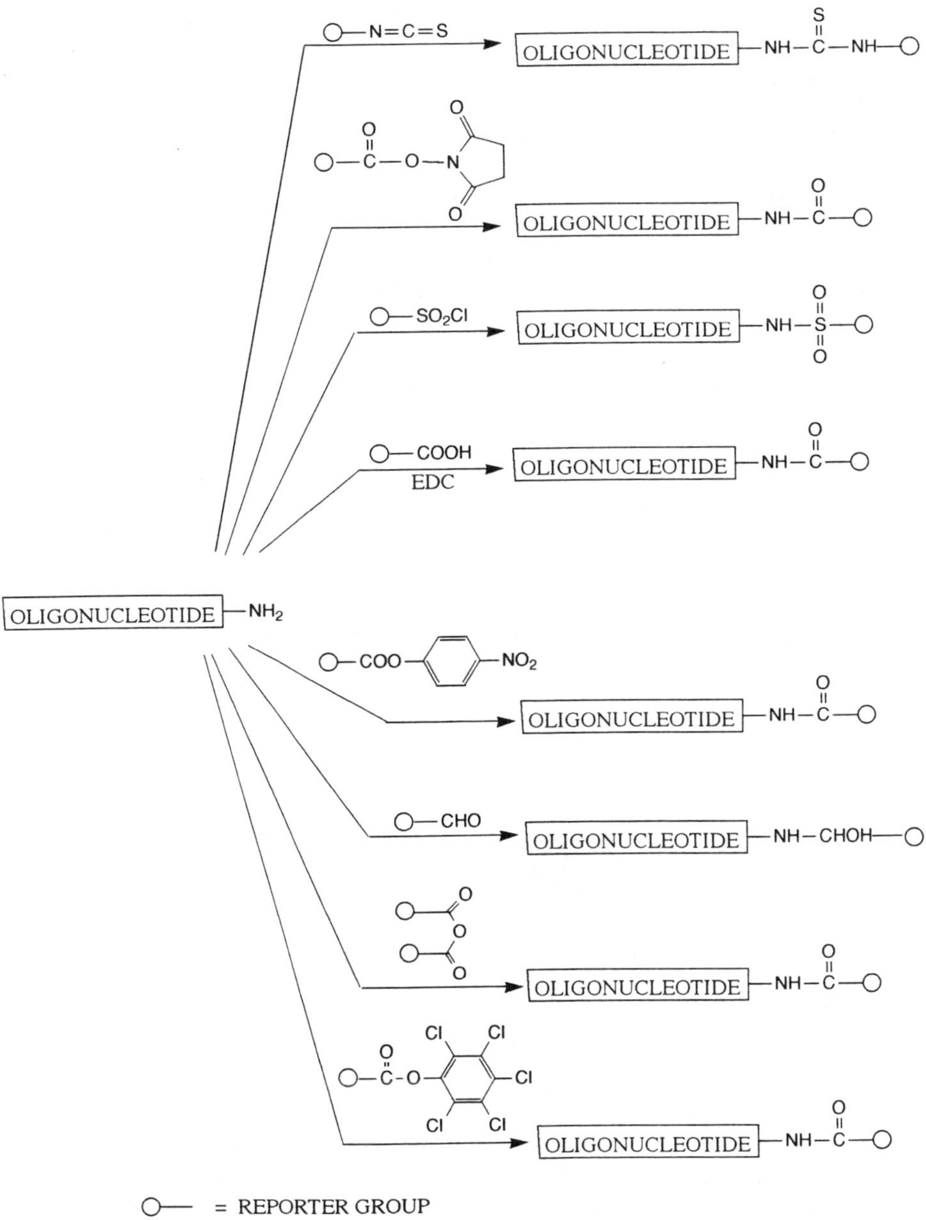

Fig. 14. Reaction of various derivatives of reporter groups to amino functionalized oligonucleotide.

References

1. Langer, P. R., Waldrop, A. A., and Ward, D. (1981) Enzymatic synthesis of biotin-labeled polynucleotides: novel nucleic acid affinity probes. *Proc. Natl. Acad. Sci. USA* **78,** 6633–6637.
2. Murasugi, A. and Wallace, R. B. (1984) Biotin-labeled oligonucleotides: enzymatic synthesis and use as hybridization probes. *DNA* **3,** 269–277.
3. Shimkus, M., Levy, J., and Herman, T. (1985) A chemically cleavable biotinylated nucleotide: usefulness in the recovery of protein-DNA complexes from avidin affinity columns. *Proc. Natl. Acad. Sci. USA* **82,** 2593–2597.
4. Kumar, A., Tchen, P., Roullet, F., and Cohen, J. (1988) Non radioactive labelling of synthetic oligonucleotide probes with terminal deoxynucleotidyl transferase. *Anal. Biochem.* **169,** 376–382.
5. Evans, R. K, Johnson, J. D., and Haley, B. E. (1986) 5-Azido-2'-deoxyuridine 5'-triphosphate: a photoaffinity-labeling reagent and tool for the enzymatic synthesis of photoactive DNA. *Proc. Natl. Acad. Sci. USA* **83,** 5382–5386.
6. Evans, R. K. and Haley, B. E. (1987) Synthesis and biological properties of 5-azido-2'-deoxyuridine 5'-triphosphate, a photoactive nucleotide suitable for making light-sensitive DNA. *Biochemistry* **26,** 269–276.
7. Trainor, G. L. and Jensen, M. A. (1988) A procedure for the preparation of fluorescence-labeled DNA with terminal deoxynucleotidyl transferase. *Nucleic Acids Res.* **16,** 11846.
8. Iverson, B. L. and Dervan, P. B. (1987) Nonenzymatic sequence-specific cleavage of single-stranded DNA to nucleotide resolution. DNA methyl thioether probes. *J. Am. Chem. Soc.* **109,** 1241–1243.
9. Gillam, I. C. and Tener, G. M. (1986) N4-(6-Aminohexyl) cytidine and deoxycytidine nucleotides can be used to label DNA. *Anal. Biochem.* **157,** 199–207.
10. Eshaghpour, H., Soll, D., and Corthers, D. M. (1979) Specific chemical labeling of DNA fragments. *Nucleic Acids Res.* **7,** 1485–1495.
11. Stade, K, Rinke-Appel, J., and Brimacombe, R. (1989) Site-directed cross-linking of mRNA analogues to the *Esherichia coli* ribosome; identification of 30S ribsomal components that can be cross-linked to the mRNA at various points 5' with respect to the decoding site. *Nucleic Acids Res.* **17,** 9889–9908.
12. Vincent, C., Tchen, P., Cohen-sodal, M., and Kourilsky, P. (1982) Synthesis of 8-(2,4 dinitrophenyl 2,6 aminohexyl) amino-adenosine 5' triphosphate: biological properties and potential uses. *Nucleic Acids Res.* **10,** 6787–6796.
13. Chu, B. C. F., Wahl, G. M., and Orgel, L. E. (1983) Derivatization of unprotected polynucleotides. *Nucleic Acids Res.* **11,** 6513–6529.
14. Chollet, A. and Kawashima, E. H. (1985) Biotin-labeled synthetic oligodeoxyribonucleotides: chemical synthesis and uses as hybridization probes. *Nucleic Acids Res.* **13,** 1529–1541.
15. Chu, B. C. F. and Orgel, L. E. (1988) Ligation of oligonucleotides to nucleic acids or proteins via disulfide bonds. *Nucleic Acids Res.* **16,** 3671–3691.

16. Bischoff, R., Coull, J. M., and Regnier, F. E. (1987) Introduction of 5'-terminal functional groups into synthetic oligonucleotides for selective immobilization. *Anal. Biochem.* **164,** 336–344.
17. Kremsky, J. N., Wooters, J. L., Dougherty, J. P., Meyer, R. E., Collins, M., and Brown, E. L. (1987) Immobilization of DNA via oligonucleotides containing an aldehyde or carboxylic acid group at the 5' terminus. *Nucleic Acids Res.* **15,** 2891–2909.
18. Ghosh, S. S., Kao, P. M., and Kwoh, D. Y. (1989) Synthesis of 5'-oligonucleotide hydrazide derivatives and their use in preparation of enzyme-nucleic acid hybridization probes. *Anal. Biochem.* **178,** 43–51.
19. Zuckermann, R., Corey, D., and Schultz, P. (1987) Efficient methods for attachment to thiol specific probes to the 3'-ends of synthetic oligodeoxyribonucleotides. *Nucleic Acids Res.* 5305–5321.
20. Ruth, J. L. (1984) Chemical synthesis of nonradioactively-labeled DNA hybridization probes. *DNA* **3,** 123.
21. Jablonski, E., Moomaw, E. W., Tullis, R. H., and Ruth, J. L. (1986) Preparation of oligodeoxyribonucleotide-alkaline phosphatase conjugates and their use as hybridization probes. *Nucleic Acids Res.* **14,** 6115–6128.
22. Forster, A. C., McInnes, J. L., Skingle, D. C., and Symons, R. H. (1985) Nonradioactive hybridization probes prepared by the chemical labeling of DNA and RNA with a novel reagent, photobiotin. *Nucleic Acids Res.* **13,** 745–761.
23. Gibson, K. J. and Benkovic, S. J. (1987) Synthesis and application of derivatizable oligonucleotides. *Nucleic Acids Res.* **15,** 6455–6467.
24. Sproat, B. S., Lamond, A. I., Beijer, B., Neuner, P., and Rider, U. (1989) Highly efficient chemical synthesis of 2'-O-methyl oligoribonucleotides and tetrabiotinylated derivatives; Novel probes that are resistant to degradation by RNA or DNA specific nuleases. *Nucleic Acids Res.* **17,** 3373–3386.
25. Prober, J. M., Trainor, G. L., Dau, R. J., Hobbs, F. W., Robertson, C. W., Zagurski, R. J., Cocuzza, A. J., Jensen, M. A., and Baumeister, K. (1987) A system for rapid DNA sequencing with fluorescent chain-terminating dideoxynucleotides. *Science* **238,** 336–341.
26. Haralambidis, J., Chai, M., and Chollet, A. (1987) Preparation of base-modified nucleosides suitable for nonradioactive label attachment and their incorporation into synthetic oligodoexyribonucleotides. *Nucleic Acids Res.* **15,** 4857–4876.
27. Roduit, J. P., Shaw J., and Chollet, A. (1988) Synthesis of oligodeoxyribonucleotides containing an aliphatic amino linker arm at selected adenine bases and derivatization with biotin. *Nucleosides Nucleotides* **6,** 349–352.
28. Urdea, M. S., Warner, B. D., Running, J. A., Stempien, M., Clyne, J., and Horn, T. (1988) A comparison of nonradioisotopic hybridization assay methods using fluorescent, chemiluminescent, and enzyme-labeled synthetic oligodeoxyribonucleotide probes. *Nucleic Acids Res.* **16,** 4937–4956.
29. Spaltenstein, A., Robinson, B. H., and Hopkins, P. B. (1988) A rigid and nonperturbing probe for duplex DNA motion. *J. Am. Chem. Soc.* **110,** 1299–1301.

30. Allen, D. J., Darke, P. L., and Benkovic, S. J. (1989) Fluorescent oligonucleotides and deoxynucleotides triphosphate: preparation and their interaction with the large (Klenow) fragment of E Coli DNA. *Biochemistry* **28,** 4601–4607.
31. Telser, J., Cruickshank, K. A., Morrison, L. E., and Netzel, T. L. (1989) Synthesis and characterization of DNA oligomers and duplexes containing covalently attached molecular labels: comparison of biotin, fluorescein, and pyrene labels by thermodynamic and optical spectroscopic measurements. *J. Am. Chem. Soc.* **111,** 6966–6976.
32. Telser, J., Cruickshank, K. A., Schanze, K. S., and Netzel, T. L. (1989) DNA oligomers and duplexes containing a covalently attached derivative of tris (2,2'-bipyridine) ruthenium(II): synthesis and characterization by thermodynamic and optical spectroscopic measurements. *J. Am. Chem. Soc.* **111,** 7221–7226.
33. Smith, L. M., Fung, S., Hunkapiller, M. W., Hunkapiller, T. J., and Hood, L. E. (1985) The synthesis of oligonucleotides containing an aliphatic amino group at the 5'-terminus: synthesis of fluorescent DNA primers for use in DNA sequence analysis. *Nucleic Acids Res.* **13,** 2399–2419.
34. Smith, L. M., Sanders, J. Z., Kaiser, R. J., Hughes, P., Dodd, C., Connell, C. R., Heiner, C., Kent, S., and Hood, L. E. (1986) Fluorescence detection in automated DNA sequence analysis. *Nature* **321,** 674–679.
35. Sproat, B. S., Beijer, B., and Rider, P. (1987) The synthesis of protected 5'-amino-2',5'-dideoxyribonucleoside-3'-phosphoramidites; applications of 5'-amino-oligodeoxyribonucleotides. *Nucleic Acids Res.* **15,** 6181–6197.
36. Sproat, B. S., Beijer, B., Rider, P., and Neuner, P. (1987) The synthesis of protected 5'-mercapto-2',5'-dideoxyribonucleoside-3'-O-phosphoramidites; uses of 5'-mercaptoligodeoxyribonucleotides. *Nucleic Acids Res.* **15,** 4837–4348.
37. Agrawal, S., Christodoulou, C., and Gait, M. J. (1986) Efficient methods for attaching non-radioactive labels to the 5'-end of synthetic oligodeoxyribonucleotides. *Nucleic Acids Res.* **14,** 6227–6245.
38. Emson, P. C., Arai, H., Agrawal, S., Christodoulou, C., and Gait, M. J. (1989) Non-radioactive methods of *in situ* hybridization—Visualization of neuroendocrine mRNA. *Meth. Enymol.* **168,** 753–761.
39. Arai, H., Emson, P. C., Agrawal, S., Christodoulou, C., and Gait, M. J. (1988) *In situ* hybridization histochemistry: Localization of vasopressin mRNA in rat brain using a biotinylated oligonucleotide. *Mol. Brain Res.* **4,** 63–69.
40. Coull, J. M., Weith, H. L., and Bischoff, R. (1986) A novel method for the introduction of an aliphatic primary amino group at the terminus of synthetic oligonucleotides. *Tetrahedron* **27,** 3991–3994.
41. Tanaka, T., Sakata, T., Fujimoto, K., and Ikehara, M. (1987) Synthesis of oligodeoxyribonucleotides with aliphatic amino or phosphate group at the 5' end by the phosphotriester method on a polystyrene support. *Nucleic Acids Res.* **15,** 6209–6224.
42. Connolly, B. A. (1987) The synthesis of oligonucleotides containing a primary amino group at the 5'-terminus. *Nucleic Acids Res.* **15,** 3131–3139.
43. Sinha, N. D. and Cook, R. M. (1988) The preparation and application of functionalized synthetic oligonucleotides: III. Use of H-phosphonate deriva-

tives of protected amino-hexanol and mercapto-propanol or hexanol. *Nucleic Acids Res.* **16,** 2659–2669.

44. Connell, C., Fung, S., Heiner, C., Brigham, J., Chakerian, V., Heron, E., Jones, B., Menchen, S., Mordan, W., Raff, M., Recknor, M., Smith, L., Springer, J., Woo, S., and Hunkapiller, M. (1987) Automated DNA sequence analysis. *BioTechniques* **5,** 342–348.
45. Connolly, B. A. and Rider, P. (1985) Chemical synthesis of oligonucleotides containing a free sulfydryl group and subsequent attachment of thiol-specific probes. *Nucleic Acids Res.* **13,** 4485–4502.
46. Letsinger, R. L., Zhang, G., Sun, D. K., Ikeuchi, T., and Sarin, P. S. (1989) Cholesteryl conjugated oligonucleotides: synthesis, properties and activity as inhibitors of replication of HIV in cell culture. *Proc. Natl. Acad. Sci. USA* **86,** 6553–6556.
47. Agrawal, S., Mayrand, S., Zamecnik, P. C., and Pederson, T. (1990) Site specific excision of RNA by RNase H and mixed phosphate backbone oligodeoxynucleotides. *Proc. Natl. Acad. Sci. USA* **87,** 1401–1405.
48. Findanza, J. A. and McLaughlin, L. W. (1989) Introduction of reporter groups at specific sites in DNA containing phosphorothioate diesters. *J. Am. Chem. Soc.* **111,** 9117–9119.
49. Agrawal, S. and Zamecnik, P. C. (1990) Site specific functionalization of oligonucleotides for attaching two different reporter groups. *Nucleic Acids Res.* **18,** 5419–5423.
50. Mathews, J. A. and Kricka, L. J. (1988) Analytical strategies for the use of DNA probes. *Anal. Biochem.* **169,** 1–25.
51. Landegren, U., Kaiser, R., Caskey, C. T., and Hood, L. (1988) DNA-diagnostics-molecular techniques and automation. *Science* **242,** 229–237.
52. Ansorge, W., Sproat, B. S., Stegeman, J., Schwager, C., and Zenke, M. (1987) Automated DNA sequencing: ultrasensitive detection of fluorescent bands during electrophoresis. *Nucleic Acids Res.* **17,** 4593–4603.
53. Brumbaugh, J. A., Middendorf, L. R., Grove, D. L., and Ruth, J. L. (1988) Continuous, on-line DNA sequencing using oligodeoxynucleotide primers with multiple fluorophores. *Proc. Natl. Acad. Sci. USA* **85,** 5610–5614.
54. Beck, S. and Pohl, F. M. (1984) DNA sequencing with direct blotting electrophoresis. *EMBO J.* **3,** 2905–2909.
55. Kaiser, R. J., Mackeller, S. L., Vinayak, R. S., Sanders, J. Z., Saavedra, R. A., and Hood, L. E. (1989) Specific primer-directed DNA sequencing using automated fluorescence detection. *Nucleic Acids Res.* **17,** 6087–6102.
56. Agrawal, S., Sarin, P. S., Zamecnik, M., and Zamecnik, P. C. (1992) Cellular uptake and anti-HIV activity of oligonucleotides and their analogues, in *Gene Regulation: Biology of Anti-sense RNA and DNA* (Erickson, R. P. and Izant, J. G., ed.), Raven, New York, pp. 273–283.
57. Loke, S. L., Stein, C. A., Zhang, X. H., Mori, K., Nakanishi, M., Subasingha, C., Cohen, J. S., and Neckers, L. M. (1989) Characterization of oligonucleotide transport into living cells. *Proc. Natl. Acad. Sci. USA* **86,** 3474–3478.
58. Lamond, A. I., Sproat, B. S., Ryder, U., and Hamm, J. (1989) Probing the structure and function of U2 snRNP with antisense oligonucleotides made of 2'-OMe RNA. *Cell* **58,** 383–390.

59. Temsamani J., Agrawal, S., and Pederson, T. (1991) Biotinylated antisense methylphosphonate oligodeoxynucleotides: inhibition of spliceosome assembly and affinity selection of U1 and U2 small nuclear RNPs. *J. Biol. Chem.* **266,** 468–472.
60. Cardullo, R. A., Agrawal, S., Flores, C., Zamecnik, P. C., and Wolf, D. E. (1988) Detection of nucleic acid hybridization by non radiative fluorescence energy transfer. *Proc. Natl. Acad. Sci. USA* **85,** 8790–8794.
61. Sinha, N. D. and Cook, R. M. (1988) The preparation and application of functionalized synthetic oligonucleotides: III. use of H-phosphonate derivatives of protected aminohexanol and mercapto-propanol or -hexanol. *Nucleic Acids Res.* **16,** 2659–2669.
62. Agrawal, S. (1989) Preparation of functionalized oligonucleoside methylphosphonate suitable for non-radioactive label attachment. *Tet. Lett.* **30,** 7025–7028.
63. Misiura, K., Durrant, I., Evans, M. R., and Gait, M. J. (1990) Biotinyl and phosphotyrosinyl phosphoramidite derivatives useful in the incorporation of multiple reporter groups on synthetic oligonucleotides. *Nucleic Acids Res.* **18,** 4345–4354.
64. Nelson, P. S., Frye, R. A., and Liu, E. (1989) Bifunctional oligonucleotide probes synthesized using a novel CPG support are able to detect single base pair mutations. *Nucleic Acids Res.* **17,** 7187–7194.
65. Petrie, C. R., Reed, M. W., Adams, A. D. and Meyer, R. B. (1992) An improved CPG support for the synthesis of 3'-Amine tailed oligonucleotides. *Bioconj. Chem.* **3,** 85–87.
66. Nelson, P. S., Shreman-Gold, R., and Leon R. (1989) A new and versatile reagent for incorporating multiple primary aliphatic amines into synthetic oligonucleotides. *Nucleic Acids Res.* **17,** 7179–7186.

Chapter 4

Functionalization of Oligonucleotides by the Incorporation of Thio-Specific Reporter Groups

Jacqueline A. Fidanza, Hiroaki Ozaki, and Larry W. McLaughlin

1. Introduction

The attachment of reporter groups, drug derivatives, or chemically reactive species to DNA in a sequence-specific manner has the potential to provide new materials for detailed spectroscopic and structural analyses as well as new classes of DNA therapeutics and diagnostics. Sequence specific labeling of a nucleic acid sequence can be achieved by a number of procedures that employ either the nucleobase, the carbohydrate or the phosphate residue as a site for attachment (for a recent review *see* ref. *1*). Specific functional groups (such as terminal phosphomonoesters, *see* Chapter 3) or the enhanced reactivity of selected sites on the purine or pyrimidine building blocks (such as the C5-position of pyrimidines, *see* Chapter 2) are often used in order to covalently attach an appropriate linker or the desired reporter group. The manner in which the nucleic acid is labeled may be dictated by the specific study involved, but in general the principles of simplicity and versatility are best employed to guide the choice of labeling procedure.

Terminal labeling may be advantageous when minimal structural perturbation is desired. For example, the addition of an agent or label to a terminal phosphomonoester is not likely to alter the structure or stability of a double-stranded or even triple-stranded complex. However, the addition of the label to an internal site (by attachment to a specific base/

From: *Methods in Molecular Biology, Vol. 26: Protocols for Oligonucleotide Conjugates*
Edited by: S. Agrawal Copyright ©1994 Humana Press Inc., Totowa, NJ

carbohydrate residue or backbone phosphate) may be more advantageous if the procedure involves the formation of an interstrand covalent crosslink or uses other agents designed to react/interact with the product complex. Site specific labeling within a sequence has generally relied on the synthesis of a modified nucleoside building block in which a linker has been incorporated for attachment of the label. However, in some cases this can destabilize the duplex structure. For example, after reaction at the exocyclic amino group of the cytosine residue, amino-labeled helices exhibited biphasic melting curves *(2–4)* suggesting local or even global structure modulation.

In the present chapter we have described a series of experiments that allow site-specific labeling of the DNA phosphate backbone using a variety of reporter groups developed for thioselective reactions.

1.1. Labeling Techniques Directed Toward the Phosphodiester Linkage

The attachment of labels or other agents to the DNA backbone offers a number of advantages over the modification of a terminal phosphomonoesters or the modification of a base residue of a nucleoside building block. By using internal phosphodiesters instead of terminal phosphomonoesters to attach the label or agent, virtually any site within the sequence is amenable to introduction of the desired functionality. The phosphate residues are not involved in interstrand base pairing so the attachment of a linker or label at such sites should not drastically alter the stability of the nucleic acid complex (unless the agent of interest is itself designed to react/interact with the complex).

Additionally, the modification of the prochiral phosphodiester residue with a single residue creates a chiral site and two phosphorus diastereoisomers (Rp and Sp). This can be advantageous, particularly for agents designed to bind or react at specific sites within a nucleic acid sequence. With double-stranded DNA (RNA) structures, one phosphorus stereoisomer directs the covalently attached derivative more toward the major groove of an A or B form helix while the second diastereoisomer directs the agent more toward the minor groove. If the desired agent binds preferentially in one of the helix grooves or reacts selectively with a functional group located in a specific groove, stereochemical labeling of the backbone can assist in enhancing the desired reactivity.

Presently two approaches have been employed to introduce labels into the backbone of DNA sequences. Oxidation of an H-phosphonate with a primary amine produces the N-substituted phosphoramidate *(5–8)* with the desired agent tethered via the nitrogen *(9–14)*. Reaction with the substituted amine takes place immediately after the introduction of the H-phosphonate linkage. The phosphoramidate formed in this manner is stable to the chemical DNA synthesis and deprotection conditions and yields the desired modified sequence. Phosphorothioates *(15–20)* can also be used as sites for alkylation by the label of interest. In this case a phosphoramidite coupling is followed by oxidation with a sulfurizing derivative *(21–24)* to generate the phosphorothioate triester. After completion of the synthesis, deprotection and isolation of the fragment, the phosphorothioate diester generated is amenable to alkylation (at sulfur) by a variety of functional groups. Both of these procedures result in DNA fragments carrying the agent of interest covalently bound to a specific internucleotidic phosphate residue.

In the remainder of this chapter we have discussed the procedures necessary to modify the DNA backbone at specific sites using one of the methods described above.

2. Materials

2.1. DNA Synthesis Reagents

1. The phosphoramidite building blocks and the solutions necessary for automated DNA synthesis were obtained from Cruachem, distributed by Fisher (Pittsburgh, PA).
2. Gold label sulfur, cystamine, anhydrous pyridine, adamantane carbonyl chloride, and most other solvents and general chemicals were from the Aldrich Chemical Co. (Milwaukee, WI).
3. Appropriately protected H-phosphonate nucleoside derivatives were products of Applied Biosystems, Inc. (Foster City, CA).

2.2. Fluorophores and Spin Labels

1. All fluorophores were obtained either from Fluka (Buchs, Switzerland) or from Molecular Probes (Eugene, OR).
2. The PROXYL spin label was a product of the Aldrich Chemical Co.

2.3. HPLC Reagents and Materials

1. HPLC grade solvents (acetonitrile and methanol) were obtained from the Aldrich Chemical Co.

2. Reversed-phase chromatographic supports (ODS-Hypersil or MOS-Hypersil) were obtained from Keystone Scientific (Bellefonte, PA) and packed into 4.6 × 250 mm or 9.4 × 250 mm columns using previously described procedures *(25)*.
3. All analyses and isolations were performed on a Beckmann (Fullerton, CA) gradient HPLC employing a variable wavelength detector.
4. Buffers:
   a. 50 m*M* triethylammonium acetate, pH 7.0 (gradient of acetonitrile).
   b. 20 m*M* KH_2PO_4 pH 5.5 (gradient of methanol).

3. Methods

Two methods are presented that allow the site-specific incorporation of labels into the DNA backbone. Both procedures involve the introduction of a sulfur residue into the sequence; in one case it is present as a phosphorothioate diester while in the second approach it exists as a simple thiol residue tethered to the DNA backbone.

3.1. The Synthesis of Phosphorothioate-Containing Oligodeoxynucleotides

The synthesis of an oligodeoxynucleotide containing phosphorothioate diesters can be accomplished by both chemical *(21–24,26–30;* discussed in detail in vol. 20 in this series) and enzymatic (for a review, *see* ref. *31*) means, but the chemical approach is more versatile for the specific placement of a single residue within a relatively short sequence.

3.1.1. Synthesis of an Oligodeoxynucleotide Containing a Single Phosphorothioate Diester

Phosphorothioate triesters are introduced at specific sites during the assembly of the chemically synthesized oligodeoxynucleotide by oxidation of the intermediate phosphite triester obtained during standard phosphoramidite chemistry. Oxidation with elemental sulfur in carbon disulfide/pyridine *(27)* generates a pentavalent phosphorus in which the phosphorothioate triester exists as a thione:

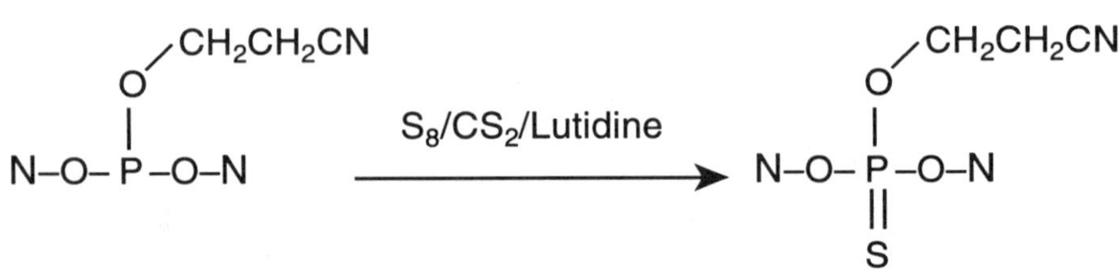

Recent work *(23,24)* has also indicated that oxidation to the phosphorothioate triester proceeds rapidly with other reagents that generally exhibit greater solubility in organic solvents than does elemental sulfur.* The thione formed in this manner is stable to the subsequent oxidation steps necessary to generate native phosphate linkages (as judged from ^{31}P-NMR experiments, unpublished results) in the remainder of the oligodeoxynucleotide. After assembly, deprotection of the oligodeoxynucleotide removes the phosphorus protecting group and generates the phosphorothioate diester.

Phosphorothioates can also be formed from the reaction of an internucleotidic H-phosphonate with elemental sulfur *(6,28–30)*:

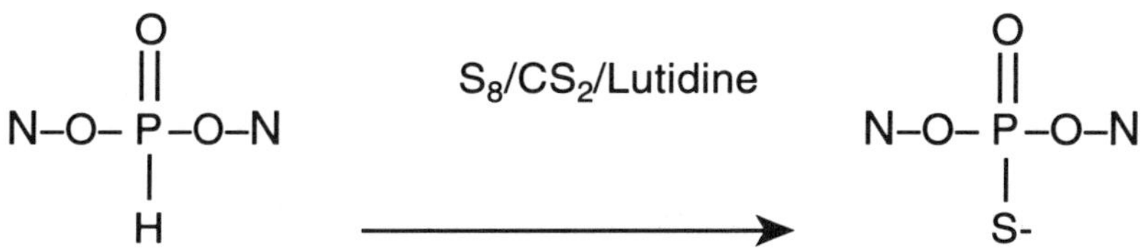

However in this approach the H-phosphonate is oxidized directly to the phosphorothioate *diester*. This species is not compatible with subsequent oxidations employing H_2O/I_2/THF/Lutidine; the phosphorothioate diesters react quickly with iodine resulting in desulfurization and conversion to the phosphate diester. The H-phosphonate method has been employed efficiently to introduce phosphorothioates at all positions within a sequence but is less useful for the introduction of a single internal phosphorothioate diester.

3.1.2. Synthesis of a Stereochemically Pure Phosphorothioate Diester

The chemical oxidation of the intermediate phosphite triester with sulfur results in roughly an equal mixture of the Rp and Sp diastereoisomers. Reaction of this mixture with the desired label or agent yields a corresponding isomeric mixture of labeled DNA fragments *(16–19)*. With short fragments it is sometimes possible to chromato-

*The poor solubility of elemental sulfur in carbon disulfide/pyridine is such that precipitation will clog the lines of an automated DNA synthesizer. Oxidations with sulfur are generally performed by interrupting the synthesis program and performing the sulfur oxidation manually. The recently developed tetraethylthiuram disulfide or 3'H-1,2-benzodithiol-3-one 1,1-dioxide derivatives eliminate this problem (the former derivative is now commercially available from Applied Biosystems while the latter can be obtained from Glen Research).

graphically resolve the two diastereoisomers, but this is not an effective general approach; with longer fragments there is often no detectable difference in retention between the labeled Rp and labeled Sp isomer. In order to effectively generate stereochemically pure DNA labeled at the phosphorothioate, it is most efficient to first generate the isomerically pure phosphorothioate dinucleotide building block with the techniques first reported from the laboratory of Fritz Eckstein *(21,32)*. In this procedure the desired nucleoside dimer containing a phosphorothioate diester is prepared by solution synthesis and is followed by separation of the two resulting diastereoisomers. After conversion of each diastereoisomer into the appropriate phosphoramidite building block, the dimer of known chirality can be incorporated into the DNA fragment. In the following example, the preparation of two complementary fully protected chiral dimer building blocks is described.

3.1.2.1. Synthesis of Rp and Sp Diastereoisomers of 5'-O-(N^6-Benzoyl-2'-deoxyadenosyl) 3'-O-[5'-O-(dimethoxytrityl)-thymidine] O-methyl Phosphorothioate

Add a solution of 5'-O-dimethoxytritylthymidine 3'-O-(diisopropylamino methoxy phosphine (882 mg, 1.25 mmol) in dry CH_3CN (5.0 mL) dropwise over a period of 10 min to a suspension of N^6-Benzoyl-3'-methoxyacetyl-2'-deoxyadenosine (427 mg, 1.0 mmol) and tetrazole (280 mg, 4.0 mmol) in dry CH_3CN (5.0 mL) under Argon gas at room temperature. After stirring for 75 min, add suspension of sulfur (320 mg, 10 mmol) in dry pyridine (15 mL) to the reaction solution and stir the mixture an additional 60 min. Remove excess sulfur by filtration and evaporate the filtrate to dryness Dissolve the residue in 50 mL of $CHCl_3$ and wash the solution with sat. $NaHCO_3$ and sat. NaCl. Dry the organic layer over Na_2SO_4 and remove the solvent by evaporation. The crude mixture of this fully protected dinucleoside phosphorothioate is purified by flash chromatography on silica gel column (Silica gel 60, ca. 100g) packed in chloroform. The diastereoisomers are eluted by using a gradient of methanol in chloroform (beginning with 1% methanol in chloroform and using 0.5% steps). Three fractions are collected to give a pure "fast" moving fraction, a "slow" moving fraction, and a larger fraction containing an isomeric mixture . The mixture is further purified by preparative tlc using silica gel plates with a concentration zone and employing $CHCl_3$-MeOH-Et_2O-H_2O (200:20:60:1, v/v) as an eluent. After consecutive development, three zones are extracted with acetone to give a pure "fast" moving zone, a "slow" moving zone, and an isomeric mixture zone. Each of the pure isomers is combined with each pure isomeric fraction obtained previously by flash chromatography. Purity analyses are carried out by tlc on silica gel plates with concentrating zones. After evaporation, each fraction is treated with dioxane (8 mL) and 28% aq. ammonia (2 mL) and incubated at ambient temperature for 1–2 h to remove the 3' protecting group. The resulting dimers are purified by silica-gel columns. Yield: "Fast" eluting isomer, 178 mg (18%); "Slow" eluting isomer, 221 mg (22%). ^{31}P-NMR ($CDCl_3$, ext H_3PO_4) "fast" isomer, 67.5 ppm; "slow" isomer, 66.9 ppm.

3.1.2.2. Analysis of the Partially Protected *R*P-dTp(s)A and *S*P-Tp(s)A Derivatives

Deprotection (by standard conditions) of a small amount of each dimer allowed determination of their nuclease sensitivity. In this case, the fast isomer is a substrate for nuclease P1 but not for snake venom phosphodiesterase, whereas the slow isomer is not cleaved by nuclease P1 but is a substrate for snake venom phosphodiesterase. These results indicate that the fast isomer is the Sp diastereomer and the slow isomer has the Rp configuration *(21,32,34)*. ^{31}P-NMR (D_2O): Sp diester, 57.6 ppm; Rp diester, 58.4 ppm.

3.1.2.3. Preparation of the *R*P-dTp(s)A and *S*P-Tp(s)A Phosphoramidite Building Blocks

Each dimer is converted to the corresponding phosphoramidite by reaction with diisopropylaminomethoxychlorophosphine using standard protocols. After isolation of the product by silica gel chromatography and its precipitation in petroleum ether, the phosphitylated dimer is recovered in yields that vary from a low of 26% to a high of 65%. Dimer purity is judged from tlc and ^{31}P-NMR spectra (Fig. 1). ^{31}P-NMR ($CDCl_3$, ext. H_3PO_4) $\delta = 67.4$, 147.1, 147.5 ppm (Sp diastereoisomer); 66.9, 147.1, 147.4 ppm (Rp diastereoisomer).

3.1.2.4. Synthesis and Analysis of the Stereochemically Pure Sequence

The Sp or Rp phosphitylated dimer block is incorporated into the appropriate position by the same procedure used for simple phosphoramidites. Deprotection and HPLC isolation of the sequence occurred by standard methods. Enzymatic analysis is then performed:

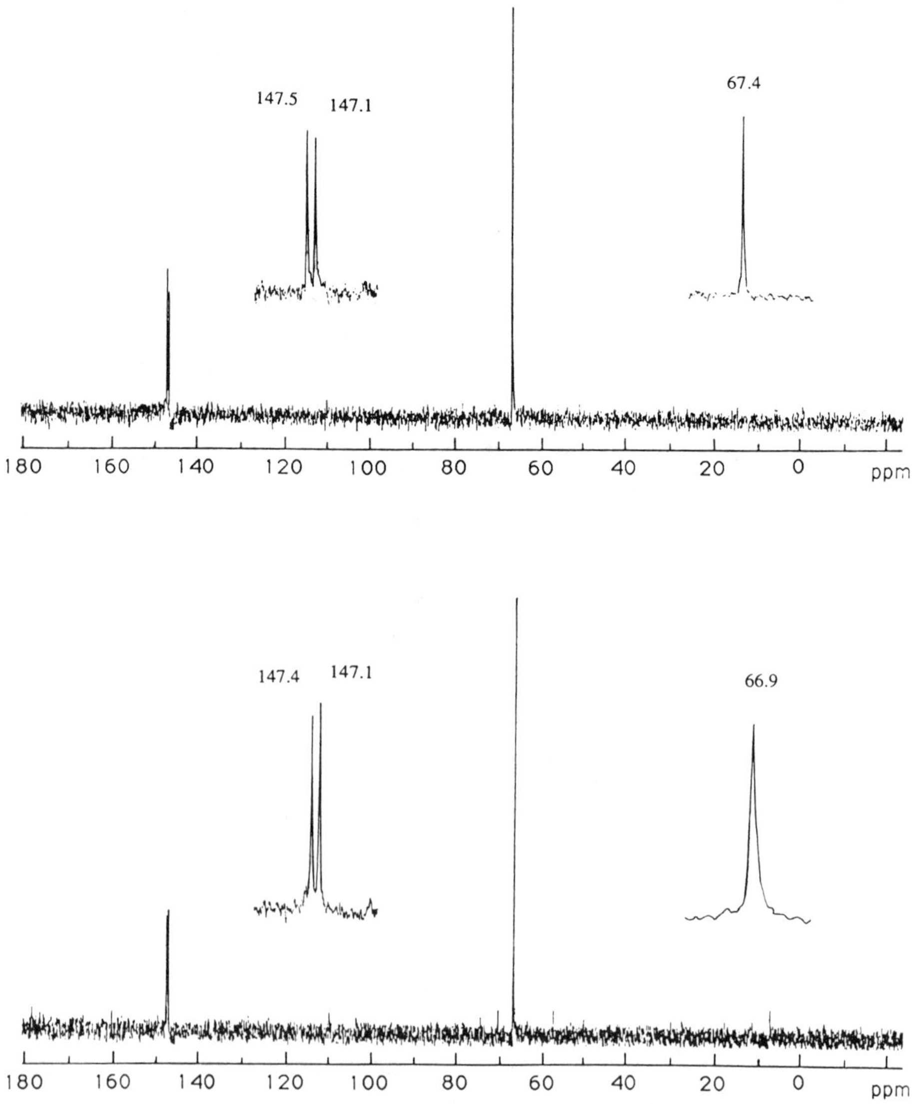

Fig. 1. ^{31}P-NMR spectra for the *Sp dTp*(s)A (above) and *Rp Tp*(s)A (below) phosphoramidite building blocks.

To 0.5 A_{260} U of 24-mer in 30 µL of 20 m*M* Tris-HCl, (pH 8.0) containing 40 m*M* $MgCl_2$ is added 9×10^{-3} U of snake venom phosphodiesterase and 2 U of bacterial alkaline phosphatase and the mixture is incubated at 37°C overnight.

HPLC analysis: (Column: ODS-Hypersil, Buffer: 50 m*M* triethylammonium acetate, pH 7.0, Gradient: 0–21% acetonitrile in 42 min). Under these conditions, the latter analysis of the Sp 24-mer results in peaks at 8.5 min (dC), 13.5 min (dG), 14.6 min (dT), 17.8 min (dA) and 28.0 [Sp-dTp(s)A]. The Rp 24-mer produces only the peaks for dC, dG, dT, and dA.

3.2. Labeling of Internucleotidic Phosphorothioate Diesters

Oligodeoxynucleotides containing a uniquely placed covalently bound reporter group tethered to the DNA backbone by alkylation of a phosphorothioate diester are easily obtained by incubation of phosphorothioate containing oligodeoxynucleotide with the probe of choice in aqueous or largely aqueous solution within a pH range of 5–8 and from 25 to 50°C.

3.2.1. Reactivity of the Phosphorothioate Diesters

Oligodeoxynucleotides containing a single phosphorothioate diester can be efficiently labeled over a wide pH range *(5–8)*. Below pH 5, competing acid catalyzed depurination begins to introduce impurities into the reaction mixture and above pH 8 base catalyzed hydrolysis of the product phosphorothioate triester reduces overall yields. Some labeling will occur at ambient temperature, but it is more efficient at 50°C. Higher temperatures appear to further enhance yields by removing secondary structure (particularly for self-complementary sequences). Under these temperature and pH conditions the labeling reactions typically proceed with greater than 80% yields after reaction time of 18 h (overnight). However, longer sequences (>20 residues) can result in somewhat reduced yields *(see below)*.

Three functional groups can be employed to covalently alkylate the phosphorothioate diester. Reagents containing γ-bromo-α,β-unsaturated carbonyls, iodo or bromo acetamides, or aziridinylsulfonamides, function effectively to label the sulfur residue and produce the corresponding phosphorothioate triester:

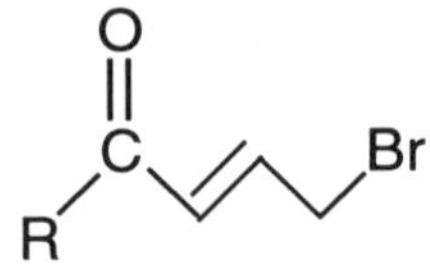

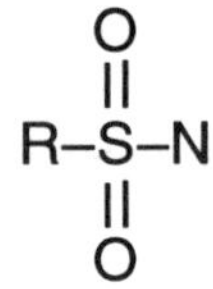

X = Br or I

Both the Rp and Sp diastereoisomers exhibit equal reactivity in all reactions monitored to date.

The concentration and amount of label required for rapid and efficient reaction with the phosphorothioate diester varies significantly depending on the functional group employed for the alkylation since the rates of competing hydrolytic reactions of the agents themselves also vary. Additionally the aqueous character of the reaction mixture will depend largely on the solubility of the label as well as the oligodeoxynucleotide. Aqueous/dimethyl formamide solutions can be commonly employed to solvate both the oligodeoxynucleotide and the label or agent of interest. The amount of DMF required in a given reaction is dependent on the solubility of the appropriate reagent. However, with DMF concentrations greater than 55%, precipitation of the oligodeoxynucleotide may become problematic. A typical labeling reaction is described below:

3.2.1.1. Labeling of a 30-mer with 5-Iodoacetamidoeosin (5-IAE)

A 100 µL reaction mixture containing 0.323 m*M* of the phosphorothioate-containing 30-mer, d[CCCGTCCTAGCAAGCCGCTGCTACCGG(s)AGG], 6.67 m*M* of 5-IAE in 20 m*M* potassium phosphate buffer (pH 8.0) and 13% DMF is incubated for 30 h at 50°C. After 30 h, HPLC analysis indicates that the reaction is approx 70% complete. The product is isolated by HPLC as described below.

3.2.2. Analysis and Purification of the Labeled Oligodeoxynucleotides

Alkylation of the phosphorothioate diester converts it to the corresponding triester. In addition to the presence of the covalently bound reporter group (often itself hydrophobic in character), the labeled product contains one less negatively charged phosphorus residue. Both of these characteristic typically make the product, triester-containing sequence, more hydrophobic than the unlabeled diester-containing sequence and the reaction mixture is amenable to separation and purification by HPLC employing reversed-phase chromatographic matrices:

3.2.2.1. Isolation of a Labeled Dodecamer Using Reversed-Phase HPLC

Prepare two buffers, Buffer A: 20 m*M* KH_2PO_4 pH 5.5, Buffer B: 20 m*M* KH_2PO_4 pH 5.5, 70% methanol and filter both buffers through a 0.45 µm membrane filter to remove particulates. Attach a 4.6 × 250 mm or 9.4 × 250 mm column of 5 µ*M* ODS-Hypersil (or other suitable reversed-phase support), wash the column with at least 10 column volumes of buffer B, and then re-equilibrate with buffer A. Program the HPLC system to elute the column under the following conditions: Flow rate: 1.5 mL/min (small column) or 3.0 mL (large column), Linear gradient: 0–100% buffer B in 60 min. Isolate the desired peak directly from the detector outflow, and remove the methanol by evaporation and then lyophilize to dryness. Desalt the labeled materials by one of two methods:

1. Prepare a Sep-Pak(C_{18}) (or Nensorb) column in the following manner: Attach a 10 mL plastic syringe cylinder to the Sep-Pak column as a solvent reservoir. Wash the column with 20 mL of methanol followed by 20 mL of double-distilled water. Dissolve the residue containing the labeled fragment in distilled water and add this solution to the Sep-Pak column. Wash the Sep-Pak column with 10 mL of water to remove the phosphate buffer followed by 10 mL of 50% aqueous methanol (higher concentrations may be required in some instances depending on the hydrophobic character of the bound label) to elute the labeled DNA fragment. Collect the appropriate fractions and lyophilize to dryness.
2. Reduce the volume if necessary and add the mixture to a column of Sephadex G-10 previously swollen and packed in distilled water. Collect the excluded volume and lyophilize it to dryness.

The ability to resolve the unlabeled sequence from the labeled product will ultimately depend both on the hydrophobic character of the label and the length of the sequence. An increase in the length of the sequence or a decrease in the hydrophobic character of the label will reduce the observed resolution. Nevertheless, HPLC resolution (and purification) functions well in many cases as illustrated for the 30-mer partially labeled with 5-iodoacetamidoeosin (Fig. 2).

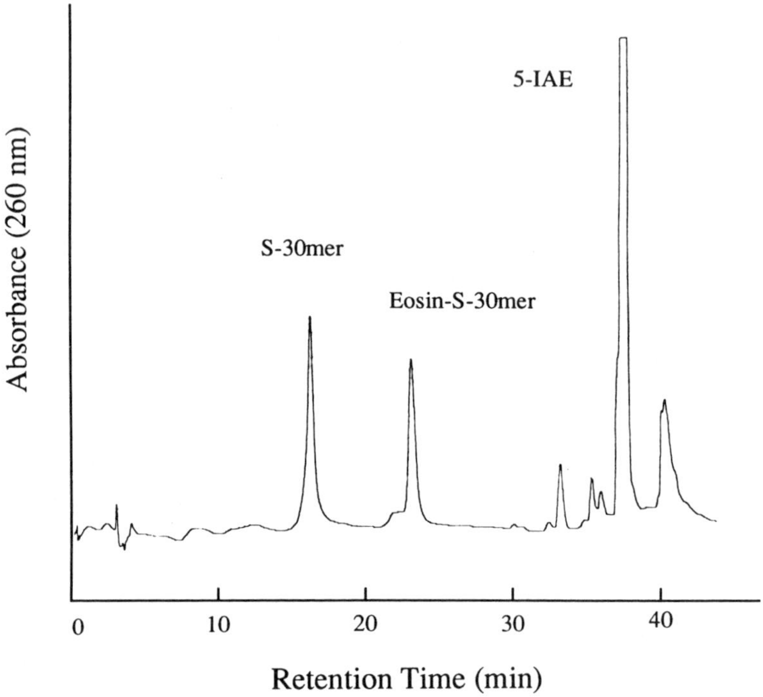

Fig. 2. HPLC analysis (after 6 h at 50°C) of the mixture resulting from the reaction between the phosphorothioate-containing 30 nucleotide fragment d[CCCGTCCTAGC AAGCCGCTGCTACCGG(s)AGG] (S-30mer) with 5-iodoacetamidoeosin (5-IAE) to produce the 30-mer carrying the label at the site of the phosphorothioate (Eosin-S-30mer). HPLC conditions are described in the text.

If the spectral characteristics of the label differ significantly from those of the nucleic acid, UV-Visible spectroscopy can be used to confirm the presence of a single label within the sequence. The hydrolytic characteristics *(see below)* of the phosphorothioate triester can also be useful to confirm the presence of the label. Incubation of the labeled oligodeoxynucleotide at pH 10 and ambient temperature for 60–120 min results in loss of the label predominately by P-S bond cleavage and generation of the native phosphate-containing sequence. HPLC analysis with the appropriate standard can be used to confirm this transformation. S1 nuclease can be used to hydrolyze the labeled oligodeoxynucleotide and produce the corresponding 2'-deoxynucleoside-5'-monophosphates in addition to the dinucleoside phosphorothioate triester-5'-phosphate carrying the label (the triester is resistant to nuclease hydrolysis). The resulting mixture can be analyzed by HPLC. With labels of high hydrophobic character, the labeled dimer may not be effectively eluted from the reversed-phase column. In such cases, heat denaturation to inactivate the nuclease, followed by raising the pH to approx 10, will hydrolyze the triester and generate the native dimer.

3.3. Properties of Labeled Oligodeoxynucleotides

The properties of the modified oligodeoxynucleotides will often depend on the nature of the labeling agent. Intercalators or other agents designed to interact with the nucleic acid helix can be expected to alter the stability and other properties of these materials. In the present section only two properties will be discussed:

1. The hydrolytic stability of the phosphorothioate triesters; and
2. The helix stability of oligodeoxynucleotides carrying labels that should not interact with the duplex DNA.

3.3.1. Hydrolytic Stability

The phosphorothioate triester produced upon alkylation of the corresponding diester is sensitive to base catalyzed hydrolysis as would be expected. In general the triesters are stable for long periods of time at ambient temperature in mildly acidic and neutral pH solutions. As the pH increases above 8 the triester undergoes hydrolysis with loss of the label. Small amounts of unidentified hydrolysis products are observed (presumably related to strand cleavage), but the major product of the hydrolysis is the phosphodiester-containing sequence and the free label resulting from P-S bond cleavage. Single-stranded sequences undergo hydrolysis more readily than the corresponding labeled duplex. An example of the stability of a labeled single-stranded oligodeoxynucleotide at different pH values is illustrated in Fig. 3.

3.3.2. Effects on Helix Stability

One of the advantages of modifying the phosphorus backbone is that the label does not interfere with the functional groups involved in interstrand hydrogen bonding. In nearly all examples examined to date, there is little to no change in the T_m value between the labeled and unlabeled duplex provided that the label is not an intercalator, groove binder, and so on.

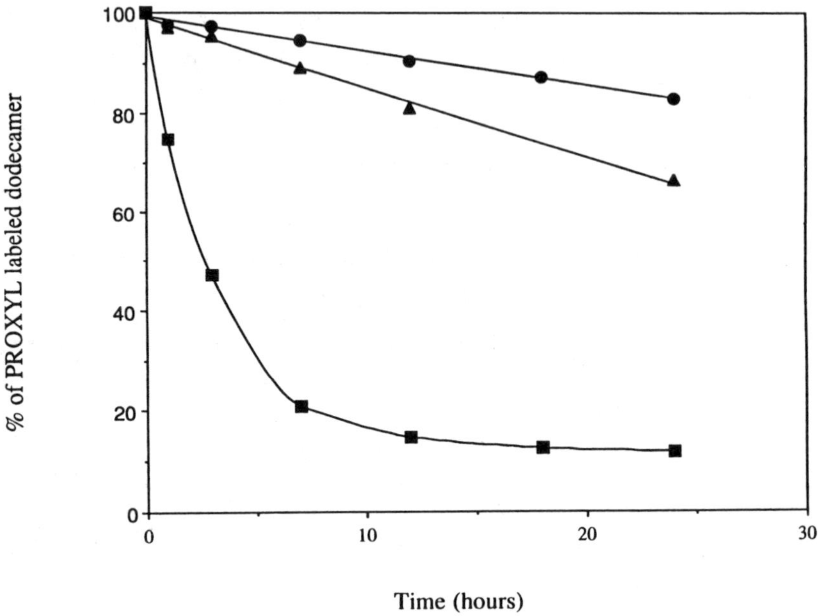

Fig. 3. Hydrolysis of the PROXYL-labeled dodecamer d[CGCA(s)AAAAAGCG] at ambient temperature and pH values of 7.0 (●), 8.0 (▲), and 10.0 (■).

3.4. The Synthesis of DNA Containing a Tethered Thiol Residue

The thioester approach described above works well for a number of labels carrying haloacetamido, γ-bromo-α,β-unsaturated carbonyl or aziridine sulfonamide functionalization. However, introducing a tethered thiol residue to the DNA backbone would increase the versatility of such backbone labeling by allowing the use of additional functional groups, such as maleimides, isothiocyanates, bromomethylaromatics, and others originally developed for the labeling of cysteine residues within proteins.

3.4.1. Introduction of a Thiol Residue into a DNA Sequence

The method chosen for the incorporation of a thiol residue site specifically into an oligodeoxynucleotide relies on the amine oxidation chemistry originally developed by Todd and coworkers *(35,36)* and extended more recently by Froehler *(6)* Letsinger *(9–13)* and Agrawal *(14)*. In this approach an internucleotidic H-phosphonate linkage is oxidized with a primary amine carrying the reporter group or linker of interest:

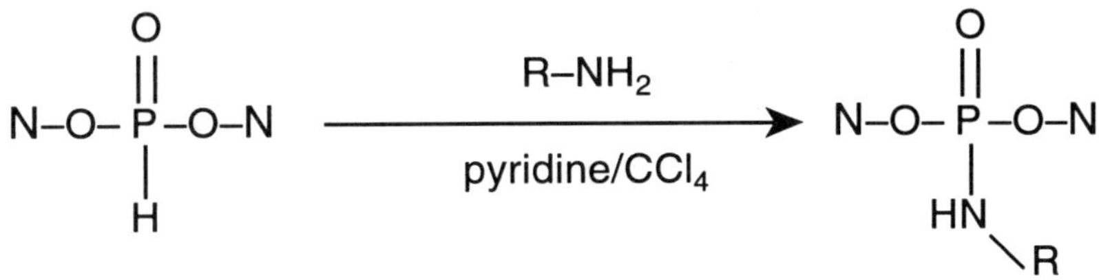

In the present case, a thiol residue can be introduced by choosing the appropriately protected aliphatic mercapto amine derivative. We have found that the simplest such derivative is cystamine, although other related derivatives should function equally well. The disulfide present in cystamine provides adequate protection of the thiol moiety during subsequent assembly of the oligodeoxynucleotide yet can be easily removed after deprotection and purification by a simple treatment with dithiothreitol (DTT). As with the phosphorothioates, the introduction of the tethered thiol through a phosphoramidate linkage results in two phosphorus diastereomers.

The procedure used to introduce the tethered thiol residue is analogous to other published work *(9–14)*. The oligodeoxynucleotide is initially elongated using standard solid phase-based phosphoramidite techniques. At the site of functionalization, a single H-phosphonate coupling is performed which is then oxidized in the presence of cystamine. After oxidation, the terminal amino group is acetylated (capped)* and the sequence further elongated using standard phosphoramidite procedures.

3.4.1.1. Incorporation of a Thiol Residue During Assembly of the Oligodeoxynucleotide

Cystamine dihydrochloride (Aldrich) is dissolved in a few milliliters of 10*M* NaOH. Enough water is added to dissolve the oily emulsion formed from the deprotonated amine. HCl is added dropwise until

*The terminal amine of the cystamine linker is acetylated during the capping step to prevent any side reactions from occurring at this site. The capping solution normally employed in the synthesis cycle is commonly used for this step, and this introduces an acetyl group to the amine. The acetamide generated in this fashion is largely stable to the deprotection step using aqueous ammonia, but this is not a concern since the acetylated amine is removed from the sequence during the DTT cleavage of the disulfide (*see* Fig. 4). However, it is also possible to use a trifluoroacetic anhydride capping reagent *(14)* should the free amine at the terminus of the linker be the desired final product.

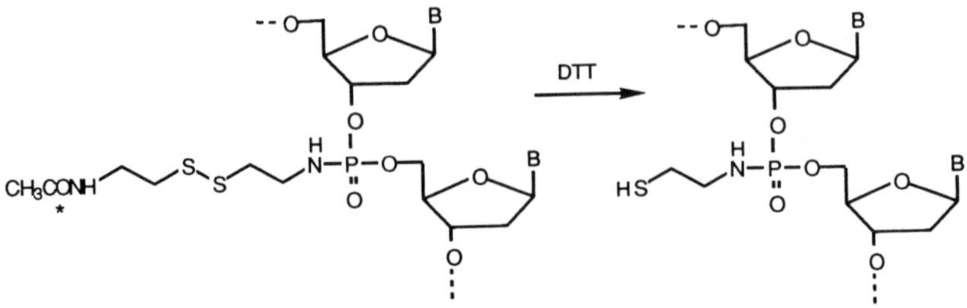

Fig. 4. The final deprotection step with dithiothreitol cleaves the disulfide and unmasks the tethered thiol or various labeling reactions. After capping with acetic anhydride and subsequent deprotection with aqueous ammonia, the cystamine linker is present partly as the free amine but largely as the acetamide derivative (*see* footnote).

the pH is in the 12–14 range. The aqueous solution is evaporated to dryness to obtain a white dry crystalline residue that is washed with dichloromethane several times to extract out the amine into the organic solvent and filtered to remove the inorganic salts. The organic solution is dried with potassium carbonate, and evaporated to yield a pale yellow oil that is dried under high vacuum overnight.

An oligodeoxynucleotide is elongated using standard phosphoramidite coupling procedures up to the residue located at the 5' side of the tether. A hydrogen phosphonate derivative (25 µmol) is then introduced at a single site by coupling for 5 min in the presence of 5 equivalents (125 µmol) of adamantane carbonyl chloride. The phosphonate diester thus formed is oxidized with 200 mg cystamine in 300 µL pyridine/carbon tetrachloride (2:1) for 1 h and then treated with acetic anhydride (or trifluoroacetic anhydride) to cap the terminal amine of the tether and any unreacted 5' hydroxyl. Oligodeoxynucleotide synthesis is then continued in the normal manner, followed by standard deprotection and HPLC isolation procedures.

HPLC analysis revealed an oligomer eluting as two closely related peaks in approximately equal proportions owing to the chirality imparted upon the oligomer by the presence of the phosphoramidate diastereoisomers (Sp and Rp) linking the disulfide-containing tether.

After completion of the synthesis, the functionalized oligodeoxynucleotide is deprotected and purified by standard procedures. After purification, the thiol residue can be unmasked by a short treatment with DTT (Fig. 4).

3.4.1.2. Deprotection of the Tethered Thiol Residue

Simple post synthesis removal of the thiol protecting group is performed by incubating the disulfide-containing oligomer in a solution containing 13 m*M* dithiothreitol and 25 m*M* Tris-HCl, pH 8.0 for 15 min at ambient temperature. The cleavage of the disulfide can be monitored by reversed-phase HPLC (Buffer A: 0.02*M* KH_2PO_4, pH 5.5, B: 0.02*M* KH_2PO_4, pH 5.5 containing 70% methanol gradient: 0–100% B in 60 min). Analysis of the reaction should indicate the presence of two closely related peaks reflecting the diastereomeric character of the oligomer. The degree of diastereomeric resolution varies with the particular sequence and the location of the tether within the strand. In a typical example, the Rp and Sp diastereomers of the dodecamer d[CGCA($NHCH_2CH_2SSCH_2CH_2NHCOCH_3$)AAAAAGCG] eluted from the column as two closely running peaks with retention times of 20.0 and 20.6 min. After 15 min of reaction, the deprotected fragments, Rp and Sp d[CGCA($NHCH_2CH_2SH$)AAAAAGCG], eluted approx 1 min earlier.

The phosphoramidate linkage is resistant to most nucleases. Treatment of a thiol containing oligodeoxynucleotide with nuclease P1, snake venom phosphodiesterase, and bacterial alkaline phosphatase produces the four common 2'-deoxynucleosides in addition to the dinucleoside phosphoramidate carrying the thiol tether. HPLC analysis of this hydrolytic mixture and comparison with the appropriate standards can be used to confirm the presence of the thiol tether (Fig. 5).

3.4.1.3. Nucleoside Analysis of d[CGCA($NHCH_2CH_2SH$)AAAAAGCG]

To 1 A_{260} U of the thiol-containing oligodeoxynucleotide is added 2 U of nuclease P1 and the mixture incubated for several hours in 40 m*M* sodium acetate, pH 5.0, 20 m*M* $MgCl_2$ at 37°C. The solution is then rebuffered to pH 8.0 using 100 m*M* Tris-HCl and 3 U of snake venom phosphodiesterase and 2 U of bacterial alkaline phosphatase are added followed by an additional incubation at 37°C. Using the HPLC conditions noted above, analysis of the resulting hydrolysate indicates the presence of five distinct species confirmed by comparison with authentic standards: 6.3 min (dC), 10.6 min, (dG), 15.0 min, (dA) and two peaks at 29.0 and 30.0 min, which correspond to the Sp and Rp diastereoisomers of d[Ap($NHCH_2CH_2SH$)A].

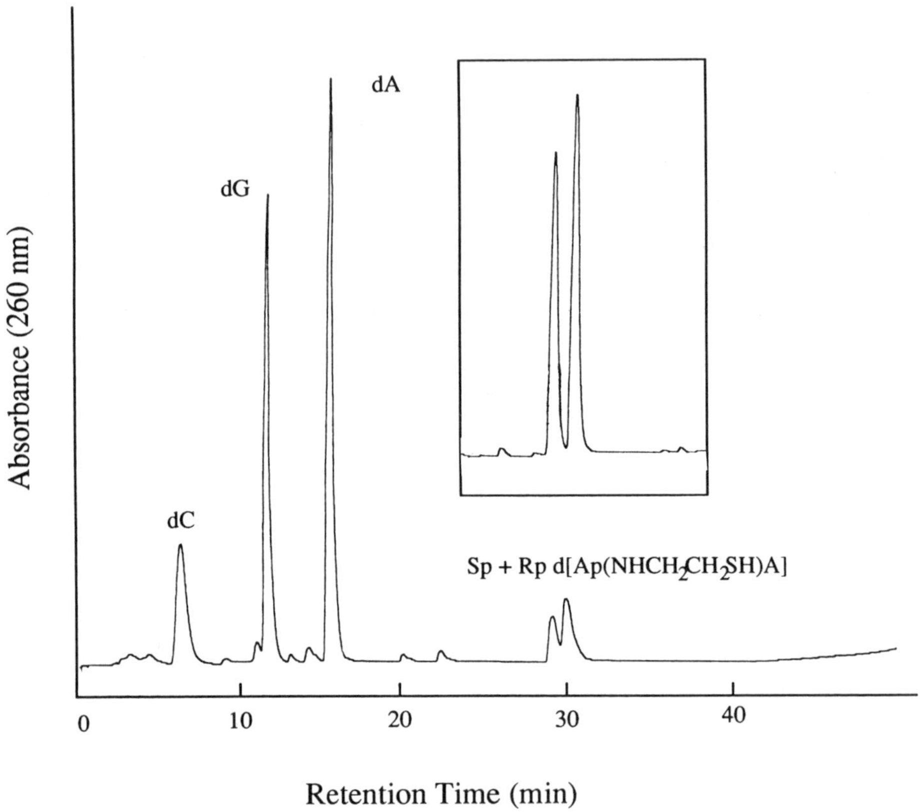

Fig. 5. Nucleoside analysis of the *Rp* and *Sp* diastereomers of the dodecamer d[CGCAp($NHCH_2CH_2SH$)AAAAAGCG] after treatment with snake venom phosphodiesterase, nuclease P1, and bacterial alkaline phosphatase. HPLC conditions are described in the text. Inset: Authentic sample of the *Rp* and *Sp* diastereomers of the dimer d[Ap($NHCH_2CH_2SH$)A].

3.5. Labeling of the Tethered Thiol Residue

The tethered thiol residue can be labeled by a variety of standard thio-specific functional groups at pH 8.0 and ambient temperature (Fig. 6). These conditions are similar to those commonly used to label cysteine residues in proteins. A typical reaction is described below.

3.5.1. Labeling of a Thiol-Containing Dodecamer with 4-Chloro-7-nitrobenz-2-oxa-1,3-diazole

A reaction mixture containing 0.06 m*M* d[CGCA($NHCH_2CH_2SH$) AAAAAGCG], 5 m*M* 4-chloro-7-nitrobenz-2-oxa-1,3-diazole (NBD-Cl) is incubated at ambient temperature in 20 m*M* Tris-HCl, pH 8.0

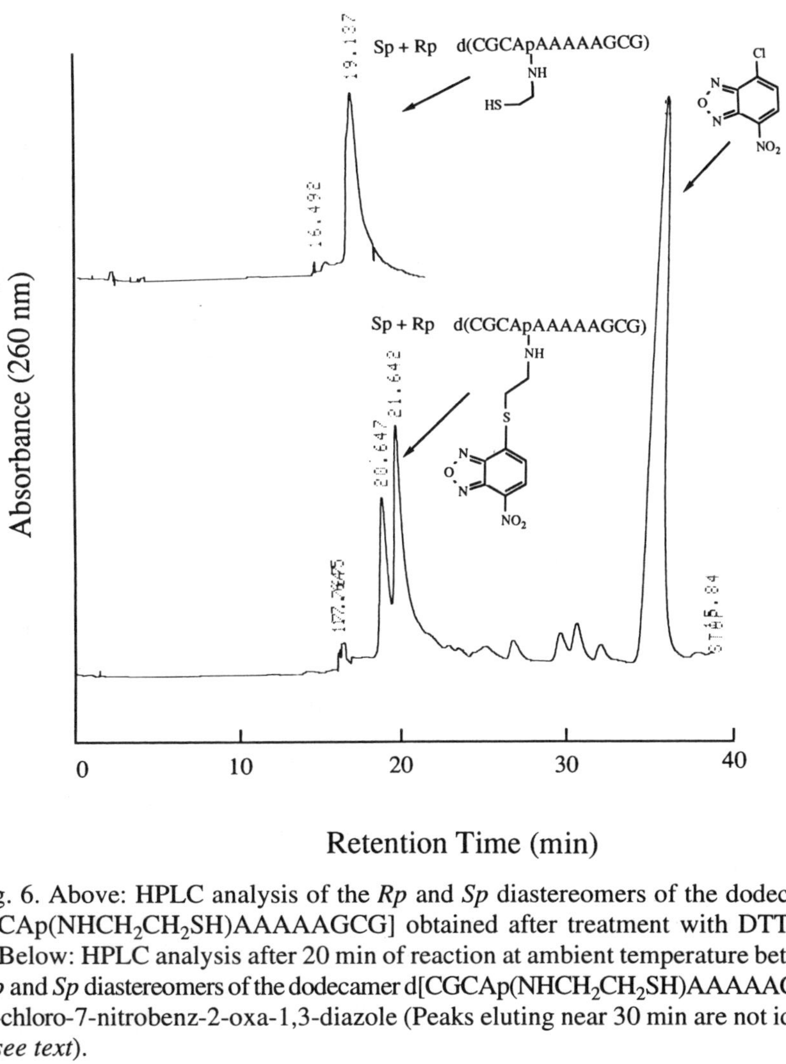

Fig. 6. Above: HPLC analysis of the *Rp* and *Sp* diastereomers of the dodecamer d[CGCAp($NHCH_2CH_2SH$)AAAAAGCG] obtained after treatment with DTT *(see text)*. Below: HPLC analysis after 20 min of reaction at ambient temperature between the *Rp* and *Sp* diastereomers of the dodecamer d[CGCAp($NHCH_2CH_2SH$)AAAAAGCG] and 4-chloro-7-nitrobenz-2-oxa-1,3-diazole (Peaks eluting near 30 min are not identified, *see text*).

50% DMF for 15 min. HPLC analysis after 15 min *(see above)* indicates that the oligodeoxynucleotide is >90% labeled (Fig. 6).

3.5.2. Reactivity, Analysis, and Purification

The thiol residue can be labeled with a variety of functional groups including benyzlbromides, maleimides, isothiocyanates, and haloacetamides. The reactions are generally complete within 20 min and are quantitative (based on HPLC analysis). It is generally not necessary to remove the excess DTT prior to the labeling reaction. Some reaction between the label and DTT may occur and produce additional small peaks in the chromatogram (*see* Fig. 6), but this reaction does not affect the overall yield of the labeling reaction or complicate chromatographic isolation of the product.

The labeled oligodeoxynucleotides are isolated by HPLC chromatography essentially as described above. In this case, resolution of the labeled and unlabeled sequences is largely dependent upon the nature of the bound label or agent since there is no difference in overall charge (unless present on the label or agent itself) after reaction with the oligodeoxynucleotide. Most of the fluorophores and other agents we have examined to date exhibit significant hydrophobic characteristics such that the labeled material elutes at a later time in the chromatographic gradient. The NBD-labeled dodecamer (*see* Fig. 6) eluted 1 to 2 min later from the column than the unlabeled fragment. By comparison, the same sequence carrying the more hydrophobic pyrene label eluted approx 8 min later than the unlabeled sequence.

The labeled oligodeoxynucleotides can be analyzed spectroscopically as described above or by HPLC analysis after incubation with nuclease P1, snake venom phosphodiesterase, and bacterial alkaline phosphatase essentially as described above.

3.6. Properties of the Thio-Labeled Oligodeoxynucleotides

DNA fragments carrying the label of interest tethered through the thiol functionality to the internal phosphoramidate are stable to hydrolysis over a wide pH range. Unlike the phosphorothioate triesters, they are very stable under alkaline conditions. Labels bound in this fashion to double-stranded DNA sequences do not appear to result in any significant change in helical stability.

4. Notes

The procedures described in this chapter allow the introduction of labels or other agents site-specifically into the phosphate backbone of DNA sequences. The procedures are simple, rapid, and in general do not require extensive synthetic expertise. In both cases the nucleophile necessary for the subsequent labeling reaction is incorporated during the assembly of the sequence and requires only moderate changes in well-known synthetic procedures. The labeled products are generally stable materials although the phosphorothioate triesters will undergo base catalyzed hydrolysis at alkaline pH values. The use of the DNA backbone as a site for labeling helps to reduce destabilization of double-stranded structures by placing the label of interest on the outer surface of the biopolymer. The ability to attach reporter groups (covalently) where desired on the DNA backbone should simplify studies involving protein binding, resonance energy transfer, structural analyses, and nucleic acid dynamics. By employing the appropriate functional group, it should additionally be possible to attach a variety of derivatives to DNA sequences including, but not limited to, peptides and proteins, antibiotics, antineoplastics, and antivirals.

Acknowledgments

This work was supported by the National Institutes of Health (grant GM-37065). L. W. M. is the recipient of an American Cancer Society Faculty Research Award.

References

1. Goodchild, J. (1990) Conjugates of oligonucleotides and modified oligonucleotides: a review of their synthesis and properties. *Bioconjugate Chem.* **1,** 165–187.
2. Telser, J., Cruickshank, K. A., Morrison, L. E., and Netzel, T. L. (1989) Synthesis and characterization of DNA oligomers and duplexes containing covalently attached molecular labels: comparison of biotin, fluorescein, and pyrene labels by thermodynamic and optical spectroscopic measurements. *J. Amer. Chem. Soc.* **111,** 6966–6976.
3. Telser, J., Cruickshank, K. A., Morrison, L. E., Netzel, T. L., and Chan, C.-K. (1989) DNA duplexes covalently labeled at two sites: synthesis and characterization by steady-state and time-resolved optical spectroscopies. *J. Amer. Chem. Soc.* **111,** 7226–7232.
4. MacMillian, A. M. and Verdine, G. L. (1990) Synthesis of functionally tethered oligodeoxynucleotides by the convertible nucleoside approach. *J. Org. Chem.* **55,** 5931–5933.
5. Nemer, M. J. and Olgilvie, K. K. (1980) Phosphoramidate analogues of diribonucleoside monophosphates. *Tetrahedron Lett.* **21,** 4153–4154.
6. Froehler, B. C. (1986) Deoxynucleoside H-phosphonate diester intermediates in the synthesis of internucleotide phosphate analogues. *Tetrahedron Lett.* **27,** 5575–5578.
7. Froehler, B. C., Ng, P. G., and Matteucci, M. D. (1986) Synthesis of DNA via deoxynucleoside H-phosphonate intermediates. *Nucleic Acids Res.* **14,** 5399–5407.
8. Froehler, B. C., Ng, P. G., and Matteucci, M. D. (1988) phosphoramidate analogues of DNA: synthesis and thermal stability of heteroduplexes. *Nucleic Acids Res.* **16,** 4831–4839.

9. Letsinger, R. L. and Schott, M. E. (1981) Selectivity in binding a phenanthridinium-dinucleotide derivative to homopolynucleotides. *J. Amer. Chem. Soc.* **103,** 7394–7398.
10. Yamana, K. and Letsinger, R. L. (1985) Synthesis and properties of oligonucleotides bearing a pendant pyrene group. *Nucleic Acids Symp. Ser.* **16,** 169–173.
11. Letsinger, R. L., Bach, S. A., and Eadie, J. S. (1986) Effects of pendant groups at phosphorus on binding properties of d-ApA analogues. *Nucleic Acids. Res.* **14,** 3487–3497.
12. Jäger, A., Levy, M. J., and Hecht, S. M. (1988) Oligonucleotide N-alkylphosphoramidates: synthesis and binding to polynucleotides. *Biochemistry* **27,** 7237–7246.
13. Letsinger, R. L., Zhang, G., Sun, D. K., Ikeuchi, T., and Sarin, P. S. (1989) Cholesteryl-conjugated oligonucleotides: synthesis, properties, and activity as inhibitors of replication of human immunodeficiency virus in cell culture. *Proc. Natl. Acad. Sci.. USA* **86,** 6553–6556.
14. Agrawal, S. and Tang, J.-Y. (1990) Site-specific functionalization of oligonucleotides for non-radioactive labeling. *Tetrahedron Lett.* **31,** 1543–1546.
15. Cosstick, R., McLaughlin, L. W., and Eckstein, F. (1984) Fluorescent labeling of tRNA and oligodeoxynucleotides using T4 RNA ligase. *Nucleic Acids Res.* **12,** 1791–1810.
16. Fidanza, J. A. and McLaughlin, L. W. (1989) Introduction of reporter groups at specific sites in DNA containing phosphorothioate diesters. *J. Amer. Chem. Soc.* **111,** 9117–9119.
17. Hodges, R., Conway, N. E., and McLaughlin L. W. (1989) "Post-assay" covalent labeling of phosphorothioate-containing nucleic acids with multiple fluorescent markers. *Biochemistry* **28,** 261–267.
18. Conway, N. E., Fidanza, J. A. ,and McLaughlin, L. W. (1989) The introduction of reporter group at multiple and/or specific sites in DNA containing phosphorothioate diesters. *Nucleic Acids Res. Symp. Ser.* **21,** 43–44.
19. Conway, N. E., Fidanza, J. A., and McLaughlin, L. W. (1990) Reaction of internucleotidic phosphorothioate diesters with reporter groups. *Phosphorus and Sulfur* **51,** 27–30.
20. Agrawal, S. and Zamecnik, P. C. (1990) Site specific functionalization of oligonucleotides for attaching two different reporter groups. *Nucleic Acids Res.* **18,** 5419–5423.
21. Connolly, B. A., Potter, B. V. L., Eckstein, F., Pingoud, A., and Grotjahn, L. (1984) Synthesis and characterization of an octanucleotide containing the *Eco*RI recognition sequence with a phosphorothioate group at the cleavage site. *Biochemistry* **23,** 3443–3453.
22. Stec, W. J., Zon, G., Egan, W., and Stec, B. (1984) Automated solid-phase synthesis, separation, and stereochemistry of phosphorothioate analogues of oligodeoxyribonucleotides *J. Amer. Chem. Soc.* **106,** 6077–6079.

23. Iyer, R. P., Egan, W., Regan, J., and Beaucage, S. L. (1990) 3*H*-1,2-benzodithiol-3-one 1,1-dioxide as an improved sulfurizing reagent in the solid-phase synthesis of oligodeoxyribonucleoside phosphorothioates. *J. Amer. Chem. Soc.* **112,** 1253–1254.
24. Iyer, R. P., Phillips, L. R., Egan, W., Regan, J. B., and Beaucage, S. L. (1990) The automated synthesis of sulfur-containing oligodeoxyribonucleotides using 3*H*-1,2-benzodithiol-3-one 1,1-dioxide as a sulfur-transfer reagent. *J. Org. Chem.* **55,** 4693–4699.
25. Piel, N. and McLaughlin, L. W. (1984) Chromatographic purification of synthetic oligonucleotides, in *Oligonucleotide Synthesis: A Practical Approach* (Gait, M. J., ed.), IRL, New York, pp. 117–134.
26. LaPlanche, L., James, T. L., Powell, C., Wilson, D. W., Uzhanski, B., Stec, W. J., Summers, M. F., and Zon, G. (1986) Phosphorothioate modified oligodeoxynucleotides III. NMR and UV spectroscopic studies of the Rp-Rp, Sp-Rp and Sp-Sp duplexes $[d(GGsAATTCC)]_2$ derived from diastereoisomeric O-ethyl phosphorothioate. *Nucleic Acids Res.* **14,** 9081–9093.
27. Ott, J. and Eckstein, F. (1987) Protection of oligonucleotide primers against degradation by DNA polymerase I. *Biochemistry* **26,** 8237–8241.
28. Stein, C. A., Subasinghe, C., Shinozuka, K., and Cohen, J. S. (1988) Physicochemical properties of phosphorothioate oligodeoxynucleotides. *Nucleic Acids Res.* **16,** 3209–3221.
29. Stein, C. A. and Cohen, J. S. (1988) Oligodeoxynucleotides as inhibitors of gene expression: a review. *Cancer Res.* **48,** 2659–2668.
30. Stein, C. A. and Cohen, J. S. (1989) Phosphorothioate oligoxynucleotide analogues, in *Oligodeoxynucleotides: Antisense Inhibitors of Gene Expression* (Cohen, J. S., ed.), CRC, Boca Raton, FL, pp. 97–118.
31. Eckstein, F. (1985) Nucleoside phosphorothioates. *Ann. Rev. Biochem.* **54,** 367–402, and references therein.
32. Cosstick, R. and Eckstein, F. (1985) Synthesis of d(GC) and d(CG) octamers containing alternating phosphorothioate linkages: effect of the phosphorothioate group on the B-Z transition. *Biochemistry* **24,** 3630–3638.
33. Potter, B. V. L., Romaniuk, P. J., and Eckstein, F. (1983) Stereochemical course of DNA hydrolysis by nuclease S1. *J. Biol. Chem.* **258,** 1758–1773.
34. Stec, W. J., Zon, G., and Uznansky, B. (1985) Reversed-phase high-performance liquid chromatographic separation of diastereomeric phosphorothioate analogues of oligodeoxyribonucleotides and other backbone-modified congeners of DNA. *J. Chromatog.* **326,** 263–280.
35. Atherton, F. R., Openshaw, H. T., and Todd, A. R. (1945) Studies on phosphorylation. Part II. The reaction of dialkyl phosphites with polyhalogen compounds in presence of bases. A new method for the phosphorylation of amines. *J. Chem. Soc.* 660–663.
36. Blackburn, G. M., Cohen, J. S., and Todd, A. R. (1966) Studies in phosphorylation. Part XXIX. The synthesis of dialkyl phosphates from monoalkyl phosphonates: direct oxidative esterification. *J. Chem. Soc.* 239–245.

CHAPTER 5

Postsynthesis Functionalization of Oligonucleotides

Barbara C. F. Chu and Leslie E. Orgel

1. Introduction

Solid-phase oligodeoxynucleotide synthesis has become a routine procedure in most molecular-biology laboratories. The reagents for the synthesis of unmodified oligomers have been available for several years, and novel commercially available reagents that permit the introduction of useful analogs into DNA are offered with increasing frequency. Solid-phase RNA synthesis is also becoming accessible to the molecular-biology community. If a suitable protected derivative is available, solid-phase synthesis is usually the preferred method of incorporating a nonstandard residue into an oligomer. However, there are still situations in which it is advantageous, or even essential, to approach the synthesis of modified oligonucleotides in a different way, namely by derivatizing unprotected oligonucleotides or their analogs.

Postsynthetic derivatization is the only method available for the modification of oligonucleotides that have been isolated from natural sources or have been synthesized enzymatically. Postsynthetic derivatization is also the most convenient method for the preparation of many 5'-[^{32}P]-labeled derivatives that have useful research applications. The development of protected derivatives of nucleoside analogs for automated DNA synthesis is usually very time-consuming, so it is often advantageous to carry out exploratory work using postsynthetic derivatization. In this chapter we discuss this approach to the preparation of modified oligonucleotides briefly, and then present in detail a few protocols developed in our laboratory. For a review of the exten-

From: *Methods in Molecular Biology, Vol. 26: Protocols for Oligonucleotide Conjugates*
Edited by: S. Agrawal Copyright ©1994 Humana Press Inc., Totowa, NJ

sive Russian literature on the postsynthetic derivatization of oligonucleotides, *see* ref. *1*.

2. Materials

The Chemical Phosphorylation Agent used for automated DNA synthesis of 5'-phosphate, 3'-phosphate, and 5'-phosphorothioate terminated oligonucleotides was obtained from Glen Research (Sterling, VA). The sulfurizing agent TETD/acetonitrile used for the automated DNA synthesis of 5'-phosphorothioate oligonucleotides was obtained from Applied Biosystems (Foster City, CA). All other reagents were obtained commercially from general chemical supply companies.

3. Postsynthetic Derivatization

Treatment of an oligonucleotide with a reagent that attacks one of the bases fails to discriminate between the different positions in the sequence at which the target base occurs. Consequently, a mixture of isomers containing one or more base modifications is obtained. This approach may be satisfactory if the position of modification is irrelevant *(2)*, but it is not useful when a particular position in the sequence must be modified. In that case it is essential to introduce at a defined position in the sequence a functional group that can be derivatized under conditions that do not permit attack on the other residues. Primary amino groups, sulfhydryl, or phosphorothioate groups and terminal phosphate residues are suitable functionalities on which to hang organic moieties.

Two general strategies are available for the introduction of suitable functional groups or linkers into oligonucleotides. One method is to use commercially available protected reagents or linkers during solid-phase DNA synthesis. The linker can subsequently be deprotected and used as an attachment site. Such linkers can be inserted at any position in the sequence. However, if they are inserted at the 5' terminus, it may no longer be possible to label the sequence using polynucleotide kinase. The other method is to introduce terminal phosphate or phosphorothioate groups into an oligomer using polynucleotide kinase and then to use simple aqueous solution chemistry to attach organic moieties to the phosphate or phosphorothioate groups. This method is compatible with 5'-[^{32}P]-labeling. In some cases both strategies are applicable, so the choice will depend on the preference of the investigator.

The position at which an oligonucleotide is derivatized may be critical for a particular application. In general, derivatization at the 3'- and 5'-termini is advantageous if the oligomer is to be hybridized to a complementary target, since many internal modifications of the bases interfere with hybridization. If several organic moieties must be attached, modification of interior residues cannot be avoided.

The reagents used in the postsynthetic derivatization of oligonucleotides and their analogs are almost always electrophiles that can attack the standard nucleotide bases. It is necessary, therefore, to establish conditions under which a sufficient yield of the desired product can be obtained without an unacceptable level of side products. In general, increasing the concentration of reagent or the time of reaction increases the extent of derivatization of the desired functional group, but also increases the yield of side products. If the reaction is allowed to continue too long, the formation of side products continues without significant increase in the yield of the desired product. It is essential, therefore, to stop the reaction soon enough. Careful control of the pH and temperature of the reaction mixture is also important.

In most of the work to be described in this chapter we have used [^{32}P]-labeled 5'-phosphomonoester groups or [^{35}S]-labeled phosphorothioate groups as attachment sites. Consequently, we have used T4 polynucleotide kinase and γ-[^{32}P]-adenosine-triphosphate or γ-[^{35}S]-adenosine-thiotriphosphate to form oligonucleotide-5'-phosphates or 5'-phosphorothioates. However, nonradioactive 5'- or 3'-terminal phosphate and phosphorothioate groups can equally well be introduced by solid phase methods (*see* Section 3.1.).

We have used water-soluble carbodiimides, which are well-established reagents for activating phosphate groups *(3–6)*, to attach a variety of molecules to the 5'- and 3'-phosphates of oligodeoxynucleotides and the 5'-phosphates of oligoribonucleotides. These procedures are not applicable to 3'-phosphates of oligomers terminated by a ribonucleotide, since on activation, they cyclize to 2',3'-cyclic phosphates. The basic chemistry that we have used to form adducts to a terminal phosphate group of an oligonucleotide is illustrated in Fig. 1. The 5'(or 3')-phosphates are first converted to the 5'(or 3')-phosphorimidazolides **(III)**. Imidazole can then be displaced by nucleophiles to form a variety of adducts *(7)*.

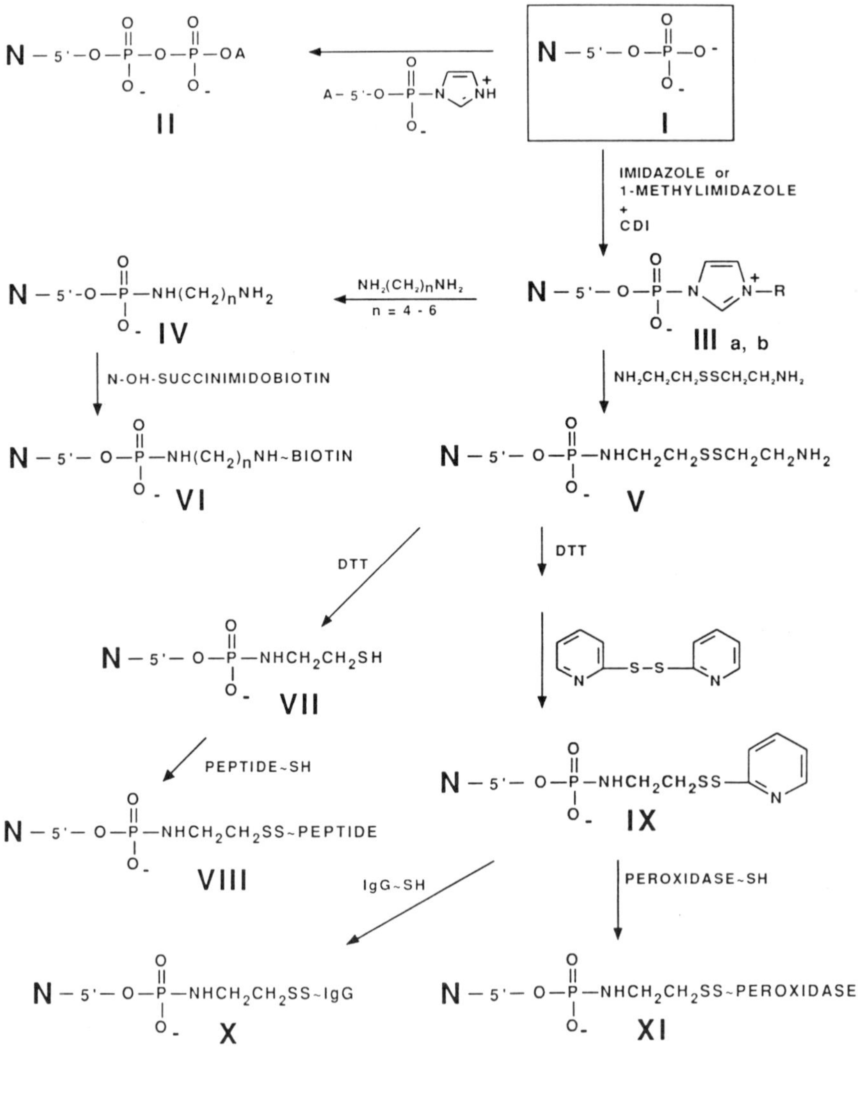

III a, R=H; III b, R=CH_3

Fig. 1. Derivatization of oligonucleotide-5'-phosphates I. The central intermediates are the phosphorimidazolides **IIIa** and **IIIb**. If it is necessary to purify the imidazolide, the reaction with carbodiimide (CDI) is carried out in the presence of imidazole to give **IIIa**. Otherwise, the reaction is carried out in the presence of 1-methylimidazole to give the labile intermediate **IIIb**, which reacts *in situ* with the amine.

Oligodeoxynucleotide-3-phosphates can be derivatized using the same chemistry. However, the activation of oligomers with a terminal ribonucleoside-3-phosphate residue results in the formation of a cyclic phosphate.

It is not usually necessary to purify the phosphorimidazolide. It can be separated from surviving carbodiimide and then immediately treated with an excess of the nucleophile. It is only if the final product cannot be separated readily from the starting phosphate that it is necessary to purify the phosphorimidazolide by HPLC. The isolation of an imidazolide intermediate can usually be avoided completely and the ligand attached to the terminal phosphate with carbodiimide in a single step by using a buffer such as 1-methyl imidazole (or lutidine) in place of imidazole. However, this method is not applicable if the ligand, for example a protein, is itself modified or polymerized by the water-soluble carbodiimide. Furthermore, the "one-pot" reaction usually requires longer exposure of the oligonucleotide to the carbodiimide and so may result in the formation of more side products.

Many useful derivatives of oligonucleotides can be conveniently prepared from commercially available activated carboxylic acids, for example biotin adducts from *N*-hydroxysuccinimidobiotin **(VI)**. In these cases, coupling to the oligonucleotide is achieved via a primary amino group *(8,9)*. Diamine linkers, such as ethylenediamine or hexamethylenediamine, are readily introduced into oligonucleotides by the method described above **(IV)**, and provide the free amino group needed for derivatization. The same coupling procedures are applicable to the acylation of amino groups introduced during solid-phase synthesis. Thiol groups can similarly be introduced by forming the cystamine adduct **(V)** and reducing it to the cystamine adduct **(VII)** with dithiothreitol *(10)*. An alternative procedure involves the introduction of a hydrazide linker that is subsequently reacted with an aldehyde *(11)*.

Internal phosphorothioate groups have been used to attach fluorescent labels to oligonucleotides *(12)*. The same procedures should work equally well for the attachment of other organic moieties, provided an appropriate organic halide is available. We have used phosphorothioate groups in a different context, to attach complexes of divalent platinum. In this way, we have developed very simple methods for crosslinking complementary oligonucleotides *(13,14)*.

Most of the examples discussed in this chapter involve short oligodeoxynucleotides. However, the methods are equally applicable to short oligoribonucleotides, to higher-mol wt nucleic acids and to synthetic oligonucleotides with modified backbones provided the derivatizing reagent does not react with groups introduced into the backbone. The only requirement is that it be possible to introduce terminal phosphate or phosphorothioate groups into the protected oligomer during solid-phase synthesis, or into the unprotected oligomer by enzymatic methods.

The procedures described above allow an almost unlimited variety of organic ligands to be attached to the 5'- or 3'-termini of oligodeoxynucleotides or nucleic acids. A terminal 3'- or 5'-phosphate group can be used as a site for the direct attachment of most water-soluble amines and phosphomonoesters, whereas carboxylic acids can be attached to terminal phosphates via diamine linkers. When a reactive halogen derivative of an organic moiety is available, a phosphorothioate group or a cystamine molecule provides a useful attachment site *(12)*. Transthiolation of a 5'-pyridyldisulfide adduct (**IX**) of an oligonucleotide or nucleic acid with a thiol-containing polypeptide or protein yields a stable nucleic acid-protein adduct **X** or **XI**, which is readily cleaved by dithiothreitol *(10)*. Derivatized oligonucleotides suitable for most applications to molecular biology or biotechnology are accessible by one or other of these ligation procedures.

3.1. Synthesis of Oligonucleotides Terminated with 5'- or 3'-Phosphate or 5'-Phosphorothioate Groups

3.1.1. Synthesis of 5'-Phosphates

A terminal 5'-phosphate can be attached to an oligonucleotide chemically during automated DNA synthesis, using a commercially available protected phosphorylating agent ("Chemical Phosphorylation Reagent," Glen Research). This is the method of choice when large amounts of a 5'-phosphorylated derivative are needed.

Unprotected oligonucleotides or nucleic acids obtained from biological sources or synthesized enzymatically must be phosphorylated using ATP and polynucleotide kinase. [^{32}P]-labeled phosphoramidites suitable for solid-phase synthesis are not available, so [^{32}P]-labeling must also be carried out enzymatically.

3.1.1.1. Chemical Phosphorylation

1. A bottle containing Chemical Phosphorylation Reagent in acetonitrile is placed in the fifth (spare) phosphoramidite reservoir position (entered as an X in the sequence on many synthesizers).

2. The desired DNA sequence is entered in the usual way with X as the first entry. The sequence 5'-pATC-3', for example, would be entered as 5'-XATC-3'.
3. Deprotection is carried out in the normal way.

3.1.1.2. Enzymatic Phosphorylation

The procedure described by Maniatis in ref. *23* gives good results. Briefly, the oligomer to be kinased (1–50 pmol) is incubated with 1–2 µ*M* γ-[^{32}P]-ATP and 20 U of polynucleotide kinase in 50 µL of buffer containing 100 m*M* Tris (pH 7.6), 20 m*M* $MgCl_2$, 10 m*M* DTT, 0.1 m*M* spermidine, and 0.1 m*M* EDTA at 37°C for 45 min. The reaction is stopped with 2 µL of 0.5*M* EDTA and the kinased oligomer is purified by HPLC or gel electrophoresis (*see* Section 3.1.4.). When larger amounts of unlabeled kinased oligomers are required, for example 1–2 µmol, the concentration of ATP is increased to 100 µ*M*.

3.1.2. Synthesis of 3'-Phosphates

Synthesis of an oligodeoxynucleotide terminated by a 3'-phosphate can be achieved by automated DNA synthesis or by adding a ribonucleotide to the 3'-end of the oligomer with terminal transferase, oxidizing with periodate, and then β-eliminating the oxidized residue (*see* Chapter 3 for details). The latter must be used to add a [^{32}P]- label to the 3' terminus of an oligonucleotide.

3.1.2.1. Chemical Phosphorylation

The strategy adopted is to synthesize an oligonucleotide in which the chemical phosphorylating agent is used to form the first addition to the resin-attached nucleoside. Then the rest of the DNA sequence is synthesized in the usual way (from the 3'-end). During standard ammonia deprotection, the linkage between the phosphate and the nucleoside support is broken by β-elimination, leaving a 3'-phosphate attached to the synthesized chain.

To synthesize the sequence 5'-ACTp-3', for example, the procedure below is followed.

1. The bottle containing Chemical Phosphorylation Reagent is placed in the fifth phosphoramidite reservoir.
2. The sequence 5'-ACTXC-3' is entered. The 3'-C is arbitrary and could be replaced by A, T, or G, since during deprotection the terminal residue is β-eliminated.
3. Deprotection is carried out in the usual way.

3.1.2.2. Enzymatic Phosphorylation

The procedures a and b described in ref. *15* a and b are used.

1. A ribonucleotide (preferably cytidine) is first attached to the substrate oligomer with terminal transferase. Thirty pmol of the oligodeoxynucleotide is incubated with 60 pmol of α-[^{32}P]-cytidine triphosphate and 16 U of terminal transferase in 20 µL of 0.1*M* cacodylate buffer (pH 7) containing 1 m*M* $CoCl_2$ at 30°C for 2 h. The oligonucleotide with a single cytidine attached is obtained in 70–80% yield. It can be separated from starting material and from a small amount of the oligodeoxynucleotide with two cytidine residues attached by gel electrophoresis.
2. The purified product is dissolved in 10 µL of water and then added to 5 µL of a 1*M* solution of cyclohexylamine in 0.5*M* di-*n*-propylmalonic acid at pH 8.2 (*see* Note 1). Five microliters of a 0.1*M* solution of sodium periodate is then added, and the reaction is left at 45°C for 90 min. The 3'-phosphate terminated oligonucleotide, usually obtained in ~90% yield, can be purified by gel electrophoresis or HPLC (*see* Note 2).

3.1.3. The Synthesis of 5'-Phosphorothioates

5'-Phosphorothioate residues can be attached to oligonucleotides either chemically during automated DNA synthesis (*see* Chapter 6 for details) or enzymatically, using adenosine-γ-thiotriphosphate and polynucleotide kinase.

3.1.3.1. Chemical Synthesis of 5'-Terminal Phosphorothioates

To obtain 5'-terminal phosphorothioates, Chemical Phosphorylation Reagent is used to form the oligonucleotide 5'-phosphite and the phosphite is oxidized with a sulfurizing agent (for example, TETD/acetonitrile available from Applied Biosystems). The following procedure is used.

1. The chemical phosphorylating agent is placed in the fifth or X phosphoramidate reservoir.
2. The sequence of the required oligomer that precedes the phosphorothioate is synthesized in the normal way.
3. The iodine bottle is removed from the synthesizer, and residual I_2 is removed from the lines by flushing with acetonitrile.
4. The sulfurizing reagent (for example, TETD/acetonitrile) is introduced into the position normally occupied by the iodine reagent.
5. The program is modified as indicated in the instructions provided with the reagent. This is necessary because the sulfurization reaction is slower than oxidation with iodine.

6. The sequence 5'-XA is entered. The A is arbitrary—it indicates to the synthesizer that the sequence already attached to the resin is to be treated as if it were the resin-attached 3'-nucleoside in a standard synthesis.
7. The product is deprotected in the normal way.

3.1.3.2. Enzymatic Synthesis of 5'-Terminal Phosphorothioates

Enzymatic synthesis of 5'-terminal phosphorothioate is carried out using the same general procedure described in Section 3.1.1. for normal phosphorylation, but with 5'-adenosine-γ-thiotriphosphate in place of 5'-adenosine-triphosphate. Higher concentrations of the triphosphate and enzyme than those used for normal phosphorylation are needed. Efficient thiophosphorylation of 3–4 nmol of a 20-mer oligodeoxynucleotide, for example, is achieved in 1 h at 37°C in the presence of 0.8 m*M* adenosine-γ-thiotriphosphate and 50 U of polynucleotide kinase.

3.1.4. Separation of Oligonucleotides from Their 5'-Phosphates and Oligonucleotide-5'-Phosphates from 5'-Phosphorothioates

It is often important to obtain a derivatized oligonucleotide uncontaminated by unmodified starting material. Otherwise, the unmodified oligomer competes for sites on a nucleic acid target, and lowers the efficiency of crosslinking, cleavage, and so forth. In such cases, it is important to start with a pure sample of the phosphorylated oligomer. If a 5'-phosphorylated oligodeoxynucleotide is prepared by solid-phase methods, this is not a problem. However, enzymatic phosphorylation is often incomplete, so enzymatically phosphorylated oligomers must be separated from unphosphorylated starting material. Oligonucleotide-5'-phosphorothioates made either enzymatically or chemically usually contain substantial amounts of the 5'-phosphate derivatives as side products. This is especially true of products made with γ-[^{35}S]-adenosine-thiotriphosphate of high specific activity. We have found that, in this case, the enzymatic reaction with polynucleotide kinase sometimes yields a product that includes more than 70% of the unlabeled 5'-phosphate derivative. Commercial, unlabeled adenosine-γ-thiotriphosphate preparations are usually considerably purer than the γ-[^{35}S]-labeled material, but even so we have usually found that about 20% of the unlabeled product made enzymatically is the phosphate.

Gel electrophoresis does not easily separate oligonucleotides from their 5'-phosphates, or oligonucleotide-5'-phosphates from the corresponding phosphorothioates, so HPLC must be used to carry out these separations. We have routinely used the reverse phase anion exchange packing material RPC-5 that we prepare in our laboratory to achieve the necessary separations for oligonucleotides up to 50 residues in length. The commercially available packing materials C18 or HEMA IEC BIO-1000 QAE 10U (Alltech, Dearfield, IL) *(16)* can sometimes be substituted for RPC-5 in work involving oligomers up to about the 20-mer.

3.2. The Synthesis of Nucleic Acids Carrying 5' or 3' Primary Amino Groups

The attachment of an amine to the 5'(or 3')-phosphate of an oligonucleotide, using a water-soluble carbodiimide such as 1-ethyl-3,3-dimethylaminopropylcarbodiimide (CDI), can be carried out in two ways. If the amine does not contain other reactive groups and the nucleic acid to be used does not have a high-mol wt, one can use a one-step procedure. The oligonucleotide together with the amine is treated with CDI in the presence of 1-methylimidazole *(7)* **(IIIb)**. This leads to the formation of a highly reactive 1-methylimidazolide adduct of the oligonucleotide. The 1-methyl imidazole is immediately displaced *in situ* by the amine to form the phosphoramidate (Fig. 1, **IV**).

Add 0.005–5 A_{260} U of the 5'(or 3')-phosphate of the oligonucleotide or nucleic acid (in 10 µL of water) to 10 µL of 0.5*M* 1-methylimidazole buffer (pH 7.2). Twenty microliters of a 2*M* solution of the amine, freshly prepared and adjusted to pH 7.2, is then added. Finally, 5 µL of a freshly prepared 1.5*M* solution of CDI is added to the reaction mixture (*see* Note 3). The mixture is kept at 50°C for 1–3 h. The final concentrations are ~0.1*M* 1-methylimidazole, 0.15*M* CDI, and ~0.2–0.4*M* amine. With most water-soluble amines, the reaction can be stopped after 2 h to obtain a 75–85% yield of the oligonucleotide phosphoramidate. With less soluble amines, such as di-*n*-hexylamine, reaction with a saturated solution of the amine for 3–4 h at 50°C is needed. Carbodiimides react with G and T residues, par-ticularly at high pHs *(17)*, so prolonged incubation at 50°C is not recommended.

The amine-derivatized oligonucleotide can usually be separated from the 5'-phosphate starting material, either by HPLC or by gel electrophoresis. If HPLC is used for purification, the entire reaction mixture can be loaded on the column. We have used RPC-5 to separate a wide variety of oligonucleotide phosphoramidates from corresponding oligonucleotide-5'-phosphates, using a perchlorate gradient at pH 12. Other commercially available anion exchange reverse phase columns such as C18 or HEMA IEC B10-1000 QAE 10 U (Alltech) could probably be used (*see also* Chapter 3). Phosphoramidates can be separated from phosphates by gel electrophoresis because they carry one less negative charge. However, it is first necessary either to remove the CDI from the reaction mixture by dialysis or to destroy the CDI by reaction with acetate before loading the reaction mixture on the gel.

When a high-mol wt nucleic acid, which cannot be exposed to CDI at high temperatures for long periods, is derivatized, or when a complex amine carrying additional functional groups that can both react with CDI (for example, an unprotected peptide) is used, the oligonucleotide 5'-phosphate is first converted to a phosphorimidazolide (Fig. 1, **IIIa**) by treating with CDI at room temperature for 1 h and the CDI is then removed rapidly by gel filtration. The phosphorimidazolide adduct, which has a limited lifetime at neutral pH, is then immediately reactedwith 0.2–0.4*M* solution of the amine for 2–4 h at 50°C. The phosphoramidates, unlike the phosphorimidazolides, can be stored indefinitely at –20°C and neutral pH.

An example of this procedure is the derivatization of an RNA that is the substrate of Qβ RNA polymerase *(18)*. Two micrograms of 5'-[^{32}P]-MDV-1 RNA (a 221-base substrate of Qβ RNA polymerase) and 16 μg of carrier yeast RNA is dissolved in 20 μL of water. Two and a half microliters of a 1.0*M* imidazole solution at pH 6 and 2.5 μL of a 1.5*M* solution of CDI is then added and the reaction is allowed to proceed at room temperature for 1 h. Seventy-five microliters of buffer containing 10 m*M* HEPES, 1 mm EDTA, and 100 m*M* NaCl (pH 7.2) is then added and the CDI removed on a Sephadex G-50 spin column *(19)* equilibrated with the same buffer. A 1.0*M* solution of cystamine (pH 7.7) is added to the recovered phosphorimidazolide (in ~0.1 mL) to give a final cystamine concentration of 0.25*M*. The reaction mixture is then incubated at 50°C for 1 h. The product is again isolated by spin chromatography and precipitated with EtOH. The product, which contains about 50–70% of the cystamine adduct, can be reduced with DTT and coupled to other thiol-containing molecules *(10)* (*see* Note 4).

3.3. Derivatization of Oligonucleotides with Pyrophosphate Linkages

3.3.1. Formation of Pyrophosphate Ligase Intermediates

Oligonucleotide 5'-phosphates (pN) react with phosphorimidazolides to form pyrophosphate adducts (Fig. 1, **II**). The oligonucleotide-5'-phosphate, for example, can be converted to the intermediate in the RNA (or DNA) ligase reaction, AppN, by reacting with adenosine 5'-phosphorimidazolide (ImpA) *(20)* (*see* Note 5).

One A_{260} U of the oligonucleotide 5'-phosphate is added to a solution containing 0.2*M* $MgCl_2$, 0.2*M* adenosine-5'-phosphorimidazolide (*see* refs. *21* and *22* for synthesis), and 0.2*M* Hepes buffer at pH 7. The reaction mixture is maintained at 50°C for 4 h to yield 80–85% of the pyrophosphate adduct. The latter can be separated from starting material by HPLC or gel electrophoresis. $Ap(pU)_6$ prepared by this method is an excellent substrate for RNA ligase. Ligase intermediates of this kind are useful when it is necessary to carry out ligations in the absence of ATP. Although we have not investigated this approach extensively, we find that it can readily be adapted to the synthesis of oligonucleotides terminated by a polyphosphate chain, for example, polynucleotide-5'-triphosphates. With minimal modification it should facilitate the synthesis of "capped" messenger RNAs and their analogs.

3.4. Attachment of a Thiol Group to the 5'(or 3')-Terminus of an Oligonucleotide

A free SH group is conveniently introduced into an oligonucleotide-5'(or 3')-phosphate via a phosphoramidate bond to the NH_2 group of cystamine (Fig. 1, **VII**). The cystamine adduct **V** of the oligomer is first formed using the one-step procedure described above. It is a stable molecule that can be stored indefinitely at –20°C. The 5'(or 3')-thioethylamino (cystamine) adduct **VII**, which is sensitive to oxidation, can be generated as needed by treatment of the 5'-cystamine adduct with 5 m*M* DTT at room temperature for 1 h. To derivatize the free thiol group *in situ*, the thiol derivatizing reagent must be in substantial excess over the surviving DTT. Derivatization *in situ* can be successfully accomplished with a reagent such as 2,2'-di-pyridyldisulfide *(see below)*. The cystamine adduct can also be reduced using DTT attached to agarose beads (Reductaryl, available from Calbiochem, La Jolla, CA). The beads can be removed after reduction is complete without leaving an excess of DTT in the solution. It is also possible to use 0.3*M* sodium borohydride as the reducing agent (*see* Notes 6 and 7).

3.5. Examples of Modification of 5'(or 3')-Amino-Terminated Oligonucleotides

3.5.1. 5'(or 3')-Biotinylated Oligonucleotides

Amino-terminated oligonucleotides react readily with *N*-hydroxysuccinimidobiotin to form biotinylated adducts of oligonucleotides. Approximately 1 mg of the activated biotin is added to a solution of the amino-terminated oligonucleotide (0.01–10 A_{260} U) in 0.2 mL of 0.2*M* HEPES buffer at pH 8 with shaking, and allowed to react at room temperature for 1 h. Excess undissolved biotin is removed by centrifugation and the biotinylated oligonucleotide **VI** separated from the starting material by HPLC or gel electrophoresis *(8,9)*.

3.5.2. Oligonucleotides Carrying Radical Generating Reagents—Fe^{2+} EDTA

EDTA can readily be attached to amino-terminated oligonucleotides. The product is a chelating agent for Fe^{2+}. When the Fe^{2+} chelates are hybridized to complementary polynucleotides in the presence of a suitable reducing agent, free radicals are generated. These radicals attack and cleave the complementary template specifically at sites close to the metal ions *(24,25)*. Phenanthroline can be attached to an oligonucleotide in a similar way, and brings about cleavage in the presence of cuprous ion and hydrogen peroxide *(26)*.

EDTA anhydride (commercially available from Aldrich Chemicals, Milwaukee, WI) reacts readily with amino-terminated oligonucleotides. For example, 1 mg of EDTA anhydride was reacted with the 5'-ethylenediamine adduct of an oligonucleotide 16-mer in 0.2 mL of 0.2*M* HEPES buffer at pH 8. After 1 h excess undissolved anhydride was removed by centrifugation and the EDTA adduct, obtained in >90% yield, was purified by HPLC *(24)*.

To cleave the complementary template, a 3 μ*M* solution of the 5'-EDTA-oligonucleotide is mixed with a 600 μ*M* solution of ferrous ammonium sulfate. The EDTA adduct in 1 μL of this solution is then hybridized to 0.015–0.05 pmol of a [^{32}P]-labeled complementary strand (5000–7000) cpm in 9 μL of buffer containing 40 m*M* Tris at pH 7.8, 0.1*M* NaCl, and 100 g of carrier DNA. The final reaction mixture (10 μL) contains 0.1*M* NaCl, 60 μ*M* Fe^{2+}, 100 μg of carrier DNA, 0.015–0.05

pmol of the [^{32}P]-labeled template, and 0.5–1.5 pmol (20–30-fold excess over template) of the EDTA oligonucleotide adduct. To initiate the cleavage reaction, 1 μL of a 50 m*M* DTT solution is added to the hybridized oligonucleotides. After 1 h, the reaction is stopped with EDTA and the cleavage products analyzed by gel electrophoresis.

3.5.3. Derivatization with Oligo-L-Lysine to Facilitate Entry of Oligonucleotides into Cells

Attachment of oligolysines has been shown to facilitate the transport of oligonucleotides into cells *(27)*. To attach oligolysines to an oligomer *(28)*, 0.2–2 nmol of the oligonucleotide-5'-phosphate is reacted with 0.2*M* oligo-lysine (pH 7) and 0.15*M* CDI in 30 μL of 0.1*M* 1-methylimidazole buffer at pH 7 for 1 h at 50°C. Add 1*M* phosphate (pH 7) to the reaction (to a final concentration of 0.1*M*) to dissociate ionically bound polylysine from the polynucleotide, and keep the mixture a further hour at 50°C. The products can be separated from starting material by denaturing gel electrophoresis.

3.5.4. Covalent Attachment of Peptides or Proteins to Nucleic Acids via Cleavable Disulfide Linkers

A nucleic acid can be joined to a protein via cleavable disulfide bonds by first converting the nucleic acid to a 5'(or 3')-2-pyridyldisulfide adduct **(IX)** (Fig. 1). The 5'(or 3')-pyridyldisulfide oligonucleotide is then reacted with a peptide or protein containing one or more sulfhydryl residues to form a nucleic acid-protein adduct linked by a cleavable disulfide bond **(X** or **XI)**. A simpler procedure is available provided the peptide or protein does not contain a disulfide linkage. The 5'(or 3')-cystamine adduct of the oligonucleotide **(V)** and the thiol-containing peptide are mixed together and reduced with DTT. The solution is then dialyzed against buffer containing 10^{-4} DTT to remove the cystamine. Further dialysis against buffer without DTT results in air oxidation of the thiols to form the oligonucleotide-peptide adduct linked by disulfide bonds **(VIII)**. The adduct is separated from the dimer of the peptide by gel electrophoresis *(10)*. Similar chemistry has been used to attach psoralen to the end terminus of an oligodeoxynucleotide *(29)* (*see* Note 8).

3.5.4.1. FORMATION OF 5'-2-PYRIDYLDISULFIDE OLIGONUCLEOTIDES

To form the pyridyldisulfide adduct **(IX)**, the nucleic acid is first converted to the cystamine derivative **V** *(10)* (*see* Section 3.2.). Treat

0.02–0.2 A_{260} U of 5'-cystamine oligonucleotide with 5 m*M* DTT in 11 µL of buffer containing 10 m*M* Tris and 1 m*M* EDTA at pH 7.2 for 1 h at room temperature to form the 5'-cystamine adduct **VII** (*see* Note 7). One and a half microliter of buffer containing 500 m*M* Tris (pH 7.2) and 1 m*M* EDTA, and 51 µL of a 3 m*M* solution of 2,2'-dipyridyldisulfide in water is then added. The final reaction mixture contains 0.8 m*M* DTT and 2.4 m*M* 2,2'-dipyridyldisulfide (an excess of 3:1 of dipyridyl-disulfide over DTT). The reaction mixture is left at room temperature for 2 h. The 5'-pyridyldisulfide adduct of the oligonucleotide is then separated from starting materials either by HPLC or by gel electrophoresis. Yields ranged from 70–80% (*see* Note 9).

3.5.4.2. Thiolation of Proteins

If free sulfhydryl groups are not present in proteins, they can be introduced by reaction with 2-iminothiolane, which attacks lysine ε-amino groups and terminal amino groups with liberation of a free SH group. We have slightly modified the procedures in refs. *30* and *31* to introduce thiol groups into peroxidase and IgG. Thirty-four micrograms (~1 nmol) of peroxidase in 40 µL of 50 m*M* sodium borate buffer, pH 8.5, is mixed with 2 mg (14 mmol) of iminothiolane (Pierce, Rockford, IL) (final concentration 0.3*M*). After 1 h at room temperature, thiolated peroxidase is separated from excess reagent by gel filtration, eluting with buffer containing 50 m*M* sodium phosphate, pH 6.9, and 1 m*M* EDTA. The resulting thiolated peroxidase, which contains approx 3 mol of thiol/mol of peroxidase, is as active as native peroxidase in the H_2O_2-diaminobenzidene assay on nitrocellulose *(32)*.

To thiolate an antibody *(31)*, 400 µg of human IgG (2.5 nmol) is dissolved in buffer containing 60 m*M* triethanolamine, 7 m*M* potassium phosphate, 100 m*M* NaCl, and 1 m*M* EDTA at pH 8. Twenty microliters of a 10 m*M* solution of iminothiolane in 1*M* triethanolamine at pH 8 is then added. The final concentration of iminothiolane in the reaction mixture is 1 m*M*. The reaction is allowed to proceed for 1 h at 0°C. The thiolated antibody is then separated from unreacted imino-thiolane by gel filtration, eluting with 5 m*M* bis-Tris-acetate buffer containing 50 m*M* NaCl, and 1 m*M* EDTA at pH 5.8. The final preparation contains about 1 mol of thiol/mol of antibody. Antibodies thiolated with this reagent retain their antibody binding properties. Thiol concentrations of each preparation are measured using the Ellman procedure *(33)*.

3.5.4.3. Formation of Nucleic Acid-Protein Adducts

To form the nucleic acid-protein adduct, 10–100 pmol of the 5' (or 3')-pyridyldisulfide oligonucleotide is added to the thiol-containing protein (5–50 µg) in 10 µL of buffer containing 10 m*M* Tris and 1 m*M* EDTA at pH 7.2, and allowed to react overnight at room temperature. If the available solution of the oligonucleotide or the protein is too dilute, the solutions are mixed and then dialyzed together against buffer containing 0.1 m*M* Tris and 0.1 m*M* E1DTA. The dialysate is then evaporated in a SpeedVac concentrator, dissolved in 10 µL of Tris-EDTA buffer and incubated overnight.

It is possible to separate protein-nucleic acid adducts from uncoupled protein or nucleic acid either by gel electrophoresis or by DEAE chromatography. Gel electrophoresis is recommended for relatively low-mol wt proteins coupled to oligonucleotides < 30 residues long. Gel electrophoresis also allows oligonucleotide adducts of protein oligomers to be separated from the desired adduct of the monomeric protein. When a 16-mer oligonucleotide probe was linked to peroxidase (mol wt 40,000) containing an average of 3 mol of thiol/mol protein, for example, electrophoresis on 6% polyacrylamide allowed separation of the 16-mer-peroxidase mono adduct from unreacted peroxidase as well as from adducts of dimerized and trimerized peroxidase *(10)*. The purified 16-mer-peroxidase adduct hybridized efficiently to its complement and retained one-third of the enzymatic activity of the native protein.

When oligonucleotides longer than 30 residues or relatively large proteins such as antibodies are used, purification by DEAE chromatography is recommended. Unreacted proteins are generally not retained on DEAE and are eluted first with low ionic buffers. Protein-nucleic acid adducts are eluted at intermediate salt concentrations and unreacted oligonucleotides or nucleic acids are eluted last at higher salt concentrations.

The choice of buffer or salts for the elution is dependent on the size of the oligonucleotide/nucleic acid used. If the oligonucleotide is < 20–30 residues long, it is convenient to use triethylammonium bicarbonate (TEAB) buffer (pH 8) because TEAB is volatile and can be removed by lyophilization. For example, when a 16-mer oligodeoxynucleotide is linked to human IgG, unreacted IgG can be eluted with 0.25*M* TEAB, the 16-mer-IgG with 0.5*M* TEAB, and unreacted 16-mer with 1.0*M* TEAB. If higher-mol wt nucleic acids are used, for example a 221-mer

MDV-1 RNA, the concentration of TEAB required for elution is too high to be practical. In this situation, unreacted IgG is eluted from DEAE with buffer containing 50 m*M* Tris pH 7.5 and 1 m*M* EDTA, the RNA-IgG adduct with the same buffer containing 0.4*M* NaCl, and unreacted RNA with the same buffer containing 0.5*M* NaCl. The RNA-IgG adduct can be removed from buffer salts and concentrated in an Amicon 30 microconcentrator *(10)*. It is advisable to add 20 µg of BSA to the solution as carrier during the concentration process.

3.6. Crosslinking of 5'-Thiol or 5'-Phosphorothioate Oligonucleotides to Their Complements

An oligonucleotide carrying a thiol or phosphorothioate residue at its 5'- or 3'-terminus can be crosslinked to a complementary oligonucleotide either with transplatinum diamminedichloride (transPt^{II}) or with K_2PtCl_4 *(13,14,28)*. The platinum complexes react rapidly with cystamine or phosphorothioate groups. The resulting adducts still contain a reactive Pt^{II}-Cl bond that attacks bases on the complementary strand, provided they are in close proximity to the attachment site of the platinum. The efficiency of crosslinking (between 35–60%) depends on the identity of the bases closest to the Pt^{II} ion. G and A residues are most readily attacked, C residues less so, and T residues only if no other bases are available.

About 0.01 pmol of 5'-[^{32}P]-labeled DNA or RNA template in 2.5 µL of 0.02*M* phosphate buffer, pH 7.7, is heated at 95–100°C for 1 min and cooled on ice. The following are then added sequentially: One microliter of buffer containing 1 m*M* phosphate (pH 7) and 0.1 m*M* EDTA, 1 µL of a solution containing 0.1 pmol of a complementary 5'-thiol or 5'-phosphorothioate-oligonucleotide, 0.5 µL of 0.5*M* $NaClO_4$, and 0.5 µL of a solution containing the required amount of transPt^{II} (*see* Note 10). The final volume is 5.5 µL After incubation overnight at room temperature, the reaction mixtures are made up to 10 µL with buffer and the cross-product separated from the uncrosslinked [^{32}P]-template by electrophoresis on a 20% denaturing gel (*see* Note 11).

4. Notes

1. The buffered cyclohexylamine solution used above is made by dissolving 5 mmol of di-*n*-propylmalonic acid in 5 mL of water and adjusting the pH to 8 with NaOH. Ten millimoles of cyclohexylamine and sufficient water to bring the final volume to 10 mL is then added.

2. When a [^{32}P]-labeled oligonucleotide terminated with phosphate at both the 5'- and 3'-termini is required, the 5'-phosphate group must be introduced first. Polynucleotide kinase contains a 3'-phosphodiesterase activity that removes phosphate groups from oligonucleotide-3'-phosphates.
3. Carbodiimides deteriorate rapidly. To insure the highest yield of product in these reactions, carbodiimides should be stored desiccated at –20°C. Fresh solutions of carbodiimide should be prepared just before use.
4. It is not possible to separate the 5'-cystamine adduct of high-mol wt nucleic acids such as MDV-1 RNA from the 5'-phosphate derivative by gel electrophoresis. However, it is possible to biotinylate the cystamine adduct, react it with avidin, and then to purify the avidin adduct of the biotinylated RNA. Treatment with dithiothreitol liberates the purified cystamine adduct of the RNA *(18)*.
5. The reaction of an oligonucleotide 5'-phosphorimidazolide (ImpN) with adenosine-5'-phosphate (pA) to form the pyrophosphate AppN is not efficient, presumably because the hydrolysis of oligonucleotide 5'-phosphorimidazolide to 5'-phosphate competes effectively with pyrophosphate formation.
6. The removal of salts and other reagents from sulfur-containing oligonucleotide analogs that have been purified by HPLC or gel electrophoresis can be achieved with nucleic acid purification cartridges or by dialysis. If the latter method is used it is important that the dialysis tubing first be treated with acetic anhydride/EDTA *(23)* to remove metal ions and other agents that can react with sulfur-containing compounds. Alternatively, pretreated dialysis tubing can be purchased from Spectrapor (Houston, TX) (Spectrapor CE tubing).
7. Buffers containing sulfur (such as HEPES buffer) should not be used when working with sulfur-containing adducts of oligonucleotides.
8. Oligonucleotides can be linked to proteins by noncleavable linkers using chemistry related to that described above. In one procedure, imidazole is displaced by a hydrazide, and the latter is subsequently coupled to an aldehyde-containing protein *(11)*. It should also be possible to use one of the many "double-headed" reagents available from Pierce Chemical Company. Typically, one functional group of the linker is an activated ester that is first coupled to an amine-terminated oligonucleotide. The second functional group is an iodide or a maleimido residue, that subsequently is reacted with the sulfhydryl group of a protein. The nucleic acid-protein product is held together by stable amide and thioether bonds. Several suitable linkers with differing lengths and water solubilities are available. See ref. *34* for an example of a related approach.
9. Pyridyldisulfide adducts of nucleic acids are not stable. They should be used as soon as possible after preparation.

10. Just before use make 0.5 m*M* solutions of transPt^{II} or K_2PtCl_4 in buffer containing 1 m*M* PO_4 (pH 7.4) and 1 m*M* EDTA. Dilutions are made in the same buffer.
11. Yields of crosslinked products are highest when sulfur-containing oligonucleotides are separated from underivatized oligomers.

Acknowledgment

We are grateful to the National Institutes of Health for support of the work from our laboratory described in this chapter (Grant No. 2 RO1 GM 33023-09).

References

1. Knorre, D. G. and Vlassov, V. V. (1985) Complementary-addressed (sequence-specific) modification of nucleic acids, in *Progress in Nucleic Acid Research and Molecular Biology, vol. 32* (Cohn, W. E. and Moldave, K., eds.), Academic, Orlando, FL, pp. 291–319.
2. Goodchild, J. (1990) Conjugates of oligonucleotides and modified oligonucleotides: a review of their synthesis and properties. *Bioconj. Chem.* **1,** 165–187.
3. Chambers, R. W. and Moffatt, J. G. (1958) The synthesis of adenosine 5' and uridine-5' phosphoramidates. *J. Am. Chem. Soc.* **80,** 3752–3756.
4. Gilham, P. T. (1968) The synthesis of celluloses containing covalently bound nucleotides, polynucleotides, and nucleic acids. *Biochemistry* **7,** 2809–2813.
5. Moffatt, J. G. and Khorana, H. G. (1961) Nucleoside polyphosphates. X. The synthesis and some reactions of nucleoside-5'-phosphoromorpholidates and related compounds. Improved methods for the preparation of nucleoside-5'-polyphosphates. *J. Am. Chem. Soc.* **83,** 649–658.
6. Ivanovskaya, M. G., Gottikh, M. B., and Shabarova, Z. A. (1982) DNA-like duplexes containing repeats. IV. Template-directed polymerization of decadeoxyribonucleotide imidazole. *Bioorgan. Khim.* **8,** 940–944.
7. Chu, B. C. F., Wahl, G. M., and Orgel, L. E. (1983) Derivatization of unprotected polynucleotides. *Nucleic Acids Res.* **11,** 6513–6529.
8. Chu, B. C. F. and Orgel, L. E. (1985) Detection of specific DNA sequences with short biotin-labeled probes. *DNA* **4,** 327–331.
9. Cholet, A. and Kawashima, E. H. (1985) Biotin-labeled synthetic oligodeoxyribonucleotides: chemical synthesis and uses as hybridization probes. *Nucleic Acids Res.* **13,** 1529 1541.
10. Chu, B. C. F. and Orgel, L. E. (1988) Ligation of oligonucleotides to nucleic acids or proteins via disulfide bonds. *Nucleic Acids Res.* **16,** 3671–3691.
11. Ghosh, S. S., Kao, P. M., and Kwoh, D. Y. (1989) Synthesis of 5'-oligonucleotide hydrazide derivatives and their use in preparation of enzyme-nucleic acid hybridization probes. *Anal. Biochem.* **178,** 43–51.
12. Hodges, R. R., Conway, N. E., and McLaughlin, L. W. (1989) "Post-assay" covalent labeling of phosphorothioate-containing nucleic acids with multiple fluorescent markers. *Biochemistry* **28,** 261–267.

13. Chu, B. C. F. and Orgel, L. E. (1990) A simple procedure for cross-linking complementary oligonucleotides. *DNA and Cell Biol.* **9,** 71–76.
14. Chu, B. C. F. and Orgel, L. E. (1990) Optimization of the efficiency of cross-linking Pt^{II} oligonucleotide phosphorothioate complexes to complementary oligonucleotides. *Nucleic Acids Res.* **18,** 5163–5171.
15a. Weith, H. L. and Gilham, P. T. (1967) Structural analysis of polynucleotides by sequential base elimination. The sequence of the terminal decanucleotide fragment of the ribonucleic acid from bacteriophage f2. *J. Am. Chem. Soc.* **89,** 5473,5474.
15b. Deng, G.-R. and Wu, R. (1983) Terminal transferase: use in the tailing of DNA and for *in vitro* mutagenesis, in *Methods in Enzymology, vol. 100* (Wu, R., Grossman, L., and Moldave, K., eds.), Academic, New York, pp. 96–116.
16. Stribling, R. (1991) High-performance liquid chromatography of oligoguanylates at high pH. *J. Chromatog.* **538,** 474–479.
17. Gilham, P. T. (1962) An addition reaction specific for uridine and guanosine nucleotides and its application to the modification of ribonuclease action. *J. Am. Chem. Soc.* **84,** 687–688.
18. Chu, B. C. F., Kramer, F. R., and Orgel, L. E. (1986) Synthesis of an amplifiable reporter RNA for bioassays. *Nucleic Acids Res.* **14,** 5591–5603.
19. Maniatis, T., Fritsch, E. F., and Sambrook, J. (1982) *Molecular Cloning*. 2nd Ed., Cold Spring Harbor Laboratory, Cold Spring Harbor, NY, p. 5,68.
20. Chu, B. C. F. and Orgel, L. E. (1984) Preparation of ligation intermediates and related polynucleotide pyrophosphates. *Biochim. Biophys. Acta* **782,** 103–105.
21. Joyce, G. F., Inoue, T., and Orgel, L. E. (1984) Non-enzymatic template-directed synthesis on RNA random copolymers: poly(C, U) templates. *J. Mol. Biol.* **176,** 279–306.
22. Mukaiyama, T. and Hashimoto, M. (1971) Phosphorylation by oxidation-reduction condensation. Preparation of active phosphorylating reagents. *Bull. Chem. Soc. Japan* **47,** 2284.
23. Maniatis, T., Fritsch, E. F., and Sambrook, J. (1982) *Molecular Cloning*. Cold Spring Harbor Laboratory, Cold Spring Harbor, NY.
24. Chu, B. C. F. and Orgel, L. E. (1985) Nonenzymatic sequence-specific cleavage of single-stranded DNA. *Proc. Natl. Acad. Sci. USA* **82,** 963–967.
25. Dreyer, G. B. and Dervan, P. B. (1985) Sequence-specific cleavage of single-stranded DNA; Oligodeoxynucleotide-EDTA-Fe(II). *Proc. Natl. Acad. Sci. USA* **82,** 968–972.
26. Chen, C.-H. B. and Sigman, D. S. (1986) Nuclease activity of 1,10-phenanthroline-copper: sequence-specific targeting. *Proc. Natl. Acad. Sci. USA* **83,** 7147–7151.
27. LeMaitre, M., Bayard, B., and Lebleu, B. (1987) Specific antiviral activity of a poly(L-lysine)-conjugated oligodeoxyribonucleotide sequence complementary to vesicular stomatitis virus *N* protein mRNA initiation site. *Proc. Natl. Acad. Sci. USA* **84,** 648–652.
28. Chu, B. C. F. and Orgel, L. E. (1989) Inhibition of DNA synthesis by cross-linking the template to platinum-thiol derivatives of complementary oligodeoxy-nucleotides. *Nucleic Acids Res.* **17,** 4783–4798.

29. Teare, J. and Wollenzien, P. (1989) Specificity of site directed psoralen addition to RNA. *Nucleic Acids Res.* **17,** 3359–3372.
30. King, T. P., Li, Y., and Kochoumian, L. (1978) Preparation of protein conjugates via intermolecular disulfide bond formation. *Biochemistry* **17,** 1499–1506.
31. Blättler, W. A., Kuenzi, B. S., Lambert, J. M., and Senter, P. D. (1985) New heterobifunctional protein cross-linking reagent that forms an acid-labile link. *Biochemistry* **24,** 1517–1524.
32. Tsang, V. C. W., Peralta, J. M., and Simons, A. R. (1983) Enzyme-linked immunoelectrotransfer blot techniques (EITB) for studying the specificities of antigens and antibodies separated by gel electrophoresis, in *Methods in Enzymology, vol. 92* (Langone, J. J. and Van Vunakis, H., eds.), Academic, New York, pp. 377–391.
33. Ellman, G. I. (1959) Tissue sulfhydryl groups. *Arch. Biochem. Biophys.* **82,** 70–77.
34. Murakami, A., Tada, J., Yamagata, K., and Takano, J. (1989) Highly sensitive detection of DNA using enzyme-linked DNA-probe. 1. Colorimetric and fluorometric detection. *Nucleic Acids Res.* **17,** 5587–5595.

CHAPTER 6

Oligonucleotide-Enzyme Conjugates

Jerry L. Ruth

1. Introduction

For nucleic acid hybridizations, the choice of nonisotopic label is constrained by hybridization and wash conditions, the sensitivity needed in the assay, and the detection method. Several recent reviews *(1–4)* have summarized methods of labeling synthetic DNA probes. In spite of significant advances in the detection of fluorescent and luminescent labels, enzymes continue to be the most sensitive reporter groups. For most applications, direct enzyme labels offer the best overall performance, with highest sensitivity, least background, and rapid detection. Several enzymes, particularly alkaline phosphatase and horseradish peroxidase, are compatible with standard hybridization conditions, and allow detection by a number of endpoints, including soluble color, dye deposition, fluorescence, and luminescence. Reported methods *(5)* for the crosslinking of enzymes to cloned double-stranded DNA does not work acceptably for oligonucleotides, giving little or no hybridization.

Selective conjugation of alkaline phosphatase and HRP to synthetic oligonucleotides was initially achieved *(6)* by the incorporation of amine-modified bases (*see* Compound **1** in Fig. 1). Oligomer probes conjugated to enzymes in such a manner can be made cleanly in a 1:1 ratio, because of their well-defined nature, avoiding excess modification of the enzyme or probe. Unlike with cloned probes, hybridizations of oligonucleotides can be done rapidly, at more moderate temperatures, and in simple buffers, preserving enzyme activity and therefore

From: *Methods in Molecular Biology, Vol. 26: Protocols for Oligonucleotide Conjugates*
Edited by: S. Agrawal Copyright ©1994 Humana Press Inc., Totowa, NJ

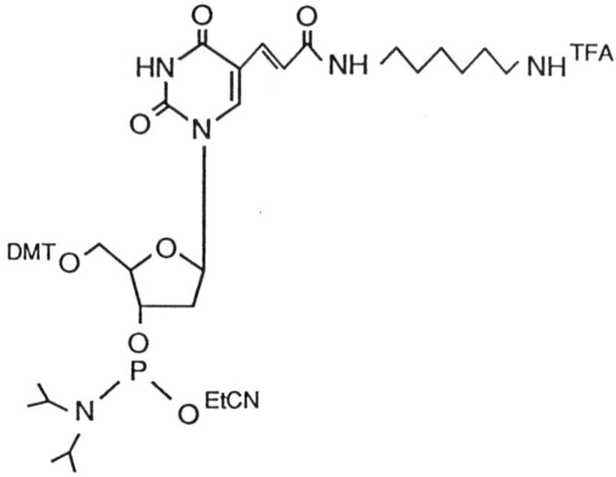

1 : C-5 Amino-dThd

Fig. 1. Functionalized thymidine base for incorporation of amines into synthetic oligonucleotides (DMT = dimethoxytrityl; TFA = trifluoroacetyl; EtCN = cyanoethyl protecting group).

maximizing signal. Surprisingly, the conjugation of enzymes to oligonucleotides has little affect on the hybridization kinetics, selectivity, or efficiency of the oligonucleotide itself. This provides nonisotopic probes that are 20–80 times more sensitive when detected by chemiluminescence than the corresponding ^{32}P label. Nonisotopic detection is also faster than the autoradiographic detection of ^{32}P (hours vs days).

Until recently, cloned probes had significant advantages in sensitivity, because of sheer number of reporter groups. This is no longer necessarily true. One oligonucleotide probe 20–30 bases in length with a single enzyme label can equal the sensitivity of a cloned probe several thousand bases in length with several hundred enzymes attached or ^{32}P labels incorporated (*see* Table 1, Section 3.2.).

The incorporation of functionalized nucleoside precursors into oligonucleotides during chemical synthesis, and subsequent attachment of the enzyme to the functional groups is summarized in Section 3. In this approach, amine "linker arm" nucleosides are incorporated into the DNA at precisely defined sites, and the subsequent attachment of the enzyme occurs only at those sites *(4,6)*. This results in the attachment of only one enzyme per oligonucleotide molecule. A number of successful varia-

Table 1
Comparison of Sensitivities for Enzymes Attached to DNA Probes*,**

Reporter	Detection method (ref)	Amount of target detection: moles	Amount of target detection: molecules	Rel. signal
Cloned probes				
HRP	DAB; color *(5)*	1×10^{-17}	6,000,000	1.3
AP	NBT/BCIP; color *(5)*	2×10^{-18}	1,000,000	8
Biotin	SA-AP; color *(2)*	2×10^{-19}	120,000	70
^{32}P	autorad, 48 h *(2)*	1×10^{-19}	60,000	138
Oligonucleotide probes†				
HRP	Color (DAB)	1×10^{-17}	8,000,000	1
Biotin	SA-AP; NBT/BCIP	1×10^{-17}	6,000,000	1.3
5'-^{32}P	autorad, 24 h	6×10^{-18}	3,000,000	3
AP	color (NBT/BCIP)	2×10^{-18}	1,000,000	8
AP	fluorescence (MUBP)	5×10^{-19}	300,000	28
AP	luminescence (AMPPD)	1×10^{-19}	60,000	138

*The results are in order of increasing hybridization sensitivity. Some reporter groups other than enzymes are included for comparison. Amount of target detected is considered to be accurate to ± twofold; hybridization was to plasmid target immobilized on membranes.

**Where "Reporter" = the signal group attached to the DNA probe; SA = streptavidin; AP = alkaline phosphatase; HRP = horseradish peroxidase; DAB = diaminobenzidine; NBT/BCIP = nitro blue tetrazolium/5-bromo-4-chloro-3-indolyl phosphate; MUBP = 4-methylumbelliferone phosphate; AMPPD = "3-adamantyl 4-methoxy 4-(2-phospho)phenyl dioxetane" (correct systematic name is disodium 3-[4-methoxyspiro(1,2-dioxetane-3,2'-tricyclo{3.3.1.1.$^{3.7}$}decan)-4-yl]phenyl phosphate).

†All nonisotopic labels (1/probe) were attached to the same sequence oligonucleotide 22-mer through an internal single linker arm dThd analog.

tions *(7–17)* have been described in Chapters 2–4 of this volume. However, the protocols contained in this chapter have intentionally been limited to published procedures using commercially available reagents. All of the reagents are designed for use on automated DNA synthesizers, allowing the routine synthesis and use of enzyme conjugates in hybridization assays.

The following chapter summarizes some useful methods for direct attachment of enzymes to functionalized synthetic oligodeoxynucleotides. The methods are based on attachment to 2'-deoxyoligomers (DNA); with care, the methods can also be adapted to RNA or RNA/DNA chimers.

2. Materials

2.1. "Linker Arm" Oligonucleotide Synthesis

1. Compound **1**, a C-5 aminoalkyl-thymidine ("Linker Arm Nucleoside") as its fully-protected phosphoramidite. Compound **1** can also be made *(4)* from 2-deoxyuridine, but the synthesis is somewhat involved and should not be undertaken casually.
2. Conventional reagents for oligonucleotide synthesis, including β-cyanoethyl phosphoramidites of the common bases, available from numerous sources.
3. Standard solutions for deprotection and purification of oligonucleotides, including concentrated ammonium hydroxide, 80% acetic acid, 2 m*M* EDTA, pH 8, and so forth.
4. Any commercially-available automated DNA synthesizer.
5. Reverse-phase HPLC for purification (optional for some uses).

For amino modification at 5', compounds such as a 5'-aminohexyl modifier *(9)* ("Aminolink") can be used in Method 3.1. without modification of the protocols; *see* Chapter 3 for more detail. One can also use 5'-thiohexyl modifiers *(10,11)*, but they require thiol-reactive cross-linkers *(11,* Chapter 5). In theory, abasic aminomodifiers *(14)* and 3'-aminoalkyl modifiers on controlled pore glass *(13)* could be used to attach enzymes to oligonucleotides, but no such reports have yet appeared. *Note:* The affect a particular modification may have on hybridization should be considered before choosing one of these reagents.

2.2. Conjugation of Oligonucleotides to Enzymes

1. Di-succinimidyl suberate ("DSS," Pierce [Rockford, IL] catalog #21555).
2. Calf intestinal alkaline phosphatase (AP; mw 141,000; Boehringer Mannheim [Indianapolis, IN] immunoassay grade cat #556602, at 10 mg/mL). (Other sources or preps may give poor conjugation.)
3. Alkaline phosphatase (AP) reaction buffer: 3*M* NaCl, 100 m*M* $NaHCO_3$, 1 m*M* $MgCl_2$, pH 8.2–8.3. Sterile filter.
4. G-25 Chromatography buffer: 1 m*M* sodium acetate, to pH 4 with HOAc. Sterile filter.
5. P-100 Chromatography buffer: 100 mM NaCl, 50 mM Tris-HCl, pH 8.5. Sterile filter.
6. Anion exchange buffers: (for MonoQ columns; others may vary). Sterile filter. Buffer A: 20 m*M* Tris-HCl, pH 8; Buffer B: 20 m*M* Tris-HCl, 1*M* NaCl, pH 8.

7. High salt buffer (HSB) (for storage of AP conjugates): 3*M* NaCl, 30 m*M* Tris, 1 m*M* $MgCl_2$, 0.1 m*M* $ZnCl_2$, 0.05% azide, pH 7.6. Sterile filter.
8. UV-Vis spectrophotometer for measuring DNA and protein concentrations.
9. Centricon 10 and 30 microconcentrators (Amicon, Beverly, MA).
10. FPLC system for anion exchange purification of the final conjugate (optional).

2.3. Hybridization of Enzyme Conjugates to Membrane-Bound Target

Abbreviations: 1X SSC = 150 m*M* sodium chloride, 15 m*M* sodium citrate, pH 7.0; SDS = sodium dodecyl sulfate; BSA = bovine serum albumin, fraction V; BCIP = 5-bromo-4-chloro-3-indolyl phosphate as its 4-toluidine salt; NBT = nitro blue tetrazolium; 4-MUP = 4-methylumbelliferone phosphate (from JBL Scientific, San Luis Obispo, CA); AMPPD = "3-adamantyl 4-methoxy 4-(2-phospho)phenyl dioxetane" (from Tropix, Inc., Bedford, MA).

1. Hybridization solution: 5X SSC, 1 w/v% SDS, 0.5 w/v% BSA.
2. Wash 1 solution: 1X SSC, 1 w/v% SDS.
3. Wash 2 solution: 1X SSC, 1 w/v% Triton X-100.
4. Wash 3 solution: 1X SSC.
5. a. Color reagent: Fresh 0.16 mg/mL BCIP, 0.17 mg/mL NBT in 100 m*M* NaCl, 100 m*M* Tris-HCl, 5 m*M* $MgCl_2$, and 0.1 m*M* $ZnCl_2$, pH 8.5–9.5.
   b. Fluorescent reagent: Fresh 30 μ*M* 4-MUP in 100 m*M* diethanolamine (highest quality available), 5 m*M* $MgCl_2$, no buffer (final pH ≈9).
   c. Luminescent reagent: 250 μ*M* AMPPD in 100 m*M* diethanolamine-HCl, 1 m*M* $MgCl_2$, pH 9.5–10.

2.4. In Situ *Hybridization of Enzyme Conjugates to Target in Immobilized Cells*

See Section 2.3. for definition of abbreviations:

1. Denaturation solution: 70% formamide, 1X SSC, 0.5% BSA, 5% methanol.
2. Hybridization solution: 5X SSC, 1 w/v% SDS, 0.5 w/v% BSA (same as for membranes).
3. Wash 1 solution: 1X SSC, 1 w/v% SDS (same as for membranes).
4. Wash 2 solution: 1X SSC, 0.5 w/v% Triton X-100, 0.075% Brij-35.
5. Wash 3 solution: 1X SSC, 0.075% Brij-35.

6. Color reagent: Same as for Section 2.3. (hybridization to membranes) + 15 w/v% polyvinyl alcohol.
7. Microscope and slides.
8. Slide processor (optional).

3. Methods

3.1. Synthesis

Reagents that allow modification at internal sites of the oligonucleotide *(4,6,7,15)*, at the 5'-terminus *(8–11)*, and at the 3'-terminus *(12,13)* have been described. Chapter 3 of this book describes modifications at 5' or 3'. This section describes the methods used to incorporate functionally-modified bases into oligodeoxynucleotide probes during synthesis *(4,6)*. No modification of synthesis cycles, postsynthesis treatment, or purification is needed. The same or similar reagents can also be used for modification of chemically-synthesized RNA or chimeric RNA/DNA oligonucleotides. Product oligomers contain a free primary amine "linker arm" base. Nonisotopic labels such as enzymes, biotin, fluorescein, or acridines can then be attached to the linker arms *(4)*; methods for attaching enzymes are summarized in Section 3.2. For attachment to enzymes, only one functional group should be incorporated per oligonucleotide. The resulting conjugates are then used as nonisotopic hybridization probes. The following protocol assumes the user has determined the desired sequence for the oligonucleotide probe.

The linker reagents are all commercially available as their β-cyanoethyl *N,N*-diisopropyl phosphoramidites. Other oligomer synthesis chemistries may require different phosphite/phosphate/phosphonate forms of the monomers.

3.1.1. Automated Synthesis of Base-Modified "Linker Arm" Oligodeoxynucleotides

1. Dissolve the linker arm phosphoramidite **1** in the designated volume of anhydrous synthesis-grade acetonitrile; final concentration is normally 100 m*M*. Insert the bottle into the appropriate amidite port on the DNA synthesizer.
2. Program the desired sequence into the synthesizer. When using the aminoalkyl thymidine analog **1**, substitute the analog in place of a single thymidine in the sequence (generally in the interior of the sequence between cytosines and/or guanines). No other modifications in the sequence or program cycle are needed.

3. For all of the analogs, the synthesis can be programmed for "Auto Deprotect" or "Manual Deprotect," and "DMT on" or "DMT off," as desired. For purification by RP-HPLC the DMT should be left on. (Best results are obtained with purified products; however, for many applications the crude can be used directly with no significant decrease in performance.) If crudes are to be used, program the synthesizer to remove the 5'-DMT protecting group by using the "DMT Off" ending program. (*Exception:* If using MMT-blocked 5'-amino compounds, do *not* remove the MMT on the synthesizer.)*
4. Begin the synthesis. After synthesis is complete, check for reproducible volume and color in the DMT releases; read the DMT releases spectroscopically, if accurate coupling yields are desired.
5. After synthesis is complete, heat the crude oligonucleotide in ammonium hydroxide as usual (8–24 h at 50–55°C) to fully deprotect the bases. This treatment will also deblock the amine linker arm functions protected by trifluoroacetyl (TFA) groups, as in compound **1**.
6. Purify the oligomer by standard HPLC, reverse phase cartridges, or gel electrophoresis, as desired.
7. Remove the 5'-DMT protecting group, if left on, by treatment of the dried oligomer with 300 µL 80% acetic acid for 40–50 min at ambient temperature. Evaporate to dryness and redissolve in 100 µL 200 m*M* sodium acetate, pH 6–7.
8. Remove all amine-containing salts (ammonia, Tris buffers, and so forth) from the oligonucleotide by desalting on a spun G-25 column equilibrated in 200 mM sodium acetate, pH 6-7. Wash column with 2 × 100 µL 200 m*M* sodium acetate, combining the load and wash eluates (≈300 µL total).
9. Ethanol precipitate by addition of 3 vol (900 µL) ethanol. Mix, cool briefly at –20° or –70°C, and pellet by centrifugation. Pipet off the supernatant (absorbance at 260 nm should be ≤0.5), and discard the supernatant. Dry the oligonucleotide pellet briefly under vacuum to remove residual ethanol.
10. Redissolve the oligonucleotide in 2 m*M* sodium EDTA, pH 7–8, for storage. *Do not add amine-containing buffers (Tris, and so on)*, or any reagents containing thiols (DTT, and so on). To determine concentration, dilute a small (5–20 µL) aliquot at least 20-fold in water, and read absorbance at 260 nm.

*When using MMT-protected amine modifiers on the 5'-terminus, do *not* program the synthesizer to remove the MMT on the machine, as this results in oligomers that are only partially (≈50%) reactive. The MMT should be left on, then removed later manually using acetic acid treatment to give fully reactive amines.

11. Calculate the oligomer concentration (C) in μ*M* by **C (μ*M*) = 107 A_{260}/*n***, where *n* is the number of bases in the oligonucleotide (this assumes 33 μg/A_{260} and an average mol wt of 308/base). Store at –20° or –70°C until use. Purity should be assessed by electrophoresis in 20% polyacrylamide gel; each linker will slow the mobility of the oligonucleotide the equivalent of 1–1.5 base U, relative to the same unmodified sequence. Useful storage life is 2–3 yr for amine linker arm oligomers.

3.1.2. Activation of Oligonucleotides and Conjugation to Alkaline Phosphatase

This section describes the method used to conjugate alkaline phosphatase or horseradish peroxidase to the amine-functionalized oligodeoxynucleotides. The approach to synthesis of enzyme-oligodeoxynucleotide conjugates is the following:

1. The linker arm oligomer is activated by reaction with homobifunctional crosslinkers;
2. The alkaline phosphatase (AP) is attached through the crosslinker; and
3. The conjugate is purified by anion exchange chromatography.

The presumed site of attachment to the enzyme is lysine residues. Conjugates prepared by this approach are 1:1 oligomer:enzyme conjugates, with only one site modified on the enzyme. If excess oligonucleotide is reacted with limiting enzyme, then 2:1 oligomer to enzyme conjugates can be isolated; in hybridization and detection, such 1:2 conjugates behave equivalent to 1:1 conjugates. Calf intestinal AP works significantly better as a conjugate than bacterial AP. *The source and composition of the AP is critical*—many commercial sources of AP with high purity and activity may react, but can lose activity and/or the ability to hybridize very quickly (within days or weeks), perhaps because of differing levels of glycosylation resulting in variations of conjugation sites. Some sources conjugate poorly, if at all. Scale of reaction is also critical. The protocols here are based on 50 nmol of oligonucleotide. The scale can be doubled or tripled by increasing the relative amounts of oligomer and enzyme; do *not* increase reaction times or volumes, or column/elution volumes. Smaller scales (1–10 nmol) give poorer yields caused by more dilute solutions and resulting slower kinetics. Since the product conjugates are more highly charged than free enzyme and much larger than free oligonucleotide, the conjugates can be easily purified to homogeneity, although the presence of free enzyme will not generally affect the use of the conjugate. However, unconjugated probe will compete nonproductively for target sites, and must be removed.

Variations of the protocols below have been tried that activate the enzyme rather than the oligomer *(9,12,16,17)*. This has invariably resulted in loss of enzyme activity and correspondingly less signal. In all reports to date, the crosslinking of alkaline phosphatase to oligonucleotides containing amine-modified bases appears to give the cleanest product with the best hybridization results. The protocol detailed in this section can also be used for 5'-amine-modified oligonucleotides. Enzymes other than alkaline phosphatase can be conjugated using similar methods, including HRP, β-galactosidase, luciferase, oxidase, nucleases, and others. For use as hybridization probes, however, most enzymes other than AP or HRP are wholly or partially inactivated by the minimum temperatures, salts, and/or detergents necessary for reasonably stringent hybridization of oligomer probes. As a result, enzymes promising superior sensitivity (such as β-galactosidase) have not been successfully used as direct probe conjugates. Both alkaline phosphatase and HRP conjugates have been used as enzyme labels, although HRP in all applications tested is 5–40-fold less sensitive than AP; *see* Table 1, Section 3.2. Reviews of synthesis and properties of modified oligodeoxynucleotides have been recently published *(3,4)*. Before attempting to attach enzymes, be sure to remove all sources of ammonium ions or primary and secondary amines (such as Tris) from the oligonucleotide by gel filtration or ethanol precipitation (*see* Section 3.1.1., steps 8 or 9). The short reaction time, low pH, and low temperature in steps 6 and 7 are important, and are needed to minimize the hydrolysis of the activated oligonucleotide. The active intermediate formed in step 5 is an NHS-activated carboxylic acid ester that hydrolyzes rapidly ($t_{1/2}$ of a few minutes) in water at alkaline pH, but is stable at pH 4. This intermediate must be kept intact until step 8.

1. Prepare 20 mg calf intestinal alkaline phosphatase (2 mL of 10 mg/mL) by dialysis with AP reaction buffer. Alternatively, put 20 mg AP directly into a Centricon-30 (Amicon) microconcentrator with 1 mL AP reaction buffer; concentrate by centrifugation at 5000*g*, maximum. Wash by refilling the Centricon with reaction buffer and reconcentrating to 100–200 mg/mL (0.7–1.4 m*M*). Measure the enzyme's concentration by conventional Bradford assay using a standard curve, and determine

specific activity of the enzyme in duplicate using *p*-nitrophenyl phosphate (pNPP) as substrate. Store the enzyme in a sterile tube at 4°C until use, up to 2 wk.

2. Prepare a gel filtration column (≈0.7 × 45 cm) of Sephadex™ G-25 in degassed 1 m*M* NaOAc, pH 4. Equilibrate at 4°C with 20–30 mL 1 m*M* NaOAc, pH 4.
3. Prepare a fresh 10 mg/mL solution of DSS crosslinker (di-succinimidyl substrate; Pierce cat #21555) by dissolving 1–2 mg DSS in 100–200 µL dry dimethyl sulfoxide (DMSO).
4. In a clean 1.5 mL Eppendorf tube, mix 50 µL 1 m*M* oligonucleotide solution (50 nmol of oligonucleotide) and 15 µL fresh 1*M* $NaHCO_3$, pH 8.5–9.
5. Add 50 µL of the 10 mg/mL DSS solution in DMSO to the oligomer. Pipet the solution up and down in the pipet to mix well.
6. *Immediately* (0.5–2 min) after mixing DSS with the oligomer, separate the DSS-activated oligomer from excess DSS by loading the reaction onto the cold G-25 column prepared in step 2. Wash on and elute with cold (4°C) 1 m*M* NaOAc, pH 4, collecting fractions of 1–2 mL/fraction. Monitor fractions by UV absorbance (254 or 260 nm wavelength). The first peak eluting will be the DSS-activated oligonucleotide, which will elute at approx 8–11 mL. *Immediately* combine appropriate fractions and go on to step 7.
7. Concentrate the combined fractions containing the activated oligomer to 60–100 µL volume using a Centricon-10 concentrator (Amicon) in a refrigerated centrifuge at 5000*g*. (This will take 1–1.5 h.) Keep cold and *immediately* go on to step 8.
8. Add a 2–3-fold molar excess of dialyzed concentrated (100–200 mg/mL) AP to the tube containing the concentrated activated oligomer from step 7 (i.e., add 14–20 mg AP/50 nmol oligonucleotide). Final volume will be ≈200 µL. Mix gently by stirring with the pipet tip (do *not* vortex the mix).
9. Allow the reaction to sit in the dark for 4–18 h.
10. Remove unconjugated oligomer by size exclusion on a 1 × 50 cm P-100 column eluting with 100 m*M* NaCl, 50 m*M* Tris-HCl, pH 8.5 buffer, at 4°C. Monitor the eluate by UV absorbance (260 nm), collecting fractions of 50 drops for about 30 fractions. (The first UV-absorbing peak is a mixture of oligonucleotide conjugate and free enzyme, and should have an $A_{260/280}$ of 0.8–1.2; unconjugated oligomer will be in later fractions, with an $A_{260/280}$ ratio of about 1.5–1.7.) Plot absorbance and combine appropriate AP-containing fractions, keeping the product cold.

(*Note:* On scales ≤50 nmol, the P-100 fractionation can be omitted if care is taken to remove unconjugated oligonucleotide during the subsequent anion exchange column in step 12.)

11. Concentrate the conjugate/enzyme mix to ≈60 µL using a Centricon 30 (Amicon) microconcentrator, refilling Centricon twice with 20 m*M* Tris-HCl, pH 8, and centrifuging in a refrigerated centrifuge.
12. Separate the desired oligonucleotide-alkaline phosphatase conjugate from free enzyme by anion exchange chromatography, preferably on a Pharmacia MonoQ-10 (0.5 × 5 cm) FPLC column. Load the cold conjugate solution on the column, washing on with 2 × 100 µL 20 m*M* Tris-HCl, pH 8. Elute a linear gradient of 0–100% B (0–1*M* NaCl in 20 m*M* Tris-HCl, pH 8) over 60 min at 1 mL/min. The free enzyme is only slightly negatively charged and elutes at about 0.15*M* NaCl, with an $A_{260/280}$ ratio of 0.5–0.9. The enzyme-probe conjugate is much more negatively charged and elutes at ≈0.45*M* NaCl (usually over 3–5 fractions at between 15–30 min); the desired conjugate has an $A_{260/280}$ ratio of 1.0–1.3, depending on oligomer length and composition. Residual unconjugated oligonucleotide ($A_{260/280}$ ratio 1.5–1.7) will elute after the conjugate, near the end of the gradient. Pool the desired fractions and keep on ice.
13. Concentrate the product on a Centricon-30 microconcentrator to approx 50 µL volume.
14. Dilute the product with sterile high salt buffer (HSB) to about 500 µL final volume. Measure protein concentration by Coomassie to determine concentration of the conjugate (generally between 5–20 µ*M*).
15. Measure enzyme activity; the specific activity should be within 10% of original, but can be lower if the enzyme has been mishandled.
16. Dilute to desired final volume in HSB (for long-term stability, final stock should be no more dilute than 0.5 µ*M*, ideally 5–10 µ*M*. (A 5 µ*M* solution is 0.7 mg/mL AP.)
17. Determine conjugate purity by electrophoresis on a nondenaturing 7.5% polyacrylamide gel. Detect bands by staining for enzyme activity using standard NBT/BCIP color reagent for alkaline phosphatase (*see* Section 2.3.); this allows strong bands with as little as 10 fmol (equivalent to 1.4 ng) enzyme.
18. Store the conjugate in sterile tubes at 2–8°C. Useful storage life should be 1.5–3 yr if handled carefully, especially if bacterial contamination is avoided.

3.1.3. Conjugation of Amine-Modified Oligonucleotides to Horseradish Peroxidase (HRP)

Use the protocol for activation and conjugation of alkaline phosphatase above (*see* Section 3.1.2.), with the following modifications:

1. Use ELA grade HRP (Boehringer Mannheim); use 4 mg HRP (mol. wt. ≈40,000) per 50 nmol oligonucleotide. Concentrate HRP in 0.1*M* $NaHCO_3$, in place of the AP reaction buffer. HRP activity does not generally need to be assayed before reaction.
2. Steps 14 and 16: use 100 m*M* potassium phosphate, pH 6.3, 0.05% azide, as storage buffer in place of HSB.

3.2. Hybridization and Nonisotopic Detection Protocols

Oligodeoxynucleotide probes can be hybridized to complementary targets in a number of formats such as *in situ,* in solution, or to target immobilized on membranes (Southern blot, northern blot, dot/slot blot). Protocols for hybridization and detection of nonisotopic oligonucleotide probes will vary somewhat depending on format, enzyme label, and modality of detection method. The most common hybridization is to target nucleic acids immobilized on membranes. Regardless of detection format, hybridization to membranes is very similar for most enzyme-labeled probes, and only detection steps will vary significantly. General protocols for hybridization of enzyme-probe conjugates are outlined below.

For use as hybridization probes, the goal of modification and labeling is not to improve natural hybridization behavior, but to have as little impact on the behavior as possible. Owing to the highly ordered nature of nucleic acid structure, helix formation, and resulting hydrogen bonding, any modification made randomly is more likely to have an adverse effect than no effect at all. This is particularly true if direct chemical modifications are made on intact DNA, such as crosslinking of enzymes *(5)*. The effect of most modifications and enzyme labels have not been studied by physicochemical methods in detail, but some modifications are known to limit general application. For example, labeled oligonucleotides containing a single internal N^4-dCyd linker arm *(7)* show anomalous two-phase melting behavior, and lack a single helix-to-coil hypochromic transition *(18)*. In contrast, modification of oligonucleotides by attachment of enzymes or other reporter groups at

C-5 of pyrimidines using compound **1** does not appear to have a measurable effect on hybridization *(4,18–20)*.

A 22-mer containing one amine thymidine linker arm **1** was conjugated to alkaline phosphatase and purified as detailed in Section 3.1.3. Subsequent hybridizations at defined probe concentrations to membrane-bound target indicated that even for oligonucleotides conjugated to relatively large reporter groups (conjugate mol wt ≈ 150,000), hybridization kinetics were about the same as those established for unmodified oligomers (mol wt ≈ 7000) in solution, with a rate constant $k \approx 5 \times 10^5$ mol^{-1} s^{-1}. In solution, the rate constant for hybridization of oligonucleotide probes was determined to be 3.6×10^5 mol^{-1} s^{-1}, which is similar to the rate constant for unmodified oligonucleotides (5.8×10^5 mol^{-1} s^{-1}) determined in parallel *(19)*. Hybridization efficiency of AP-labeled or unmodified oligonucleotide probe to either plasmids in solution or fixed on membranes was found to be nearly quantitative (≥85 ± 5% of the input target) *(19)*. Subsequent results with more than 100 oligonucleotide probes are consistent with these results. This strongly suggests that modification and attachment of even relatively large enzymes at C-5 has very little impact on the hybridization strength, selectivity, kinetics, or efficiency of oligonucleotide hybridization. No such detailed studies of other enzyme labels or 5'-labeled probes have been reported.

Choice of enzyme and detection format is highly dependent on application. However, for standard hybridizations to targets immobilized on membranes, there is a wide variation of label sensitivities (*see* Table 1), with over 100-fold difference between the least sensitive and most sensitive label. Of all the reporter groups tested, only alkaline phosphatase is more sensitive than 24 h autoradiography with ^{32}P. The most sensitive reporter group currently available for oligonucleotides is alkaline phosphatase using a luminescent dioxetane derivative *(20)* (AMPPD from Tropix, Inc.), with sensitivities about 50-fold more sensitive than ^{32}P autoradiography in one-fifth the time. In applications such as Southern blots, this allows the hybridization and detection of specific bands to be done in one working day, compared to 5–7 d needed to use a ^{32}P-cloned DNA probe for detection of the same bands, with equal sensitivities.

This general protocol can be used for Southern blots, Northern blots, "slot" blots and "dot" blots on nylon membranes. Nitrocellulose membranes can also be used. The protocol has been optimized for alkaline phosphatase-labeled oligomers of 20–30 bases in length using nylon membranes. Hybridizations with 5 n*M* probe for 10 min or 1 n*M* probe for 1 h are ≥90% complete. Alkaline phosphatase-labeled probes should not be hybridized for longer than 1 h at 60°C or 3 h at 50°C owing to inactivation of the enzyme, especially in the presence of detergents such as SDS. The same protocol can be used for hybridization of HRP- or ^{32}P-labeled oligomers by changing only the appropriate detection step. (*Note:* For detection of dot or slot blots by soluble fluorescence, each dot must be cut out before or after hybridization, and detected in separate tubes or wells.)

3.2.1. Hybridization of Enzyme Conjugates to Target Immobilized on Membranes

1. Wet the membrane with hybridization solution (50–100 µL/cc^2 membrane) for 15–20 min at hybridization temperature (*see* step 2) in a sealable bag, tube, or wash tray using a water bath as a heat source. (Some detection methods require additional blocking of the membrane before hybridization; for example, if alkaline phosphatase-labeled oligomers are to be detected with luminescent substrates such as AMPPD [from Tropix] or Lumiphos [from Lumogen], the wetting step should include 0.1-0.2 w/v% casein in 0.5X SSC, 0.1% Tween–20 for 30 min to prevent background owing to the substrate itself.)
2. Remove the solution. Wash the membrane briefly in warm hybridization buffer. Mix the probe to a final concentration of 2–5 n*M* (2 n*M* ≈ 15 ng/mL) in fresh hybridization solution and add 50–100 µL/cc^2 to the membrane. Incubate for 30 min at hybridization temperature with agitation. (For most oligomers of 18–30 bases in length, appropriate hybridization temperatures will be 50–60°C using this protocol, and optimum wash temperatures will be 10°C lower.)
3. Transfer the membrane to a wash tray or bottle containing 1.5 mL preheated wash 1 solution/cc^2 of membrane. Agitate at wash temperature (optimum hybridization minus 10°C) for 10 min.
4. Remove the wash 1 solution, and add wash 2 solution at 1.5 mL/cc^2 membrane. Agitate at wash temperature for 10 min. (The Triton X-100 detergent in wash 2 is necessary to displace SDS from the membranes when using NBT/BCIP dye deposition, since SDS causes chemical deposition of the dyes.)
5. Remove the wash 2 solution, and add wash 3 solution at 1.5 mL/cc^2 membrane. Agitate at ambient temperature for 10 min.

6. Remove wash 3, and detect immediately by either:
   a. Color: Immerse the membrane fully in color reagent, and incubate at 20–37°C until color development is adequate. Quench by washing in water. Expected detection limits are 3–9 × 10^6 molecules (5–15 × 10^{-18} mol).
   b. Soluble fluorescence: (Each membrane dot must be developed independently) To the damp dot, add 100 µL fluorescence reagent, and incubate for 30–240 min. Detect by excitation at 365 nm and emission at 450 nm. Expected detection limits are 3–9 × 10^5 molecules (5–20 × 10^{-19} mol).
   c. Luminescence: Immerse the membrane in luminescent reagent, using 25–50 µL solution/cc^2 of membrane. Expose to Kodak XAR-5 film (or the equivalent; emission maximum is at 477 nm wavelength) in a light-tight cassette for 1–24 h, then develop film as usual. Expected detection limits are 6–20 × 10^4 molecules (1–3 × 10^{-19} mol).

3.2.2. In Situ *Hybridization of Enzyme Conjugates to Target in Immobilized Cells*

The following protocol has been optimized for hybridization to free cells (cultured cells, isolated white blood cells, or vaginal swabs) using probes for human papilloma virus (HPV) or human immunodeficiency virus (HIV). Applications to other cell types or to tissues may require modification of some steps, such as pretreatment of tissue with proteinase.

1. Smear the cells onto a standard microscope slide, and fix them by dipping in methanol for 15–30 s. Remove and let dry.
2. Denature the cellular nucleic acids by covering with denaturation solution and heating at 90°C for 15 min. Remove the liquid.
3. Rinse twice in 1X SSC.
4. Add 20–50 µL hybridization solution containing 10–20 n*M* AP-labeled oligonucleotide probe. Heat at hybridization temperature (usually 45–50°C) for 2 h (unless using a humidified chamber, the sample must be covered with a coverslip to prevent excess evaporation).
5. Remove liquid from the slide. Rinse once in 1X SSC.
6. Wash twice in wash 1 solution heated to 40°C for 3 min each.
7. Wash twice in wash 2 solution heated to 40°C for 3 min each.
8. Wash once in wash 3 solution heated to 40 C for 3 min each.
9. Rinse twice briefly in color reagent.
10. Cover the sample with color reagent and incubate at 37°C for 1–4 h.
11. Counterstain with 1% Eosin B for 3 min, then rinse with water and blot/air dry.
12. Mount in crystal mount.
13. View results microscopically for purple dye deposition.

Expected detection limits are approx 20–50 copies of target per cell.

4. Notes

Although most important aspects have been mentioned during the previous discussions, some are critical and justify repeating.

4.1. Notes on Synthesis of Modified Oligonucleotides

During the chemical synthesis of the amine-modified oligonucleotide, there are no changes to conventional phosphoramidite procedures necessary except the proper programming of the linker arm nucleoside coupling step. For best results, as with any oligonucleotide synthesis, all of the reagents on the synthesizer (β-cyanoethyl phosphoramidites of the four common bases, activator, acetonitrile, capping reagents, and so on) should all be fresh, or less than a week old. During workup, the following additional precautions should be noted:

1. For enzyme conjugation, do *not* incorporate more than one amine linker per sequence. Two or more linkers may crosslink to the same enzyme molecule, resulting in adverse effects on hybridization.
2. The linker arm amidite **1** should be used within a week of dissolving the compound in acetonitrile. Coupling yields of **1** should be 98.5–99.9% initially, but may begin to fall off after a week (although product can still be isolated easily after two weeks, albeit in lower yields).
3. If 5'-amino linkers blocked with monomethoxytrityl (MMT) are used in place of **1**, the MMT group should *not* be removed on the machine, or product will be only about 50% reactive. The MMT should be removed by treatment with 80% acetic acid manually, after oligomer deprotection. If the 5'-amino linkers are protected with trifluoroacetyl (TFA) groups, this precaution is not necessary.
4. *Important:* After the linker arm oligonucleotide is purified and before enzyme conjugation, care should be taken to remove all amine-containing compounds, particularly Tris salts. Any primary or secondary amines will compete effectively for reaction with enzyme, which is present at concentrations of only about 500 μ*M* during the conjugation step. A stock storage solution of 1 m*M* linker arm oligomer in 2 m*M* EDTA, pH 8, with no other buffer works well, with a useful storage life of several years at –70°C.
5. If the product linker arm oligonucleotide is purified, calculate concentration (**C**) in μ*M* by **C** (μ*M*) = 107 A_{260}/n, where *n* is the length in base units. (This assumes 33 μg/A_{260} and mol wt 308/base for the average heteropolymer.)

4.2. Notes on Synthesis and use of Enzyme Conjugates

Unlike the oligomer synthesis, which is essentially standard chemistry substituting **1** for thymidine, the enzyme conjugation requires specific reagents and precautions at many steps:

1. Alkaline phosphatase is a better reporter group (more sensitive and more stable) than HRP, for most applications.
2. use only calf alkaline phosphatase, Boehringer Mannheim #556602 (10 mg/mL in HSB), as the enzyme source. Most other preparations, even from the same manufacturer, tend to produce storage or use anomalies in the product conjugate.
3. After reaction of DSS crosslinker with oligonucleotide, the *N*-hydroxysuccinimidyl (NHS) intermediate ester of the oligomer must be kept intact (Section 3.2.1., steps 3–5) until coupling with the enzyme. The NHS ester is hydrolyzed quickly at even slightly alkaline conditions. The initial reaction with oligomer is fast (2 min) under these conditions, so after mixing immediately bring the reaction to pH 4–5, keep cold (4°C), and do *not* expose to any amines until after conjugation with the enzyme. The NHS ester is stable for many hours under these conditions.
4. When purifying the conjugate, be sure to remove any unconjugated oligomer, which will compete effectively for hybridization sites. Many commercial sources of custom synthesis contain unconjugated oligomer. Significant amounts (5–50%) of free enzyme do not adversely affect hybridization, but make accurate quantitation of the product conjugate inaccurate, and may add to general filter backgrounds.
5. Store the product conjugate in sterile conditions in 0.5–10 μ*M* concentrations using high salt storage buffer (*see* Section 2.2.1.) at 4°C. Stocks more dilute than ≈250 n*M* tend to decompose sooner, because of low overall protein concentrations and possible dissociation of the enzyme subunits. Useful life is up to 3–4 yr if unopened or not contaminated. Most eventual failures of conjugates can be tracked to bacterial or proteinase contamination.
6. When using chemiluminescent detection of alkaline phosphatase, most background derives from the substrate, not the conjugate. To block this effectively, pretreat the filter before hybridization by soaking in 0.2% fresh casein in 0.5X SSC, 0.1% Tween-20, with agitation for 30–60 min, then rinse in hybridization buffer before adding probe.
7. Hybridization of the conjugates is complete very fast (15 min) at 2.5 nM concentrations. To minimize probe requirements, the conjugate can be diluted to 1 n*M* using 60 min hybridizations. Further dilution of the

probe is difficult since signal will be lost because of either incomplete hybridization or loss of enzyme activity caused by prolonged heating. Avoid hybridization for longer than 1 h at 60°C or 3 h at 50°C.

References

1. Matthews, J. A. and Kricka, L. J. (1988) Analytical strategies for the use of DNA probes. *Anal. Biochem.* **169,** 1–25.
2. Leary, J. J. and Ruth, J. L. (1989) Nonradioactive labeling of nucleic acid probes, in *Nucleic Acid and Monoclonal Antibody Probes: Applications in Diagnostic Microbiology* (Swaminathan, B. and Prakash, G., eds.), Marcel Dekker, New York, pp. 33–57.
3. Goodchild, J. (1990) Conjugates of oligonucleotides and modified oligonucleotides: a review of their synthesis and properties. *Bioconjugate Chem.* **1,** 165–187.
4. Ruth, J. L. (1991) Oligodeoxynucleotides with reporter groups attached to the base, in *Oligonucleotides and Analogues: A Practical Approach* (Eckstein, F., ed.), Oxford University Press, Oxford, UK, pp. 255–282.
5. Renz, M., and Kurz, C. (1984) A colorimetric method for DNA hybridization. *Nucleic Acids Res.* **12,** 3435–3444.
6. Jablonski, E., Moomaw, E., Tullis, R., and Ruth, J. (1986) Preparation of oligodeoxynucleotide-alkaline phosphatase conjugates and their use as hybridization probes. *Nucleic Acids Res.* **14,** 6115–6128.
7. Urdea, M., Warner, B., Running, J., Stempien, M., Clyne, J., and Horn, T. (1988) A comparison of non-radioisotopic hybridization assay methods using fluorescent, chemiluminescent, and enzyme-labeled synthetic oligodeoxyribonucleotide probes. *Nucleic Acids Res.* **16,** 4937–4956.
8. Connolly, B. A. (1987) The synthesis of oligonucleotides containing a primary amino group at the 5'-terminus. *Nucleic Acids Res.* **15,** 3131–3139.
9. Ghosh, S. S., Kao, P. M., and Kwoh, D. Y. (1989) Synthesis of 5'-oligonucleotide hydrazide derivatives and their use in preparation of enzyme-nucleic acid hybridization probes. *Anal. Biochem.* **178,** 43–51.
10. Connolly, B. A. (1985) Chemical synthesis of oligonucleotides containing a free sulphydryl group and subsequent attachment of thiol-specific probes. *Nucleic Acids Res.* **13,** 4485–4502.
11. Alves, A. M., Holland, D., Edge, M. D., and Carr, F. J. (1988) Hybridization detection of single nucleotide changes with enzyme labelled oligonucleotides. *Nucleic Acids Res.* **16,** 8722.
12. Nelson, P. S., Frye, R. A., and Liu, E. (1989) Bifunctional oligonucleotide probes synthesized using a novel CPG support are able to detect single base pair mutations. *Nucleic Acids Res.* **17,** 7187–7194.
13. Zuckermann, R., Corey, D., and Schultz, P. (1987) Efficient methods for attachment of thiol-specific probes to the 3'-ends of synthetic oligodeoxyribonucleotides. *Nucleic Acids Res.* **15,** 5305–5321.
14. Nelson, P. S., Sherman-Gold, R., and Leon, R. (1989) A new and versatile reagent for incorporating multiple primary aliphatic amines into synthetic oligonucleotides. *Nucleic Acids Res.* **17,** 7179–7186.

15. Li, P., Medon, P. P., Skingle, D. C., Lansern, J. A., and Symons, R. H. (1987) Enzyme-linked synthetic oligonucleotide probes: non-radioactive detection of enterotoxigenic *Escherichia coli* in faecal specimens. *Nucleic Acids Res.* **15,** 5275–5287.
16. Murakami, A., Tada, J., Yamagata, K., and Takano, J. (1989) Highly sensitive detection of DNA using enzyme-linked DNA-probe. 1. colorimetric and fluorometric detection. *Nucleic Acids Res.* **17,** 5587–5595.
17. Corey, D. R., Pei, D., and Schultz, P. G. (1989) Sequence-selective hydrolysis of duplex DNA by an oligonucleotide-directed nuclease. *J. Am. Chem. Soc.* **111,** 8523–8525.
18. Telser, J., Cruickshank, K. A., Morrison, L. E., and Netzel, T. L. (1989) Synthesis and characterization of DNA oligomers and duplexes containing covalently-attached labels: comparison of biotin, fluorescein, and pyrene labels by thermodynamic and optical spectroscopic measurements. *J. Am. Chem. Soc.* **111,** 6966–6976.
19. Podell, S., Maske, W., Ibanez, E., and Jablonski, E. (1991) Comparison of solution hybridization efficiencies using alkaline phosphatase-labelled and ^{32}P-labelled oligodeoxynucleotide probes. *Mol. Cell. probes* **5,** 117–123.
20. Bronstein, I., Voyta, J. C., and Edwards, B. (1989) A comparison of chemiluminescent and colorimetric substrates in a hepatitis B virus DNA hybridization assay. *Anal. Biochem.* **180,** 95–98.

CHAPTER 7

Oligonucleotides Containing Degenerate Bases

Synthesis and Uses

Paul Kong Thoo Lin and Daniel M. Brown

1. Introduction

Major uses of oligodeoxyribonucleotides are as hybridization probes, sequencing primers, and, more recently, as primers for the polymerization chain reaction. When a protein sequence or part thereof is known, the construction of oligomer probes and primers is complicated by the codon degeneracy. The chain multiplicity in such probes or primers may reach very large numbers, leading to often ineffective reagents or poor signals. A number of ways of mitigating this general problem have been devised *(1)*; among these is the use of hypoxanthine I as a base that may be inserted at positions of degeneracy. Although it forms basepairs with C, A, G, and T of weak and varying stabilities (in that order), its ability to sustain the dissociation temperatures of DNA duplexes containing it above those with mismatches has led to its use first in probes and more recently in primers *(2–6)*. We discuss I later, but here we describe the synthesis of monomers and oligomers containing the pyrimidine, 6*H*,8*H*-3,4-dihydropyrimido[4,5-c][1,2]oxazin-7-one, P (showing C, T degeneracy) and 2-amino-6-methoxyaminopurine, K (showing A,G degeneracy) (Fig. 1) *(7,8)*. These bases were chosen on the grounds that their amino-imino tautomeric constants were much nearer to unity than the normal bases and that, for example, the P base could form Watson-Crick basepairs with both A and G as shown in Fig. 1 (a,b). This was established by NMR spectroscopy of self-complementary

From: *Methods in Molecular Biology, Vol. 26: Protocols for Oligonucleotide Conjugates*
Edited by: S. Agrawal Copyright ©1994 Humana Press Inc., Totowa, NJ

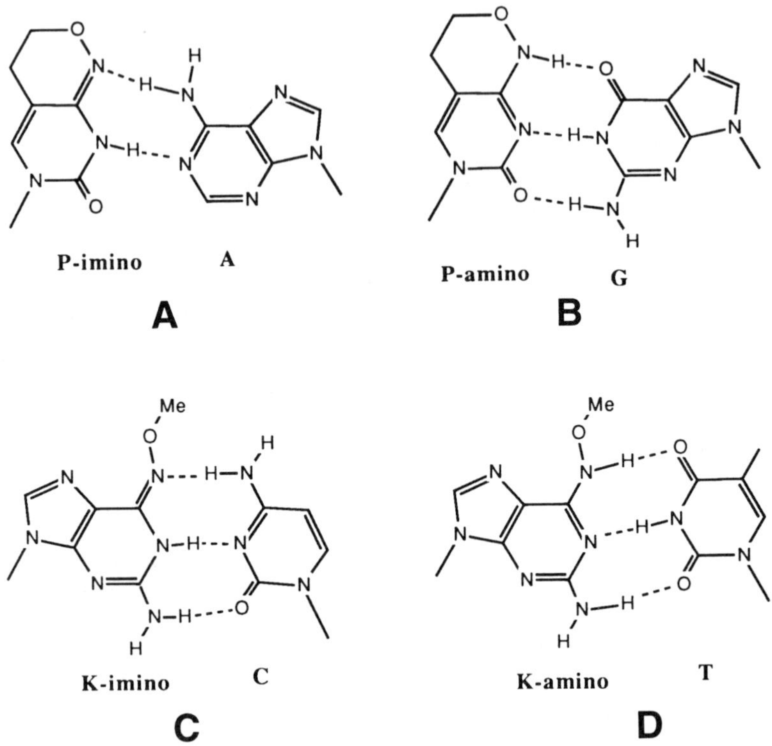

Fig. 1. Pyrimidine analog P in its imino (**A**) and amino (**B**) tautomeric forms pairing with adenine and guanine. Purine analog K in its imino (**C**) and amino (**D**) tautomeric forms pairing with cytosine and thymine. Both analogs form Watson-Crick basepairs.

8-mer duplexes *(9)*. The T_m values of duplexes containing A/P, G/P, C/K, and T/K basepairs establish their value in practice. Such melting transitions, too, give information necessary for establishing annealing temperatures in PCR experiments and in other experiments involving hybridization. Some relevant melting data is discussed later. The oligomers containing these degenerate bases are also discussed as primers for PCR amplification *(10)*.

2. Materials and Methods

2.1. Solvents

All solvents are commercially available and are dried and distilled before use except for those used in solvent extraction and column chromatography.

2.2. Chemicals

All organic reagents were obtained from either Aldrich (Gillingham, UK), Fluka (Glassop, UK), Sigma (Poole, UK), or BDH (Poole, UK). Monomers for oligonucleotides synthesis were from Cruachem (Glasgow, UK) and Applied Biosystems (Warrington, UK).

2.3. Buffers

1. 6X SSC. This buffer is used in T_m experiments and is prepared by dissolving sodium chloride (5.26 g), sodium citrate (2.65 g), in distilled water (80 mL). The pH is adjusted to 7.0 with 10*M* HCl and the solution finally made up to 100 mL with water.
2. 1X TBE. Used in gel electrophoresis, it is obtained by dilution of a stock solution of 10X TBE. The latter is made with Tris base (108 g), boric acid (55 g), EDTA (9.3 g) and made up to 1000 mL with distilled water.
3. Ion exchange HPLC buffers. Buffer A: 0.001*M* KH_2PO_4 (pH 6.3), 60% formamide: Buffer B: 0.3*M* KH_2PO_4 (pH 6.3), 60% formamide.

2.4. Biochemicals

Taq polymerase was from Promega and came with its own buffer. Agarose was from Bethesda Research Laboratories (BRL, Gaithersburg, MD), dNTPs were from Pharmacia (Piscataway, NJ), and PHI X 174 RF DNA *Hae* III Digest from Biolabs.

2.5. Chromatography

Flash column and thin layer chromatography (tlc) require Kieselgel 60H (7736) and $60F_{254}$ (Merck, Rahway, NJ), respectively, with chloroform–methanol mixtures unless otherwise stated.

For monitoring reactions and checking purity of products, the tlc plates should be viewed under shortwave UV irradiation. For detection of nucleosides the anisaldehyde reagent is used. To prepare the latter add conc H_2SO_4 (10 mL) to a solution of anisaldehyde (10 mL) in 95% ethanol (180 mL) containing 15–20 drops of acetic acid. Plates are dipped in the reagent, blotted, and heated on a hotplate. Deoxyribosyl derivatives give dark blue spots; dimethoxytrityl derivatives give orange spots in the cold.

HPLC was performed on a Waters system, using a Whatman Sax Partisphere or a Hichrom partsil 10 Sax ion exchange column.

2.6. Spectra

NMR ^{1}H and ^{31}P NMR spectra were recorded with a Bruker WM 250 MHz and AM400 MHz spectrometer, using Me_4Si and TMP as external standard respectively. Unless otherwise stated, values given are on the δ scale; singlet absorption, integration value, and assignment are in parentheses.

Ultraviolet spectra were recorded on a Beckman DU-65 spectrometer and for T_m measurements of DNA duplexes, a Perkin Elmer Lambda 2 spectrometer fitted with a Peltier block and temperature programmer was used.

2.7. Polymerase Chain Reactions

These were carried out on a Techne programmable Dri-block PHC-1 apparatus.

3. Monomer Syntheses

3.1. Synthesis of P- Monomer

3.1.1. 5-(2-Hydroxyethyl) Uracil (11)

3.1.1.1. Sodium Salt of α-Hydroxymethylene-δ-Butyrolactone

A. To a suspension of sodium hydride (12 g) in diethyl ether (500 mL) add methanol (16 g) dropwise. After refluxing for 6 h, cool the reaction vessel and add a mixture of methyl formate (40 mL, 0.65*M*) and δ-butyrolactone (38 mL, 0.5*M*) dropwise over a period of 1 h with stirring. Overnight stirring at room temperature generates a thick cream precipitate that is filtered off, washed with ether, and dried over P_2O_5 (62 g).

B. To a solution of urea (30 g) in cold 5*M* HCl (200 mL) add the sodium derivative of α-hydroxymethylene-δ-butyrolactone (34 g, 0.5*M*). After stirring overnight in the cold, collect the precipitate formed, and wash with cold water and dry over P_2O_5 (23 g, 59%). Add this, α-(1-carbamyliminoethylene)-δ-butyrolactone (13.4 g, 86 m*M*), to a solution of sodium ethoxide (2.18 g, $9.46 \times 10^{-2}M$ of Na in 250 mL of absolute-EtOH) and reflux for 6 h, during which time a solid separates. Collect the latter, dissolve in water (400 mL), heat the solution to 55°C and add Dowex 50 (H^+) to neutrality. Filter and then reduce in volume to about 200 mL. After cooling to ~5°C, the product (12.5 g; 92%) crystallizes out of solution.

3.1.1.2. 2-Deoxy-3,5-di-*O*-*p*-toluoyl-α-D-ribosyl chloride *(12)*

To a solution of deoxyribose (13.6 g, 0.1*M*) in methanol (243 mL) add 1% HCl/CH_3OH solution (27 mL). Leave the mixture stirring at room temperature for 15 min to form the methyl glycoside, add silver carbonate (5 g), filter to give a clear solution, and evaporate to dryness. Remove all the methanol by coevaporation twice with pyridine. Dissolve the resulting syrup in pyridine (80 mL) then add *p*-toluoylchloride (34 g, 0.22*M*) dropwise with cooling. Leave the mixture overnight at room temperature, after which add water. Extract the solution with diethyl ether (3 × 75 mL), wash the ether fraction with water, dil. H_2SO_4 and $KHCO_3$ solution, dry (Na_2SO_4) and evaporate to afford a golden syrup (70%). Dissolve the latter in glacial acetic acid (50 mL) in a large beaker and while keeping the temperature below 10°C, add sat-HCl in glacial acetic acid (80 mL) (20 g of dry HCl dissolved in 300 g of acid). After 10 min the product crystallizes out. Filtration of the product is followed by a cold ether wash and drying *in vacuo* over P_2O_5 and NaOH gives the product as a white powder (21 g, 55%), stable under vacuum for several weeks or at –20°C indefinitely.

3.1.1.3. 3',5-Di-*O*-*p*-Toluoyl-5-(2-Hydroxyethyl)-2'-Deoxyuridine (1)

A suspension of 5-(2-hydroxyethyl)uracil (4.0 g; 25.6 mmol) in hexamethyldisilazane (36 mL) and trimethylchlorosilane (5 mL) is heated under reflux for 4 h. Remove excess reagent from the clear solution *in vacuo* and coevaporate with xylene. To the resulting oil, in pure dry chloroform (100 mL), add α-3,5-di-*O*-*p*-toluoyl-2-deoxy-*D*-ribofuranosyl chloride (11.46 g; 29 m*M*, dried over P_2O_5 and NaOH) and stir the solution overnight at room temperature. After washing the chloroform solution with $NaHCO_3$ solution and then water and drying, the product (10 g; 79%) is isolated as a foam following silica gel chromatography. The β-nucleoside is contaminated with up to 15% of the α-anomer as measured by nmr spectroscopy but is used in the next step without further purification.

3.1.1.4. 3',5'-Di-*O*-Toluoyl-5-(2-Phthalimido-Oxyethyl)-2'-Deoxyuridine (2)

The protected hydroxyethyldeoxyuridine is used without removal of α-anomer contaminant. Dissolve it (3.13 g; 6.36 m*M*) in anhydrous THF (200 mL) and add *N*-hydroxyphthalimide (1.96 g; 12 m*M*), triphenylphosphine (3.15 g; 12 m*M*), and then diethylazodicarboxylate (2.2 mL; 14 m*M*). The exothermic reaction gives a transient deep red solution and after 1 h evaporate the yellow solution under reduced pressure. Addition of ether precipitates the colorless crystalline product (2.8 g; 70%), mp 141°C. Only traces of the α-anomer are apparent by tlc, removed by one further crystallization from ethanol. ^{1}H-NMR ($CDCl_3$) 2.33 (3H, s, CH_3-Ar), 2.43 (3H, s, CH_3-Ar), 2.47–2.77 (4H, m, H-2', H-2", 5-CH_2-), 4.17–4.36 (2H, m, -CH_2O-), 4.56 (1H, m, H-4'), 4.64–4.79 (2H, m, H-5', H-5"), 5.67 (1H, m, H-3'), 6.52 (1H, m, H-1'), 7.12–7.28 (4H, m, Ar-H), 7.70–7.83 (5H, m, phthaloyl-H, H-6), 7.90–8.00 (4H, m, Ar-H), 8.53 (1H, s, N-H).

3.1.1.5. 1-(3,5-Di-*O*-*p*-Toluoyl-2-Deoxyribofuranosyl)-4-Triazolo5-(2-Phthalimido-Oxyethyl)-^{1}H-Pyrimidin-2-One **(3)**

Prepare a solution of phosphoryl tristriazolide from triazole (1.22 g), phosphorylchloride (0.36 mL), and dry triethylamine (2.88 mL) in acetonitrile (30 mL) and add a solution of the deoxyuridine derivative (0.6 g; 0.096 mmol) in acetonitrile (5 mL). After 2 h remove the solvent and after washing the product in chloroform in the usual way pour into hexane. The precipitate (0.58 g; 90%) shows one spot on tlc. ^{1}H-NMR ($CDCl_3$) 2.26 (3H, s, CH_3), 2.43 (3H, s, CH_3), 2.47–2.58 (1H, m, H-2'), 3.08–3.28 (3H, m, 5-CH_2, H-2"), 4.19–4.42 (2H, m, -CH_2O-), 4.63–4.85 (3H, m, H-4', H-5', H-5"), 5.65 (1H, m, H-3'), 6.46 (1H, m, H-1'), 7.05–7.29 (4H, m, Ar-H), 7.70–7.83 (5H, m, H-6, phthaloyl-H), 7.94–8.00 (4H, m, Ar-H), 8.74 and 9.28 (1H each, s, triazole-H).

3.1.1.6. 6-(3,5-Di-*O*-*p*-Toluoyl-β-d-2-Deoxyribofuranosyl)-3,4-Dihydro-8H-Pyrimido[4,5-c][1,2]Oxazin-7-One **(4)**

Dissolve the triazolo-compound (1.0 g; 1.5 m*M*) in saturated ammonia-dry dioxan (100 mL) and stir the solution overnight at room temperature. Reaction is complete (tlc) and after evaporation *in vacuo*, chromatograph the crude product. It forms colorless crystals (0.64 g; 85%), mp 181°C from acetonitrile.

3.1.1.7. 6-(5-*O*-Dimethoxytrityl-β-d-2-Deoxyribofuranosyl)-3,4-Dihydro-8H-Pyrimido-[4,5-c][1,2]Oxazine-7-One **(6)**

Stir the above di-toluoyl derivative **4** (*see* Section 3.1.1.6.) overnight in NH_3-MeOH solution (20 mL). Remove solvent and triturate thoroughly with ether to give the free nucleoside **5** as a thick gum (82%), ^{1}H-NMR

(DMSO-d_6) 1.89–2.09 (2H, m, H-2', H-2"), 2.48–2.54 (2H, m, -CH_2-), 3.46–3.56 (2H, m, H-5"), 3.64–3.72 (1H, m, H-4'), 3.82 (2H, t, J = 5.6Hz, -OCH_2-), 4.16–4.21 (1H, m, H-3'), 4.95 (1H, t, J = 5.3Hz, -OH-5'), 5.19 (1H, d, J = 4.1Hz, -OH-3'), 6.12–6.18 (1H, m, H-1'), 7.00 (1H, s, H-6), 10.51 (1H, s, N-H). This (0.5 g; 1.84 m*M*) is dissolved in pyridine (10 mL) and treated with dimethoxytritylchloride (0.75 g; 2.23 mmol). After chromatography the product **6** gives colorless needles from acetonitrile, mp 182–184°C, (0.44 g; 70%). ^{1}H-NMR (DMSO-d_6) 2.01–2.22 (3H, m, H-2', H-2", -CH_2-), 3.11–3.32 (2H, m, -CH_2O-), 3.68–3.77 (2H, m, H-5', H-5"), 3.74 (6H, s, 2 × OCH_3), 3.84–3.86 (1H, m, H-4'), 4.29–4.31 (1H, m, H-3'), 5.30 (1H, d, J = 4.33Hz, 3'-OH), 6.18 (1H, t, J = 6.51Hz, H-1'), 6.89–7.40 (14H, m, H-5, Ar-H), 10.57 (1H, d, N-H).

3.1.2. P-Monomer

The above DMT-derivative **6** (2.33 g, 4.1 m*M*) is dissolved in THF (75 mL) followed by addition of *N,N*-diisopropylethylamine (16.3 m*M*, 2.85 mL) and 2-cyanoethyl *N,N*-diisopropylchlorophosphoramidite (1.3 mL, 5.69 m*M*). After 2.5 h, reaction is complete and the solution is diluted with ethyl acetate, washed with saturated solutions of sodium chloride and bicarbonate. Evaporation of the solvent followed by column chromatography affords the product **7** as a white foam (2.53 g, 81%). ^{31}P-NMR ($CDCl_3$) 148.80, 149.19.

The DMT-derivative **6** (100 mg) may also be converted to the triethylammonium salt of the 3'-*O*-hydrogenphosphonate **8** purified by two precipitations from pentane. The pale yellow powder (90 mg) has ^{31}P-NMR (C_5D_5N-CD_3CN) 2.45.

3.2. Synthesis of K Monomer

3.2.1. O-*Methylhydroxylamine* (13)

O-Methylhydroxylamine hydrochloride (8.0 g) is treated with KOH (10 g) dissolved in water (10 mL). Distill the free amine through hot KOH pellets and collect at –10°C to give the product as a colorless liquid (2.5 g, 60%) (bp 49–50°C).

*3.2.2. 2-Amino-6-Chloro-9-(2-Deoxy-3,5-Di-*O-p-*Toluoyl-β-D-Ribofuranosyl) Purine* **(9)**

A suspension of finely powdered KOH (3.2 g, 58 m*M*) and tris[2-(2-methoxyethoxy)ethyl]amine (TDA-1) (0.376 g, 1.16 m*M*) are stirred in anhydrous acetonitrile (240 mL) at room temperature under argon. After 15 min, add 2-amino-6-chloropurine (2.0 g, 11.6 m*M*) and continue stirring for 10 min. Add 2-deoxy-3,5-di-*O*-*p*-toluoyl-α-D-ribosyl chloride (4.88 g, 12.0 m*M*) and, after 40 min filter the suspension and dry. The crude product (6.0 g) is purified by flash-column chromatography, the faster-running major component is collected, and the product (3.22 g, 51%) crystallized from acetonitrile to give needles, mp 187–188°C; λ_{max} (95% EtOH) 222, 242, and 310 nm (broad); ε 4.44, 4.45, and 3.85 ^{1}H-NMR (Me_2SO): 2.37 (s, 3 H, CH_3), 2.40 (s, 3H, CH_3), 2.69–2.79 (m, 1 H, H-2'a), 3.19–3.31 (m, 1 H, H-2'b), 4.51–4.65 (m, 3 H, H-4', 5'a, 5'b), 5.73–5.76 (m, 1 H, H-3'), 6.40 (t, 1 H, *J*6.6 Hz, H-1'), 7.02 (s, 2 H, NH_2-2), 7.03–7.39 (m, 4 H, Ar), 7.82–7.95 (m, 4 H, Ar), 8.35 (s, 1 H, H-8).

The slower running, minor component from the chromatography is the 7-isomer.

*3.2.3. 2-Amino-9-(2-Deoxy-3,5-Di-*O*-*p*-Toluoyl-β-D-Ribofuranosyl)-6-Methoxyaminopurine* (10)

To a solution of **9** (0.2 g, 0.368 m*M*) in dry EtOH (2 mL) add methoxyamine (0.5 mL), and heat the sealed vessel at 90°C for 4 h. Evaporate the solvent and chromatograph a solution of the product in $CHCl_3$ to give **10** (100 mg, 49%) as a foam. ^{1}H-NMR data (Me_2SO): 2.38 (s, 3H, CH_3), 2.40 (s, 3H, CH_3), 2.62–2.70 (m, 1H, H-2'-a), 3.00–3.12 (m, 1 H, H-2'b), 3.73 (s, 3 H, $NOCH_3$), 4.46–4.64 (m, 3 H, H-4', 5'a, 5'b), 5.66–5.69 (m, 1 H, H-3'), 6.18–6.24 (m, 1 H, H-1'), 6.58 (b, 2 H, NH_2), 7.31–7.38 (m, 4 H, Ar), 7.72 (s, 1 H, H-8), 7.86–7.98 (m, 4 H, Ar), 9.84 (s, 1 H, NH).

*3.2.4. 9-[2-Deoxy-5-*O*-(4,4'-Dimethoxytrityl)-β-D-Ribofuranosyl]-2-Dimethylaminomethyleneamino-6-Methoxyaminopurine 3'-(2-Cyanoethyl* N,N*-Di-Isopropylphosphoramidite)* (13)

Compound **10** is heated at 55°C overnight with saturated NH_3/MeOH to give the free nucleoside **11** quantitatively, a solution of which (0.4 g, 1.35 m*M*) in anhydrous *N,N*-dimethylformamide (2.5 mL) and *N,N*-dimethylformamide dimethyl acetal (2.5 mL) is stirred at 50°C for 2 h. Removal of the solvent and further coevaporation of toluene and acetone from the residue *in vacuo* gives the N^2-dimethylaminomethylene derivative (one spot on tlc). Treat the crude product with 4,4'-dimethoxytrityl chloride (0.54 g, 1.62 m*M*) in pyridine at room temperature for 1.5 h, remove the solvent *in vacuo* then chromatograph the dark-blue foam with CH_2Cl_2-Me_2CO (4:1) to afford the dimethoxytrityl derivative **12** (131 mg) as a pale yellow foam. ^{1}H-NMR (Me_2SO): 3.01 (s, 3H, NCH_3), 3.10 (s, 3 H, NCH_3), 3.72 (s, 9 H, 3 OCH_3), 6.77–7.36 (m, 13 H, Ar), 7.77 (s, 1 H, H-8), 8.48 (s, 1 H, N=CHN), 8.88 (s, 1 H, NH).

Treat a solution of **12** (120 mg, 0.19 m*M*) in anhydrous tetrahydrofuran (5 mL) and *N,N*-diisopropylethylamine (0.132 mL, 7.6 m*M*), with the exclusion of moisture, with 2-cyanoethyl-*N,N*-diisopropylchlorophosphoramidite (0.66 mL, 0.285 m*M*). Reaction is complete in 1 h. Dilute the solution with ethyl acetate, wash with saturated aqueous NaCl, bicarbonate, and dry (Na_2SO_4). Chromatography of the product with ethyl acetate-CH_2Cl_2-Et_3N (45:45:10) affords the K-monomer **13** (115 mg, 73%). ^{31}P-NMR data ($CDCl_3$): 148.93 and 149.15.

3.3. Synthesis of Oligonucleotides

The above amidite monomer foams **7** and **13** dried over P_2O_5 are crushed and weighed rapidly into small vials, sealed, and stored at –20°C. They are stable indefinitely under these conditions. Before use dry acetonitrile is injected to give a final concentration of 100 m*M*. They are incorporated into oligonucleotides, using the normal program on an Applied Biosystems 380B instrument. Deprotection is complete after treatment with conc aqueous NH_3 at 55°C overnight. Purification is carried out by HPLC.

3.3.1. Synthesis of Functionalized Controlled Pore Glass (CPG) Support Carrying the 3'-O-Succinate of (5-Dimethoxytrityl-2-Deoxyribofuranosyl)-3,4-Dihydro-8H-Pyrimido [4,5-c][1,2] Oxazin-7-One

Treat the 5'-dimethoxytrityl derivative of the nucleoside P **6** (60 mg) in dry pyridine with succinic anhydride (50 mg) and 4,4-dimethylaminopyridinc (10 mg) for 4 d. Purify thc 3'-*O*-succinatc by chromatography and convert to its 4-nitrophenyl ester by reaction with 4-nitrophenol and dicyclohexylcarbodiimide. Shake the nitrophenyl ester (45 mg) and triethylamine (0.1 mL) in dimethylformamide (DMF) with vacuum-dried aminoalkyl controlled pore glass, CPG (Pierce Inc.) for 24 h. Wash the CPG with DMF, ether, and dry then treat with acetic anhydride in pyridine for 10 min, before washing and drying as before. The nucleoside loading of the functionalized CPG is 57.2 µmol/g.

3.3.2. Melting Transitions of Oligonucleotide Duplexes

Melting transitions are measured at 260 nm in 6X SSC buffer at an oligomer strand concentration of ~3 µ*M*. The temperature is increased by 1°C/min and the melting temperatures (T_m) are determined as the maxima of the differential curves, with an error of ±1°C.

3.3.3. Polymerase Chain Reaction

Each 100 µL reaction contains 1 µmol template DNA, 200 pM of each primers, 50 µm of each dNTP, 5 µ of *taq* polymerase in buffer provided (Promega) and 75 µL of paraffin oil. Thermal cycles are: denaturing temperature at 92°C for 1 min; annealing temperature at 36–44°C for 1 min; chain extension temperature at 62–70°C for 1 min for 30 cycles. Thirteen microliter of the PCR products are applied to a 2% agrose gel with 1X TBE as running buffer that contains 4 µL of ethidium bromide (10 mg/mL) with a current of 50 mA for 1.5 h. All the PCR products are run against Phi X174 RF DNA-*Hae* III Digest as marker.

4. Discussion

4.1. Synthesis of P-Monomer

The synthetic route to the 3'-cyanoethyl-*N*,*N*-diisopropylphosphoramidite **7** and the H phosphonate **8** of the *P*-nucleoside is set out in Fig. 2. Initially, condensation of α-3,5-di-*O*-*p*-toluoyl-2-deoxy-D-ribofuranosyl chloride with the fully trimethyl silylated base, 5-(2-hydroxyethyl)uracil, gives the nucleoside **1** in a synthesis modified from Griengl et al. *(14)*. The α-anomer represents a small contaminant (10–15%) but in the later stages of the synthesis, which includes several recrystallizations, the pure β-anomer is obtained. A Mitsunobu reaction leads to the crystalline phthalimidooxy derivative **2**, converted in high yield to the 4-triazolo pyrimidine **3** by the method of Reese and Skone *(15)*. Phthalimido derivatives usually require vigorous conditions for removal of the protecting group, but phthalimidooxy compounds are much more reactive. Ammonia-dioxan overnight gives rise quantitatively to the bicyclic analog **4**. The crystalline compound is deacylated quantitatively in methanolic ammonia overnight, best, at 55°C in a pressure vessel to assist dissolution. The intermediate nucleoside **5** is triturated with ether and the solid product directly converted to the 5'-*O*-dimethoxytrityl derivative **6**. After chromatographic purification it is converted to the phos-

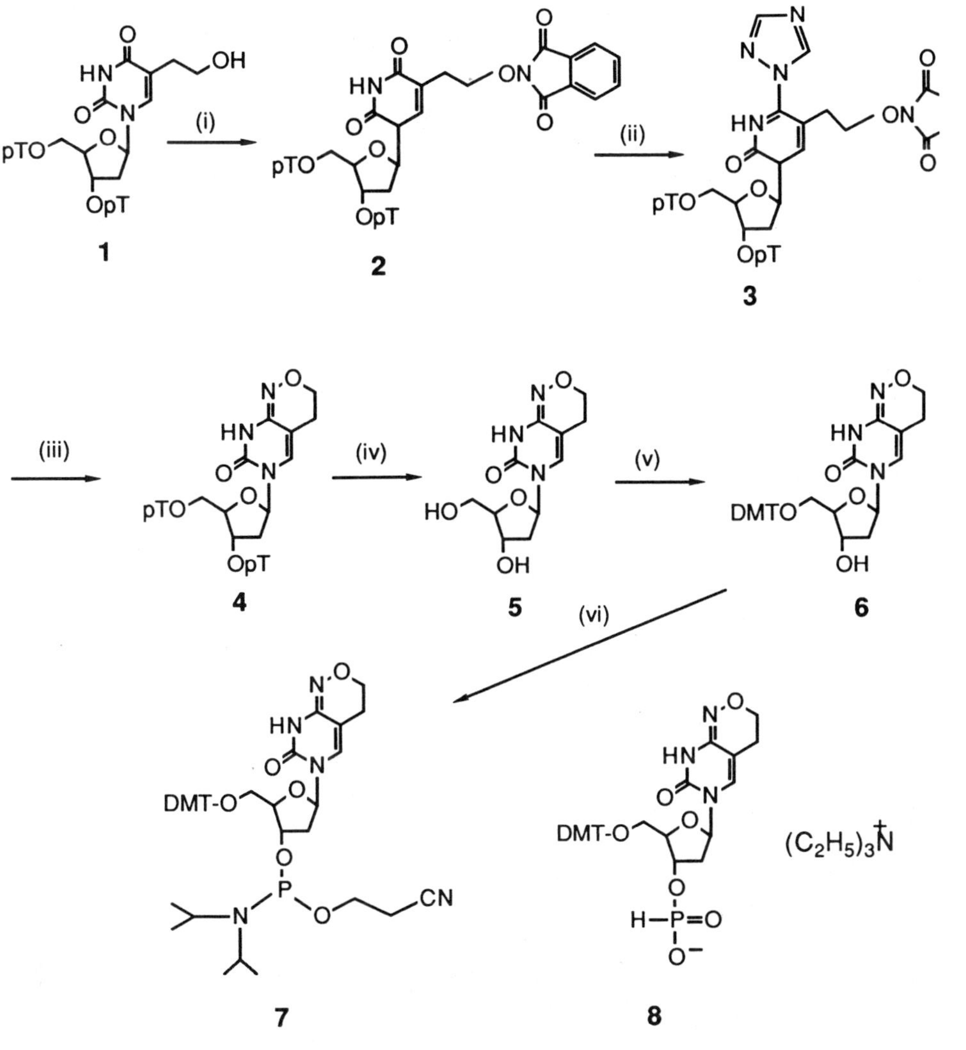

pT = pMe C_6H_4CO.   DMT = Dimethoxytrityl

Fig. 2. (i) Triphenyl phosphine, *N*-hydroxyphthalimide, DEAD in THF (ii) Triazole, $POCl_3$, Et_3N in CH_3CN (iii) NH_3/dioxan (iv) NH_3/methanol (v) Dimethoxytrityl chloride/pyridine at 55°, 2 h (vi) 2-cyanoethyl *N,N*-diisopropylphosphoramidite, *N,N*- diisopropylethylamine in THF, 2.5 h.

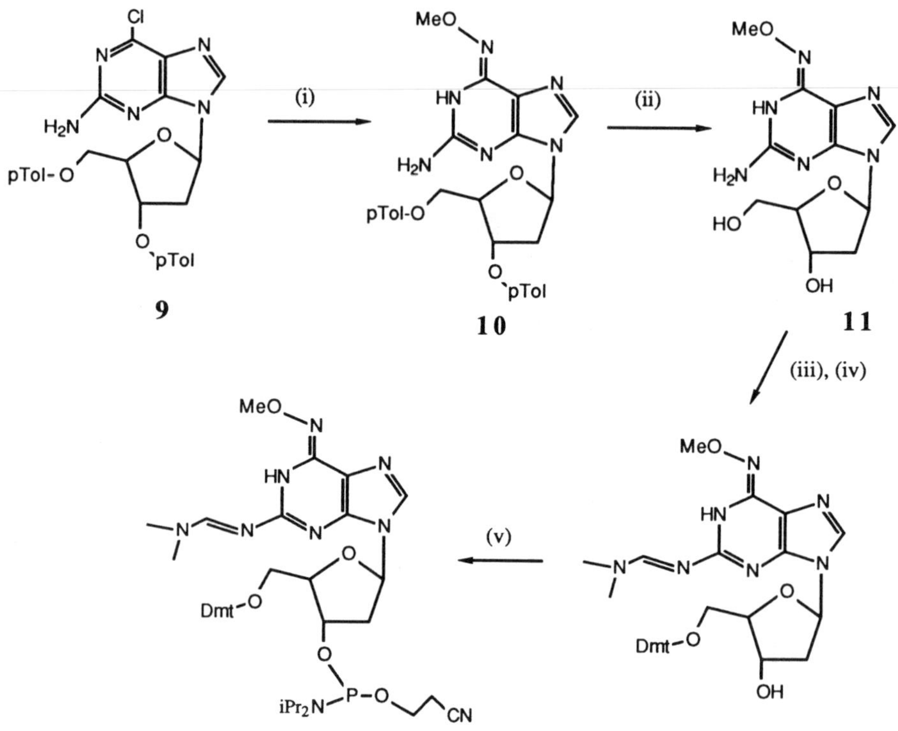

Fig. 3. (i) Methoxyamine, ethanol, o/n 90° (ii) NH_3/methanol (iii) *N,N*-dimethylformamide dimethylacetal in DMF at 50° for 2 h (iv) Dimethoxytrityl chloride/pyridine at 55°, 2 h (v) 2-cyanoethyl *N,N*-diisopropylphosphoramidite *N,N*-diisopropylethylamine in THF, 2.5 h.

phoramidite **7** and H-phosphonate **8** monomers in the usual way. Both couple in machine oligomer synthesis, using the normal cycles.

The DMT derivative **6** reacts with succinic anhydride, best in excess, to give the 3'-*O*-succinate that may then be coupled via its nitrophenylester to aminoalkyl controlled pore glass (CPG) support beads in the usual way. The latter is used in oligomer synthesis.

4.2. Synthesis of K Monomer

The synthesis of monomer **13** derived from the nucleoside 9-(β-D-2-deoxyribofuranosyl)-N^6-methoxy-2,6-diaminopurine **11** is set out in Fig. 3. Condensation of 2-amino-6-chloropurine (Aldrich Chemical

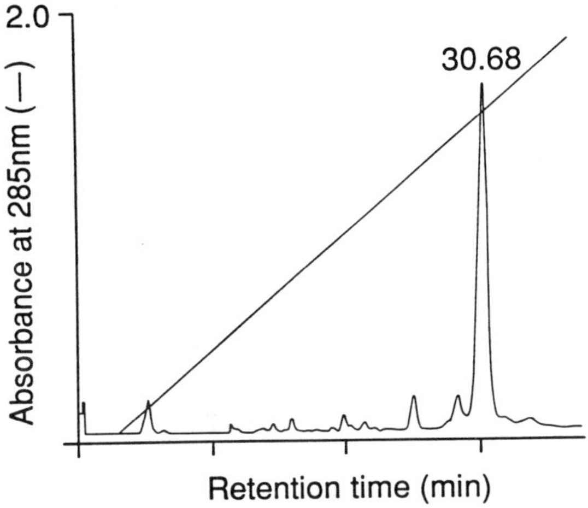

Fig. 4. Purification of crude product containing the 17-mer d(ACTTG**K**CC**K**C**P**ATT**P**TG) on a 10 × 250 mm Hichrom Partisil 10 Sax column. Gradient: 0–100% buffer B in 35 min. Flow rate 3.0 mL/min at room temperature.

Co.) with the α-3,5-di-*O*-p-toluoyl-2-deoxy-D-ribofuranosyl chloride under phase transfer conditions using the method of Seela and coworkers *(16)* gives **9** in good yield. Contaminating 7-isomer is readily removed by chromatography. The conversion of **9** to the 6-methoxyamino derivative **10** is best effected by free methoxyamine in ethanol *(17)*, whence deacylation by methanolic ammonia affords **11**. N^2-Acyl derivatives of **11** are found to be very difficult to deprotect in subsequent steps whereas the N^2-dimethylaminomethylene derivative is easily formed and deprotected *(18)*. Its dimethoxytrityl derivative **12** is converted to the 3'-(*N,N*-diisopropyl)-cyanoethylphosphoramidite **13** (K-monomer) in the usual way.

4.3. Melting Transitions of Duplexes Containing Degenerate Oligonucleotides

The phosphoramidites derived from the deoxyribonucleosides P and K are used in solid phase machine synthesis in the usual way, as regards concentrations and coupling times and purified by ion-exchange HPLC. An HPLC analysis of a crude heptadecamer containing P and K residues is shown in Fig. 4. Table 1 gives the melting transitions (T_m) of a set of related 17-mer duplexes containing one or more P residues

Table 1
Melting Temperature (T_m) of Heptadecamer Duplexes in 6X SSC Buffer at pH 7.0

	T_m (°C)		T_m (°C)
1. ACTTGGCC**A**CCATTTTG TGAACCGG**T**GGTAAAAC	72	11. ACTTGGCC**G**CCATTTTG ----------------**T**--------------	70
2. ACTTGGCC**G**CCATTTTG TGAACCGG**C**GGTAAAAC	75	12. ACTTGGCC**A**CCATTTTG ----------------**C**---------------	64
3. ACTTGGCC**A**CCATTTTG ----------------**P**--------------	72	13. ACTT**G**GCC**A**CC**A**TTTTG --------**T**------**C**----**C**---------	43
4. ACTTGGCC**G**CCATTTTG ----------------**P**--------------	73	14. ACTTGGCC**K**CCATTTTG ----------------**T**---------------	67
5. ACTTGGCC**A**CC**A**TTTTG ----------------**P**----**P**---------	71	15. ACTTGGCC**K**CCATTTTG ----------------**C**---------------	66
6. ACTTGGCC**G**CC**A**TTTTG ----------------**P**----**P**---------	70	16. ACTTGGCC**K**CC**K**TTTTG ----------------**T**-----**T**---------	64
7. ACTT**G**GCCACC**A**TTTTG --------**P**------------**P**---------	66	17. ACTTGGCC**K**CC**K**TTTTG ----------------**C**-----**T**---------	62
8. ACTT**G**GCC**A**CC**A**TTTTG --------**P**-------**P**----**P**---------	66	18. ACTTG**K**CC**K**CCATTTTG ----------**C**----**T**---------------	58
9. ACTT**G**GCC**G**CC**A**TTTTG --------**P**-------**P**----**P**---------	64	19. ACTTG**K**CC**K**CCATTTTG ----------**C**----**C**---------------	59
10. ACTT**G**GCC**A**CC**A**TTTT**G** --------**P**-------**P**----**P**--------**P**	63	20. ACTTG**K**CC**K**CC**K**TTTTG ----------**C**----**T**-----**T**---------	58
		21. ACTTG**K**CC**K**CC**K**TTTTG ----------**C**----**C**-----**T**---------	57

(paired with A and G) and a related set containing K (paired with C and T). The values have been discussed in more detail elsewhere *(7,8)*, but it is noteworthy that these 17-mers with P/A and P/G basepairs have essentially the same stabilities and that even with four such residues the T_m difference in relation to the fully complementary duplexes is only 9°C. The same observations are made with K-containing oligomers except that the T_m values are somewhat lower. Table 2 shows a comparison of fully complementary oligomers (entry 1,2) with a triple

Table 2
Melting Temperatures (T_m) of Heptadecamer Duplexes in 6X SSC Buffer at pH 7.0

	T_m (°C)		T_m (°C)
1. ACTTGGCCACCATTTTG TGAACCGGTGGTAAAAC	72	7. ACTTGICCICIATTITG ----------C---T-G-----A---	50
2. ACTTGGCCGCCATTTTG TGAACCGGCGGTAAAAC	75	8. ACTTGKCCKCPATTPTG ----------C----T--G------A---	55
3. ACTTGGCCACCATTTTG --------T-------C----C---------	43	9. ACTTGICCICPATTPTG ---------C----T-G-----A----	59
4. ACITGICCACCITTITG ---A----C---------T---A---	57	10. ACTTGICCICIATTITG ---------C----C-G----A---	53
5. ACPTGKCCACCKTTPTG ----A----C----------T----A----	60	11. ACTTGKCCKCPATTPTG ----------C----C--G-----A----	55
6. ACPTGICCACCITTPTG ----A---C---------T---A---	60	12. ACTTGICCICPATTPTG ---------C---C--A-----A----	63

mismatched oligomer (entry 3) and a set of oligomers containing P and K (entries 5,8,16). These are compared, too, with others containing hypoxanthine (I) alone or with P. In these latter oligomers (entries 6, 7, 9, 10, 12), I is considered solely as a purine and not as a "universal" base *(4)*. With regard to this limited set of observations, oligomers containing P and K are essentially similar to those containing P and I, and advantageous over those containing I alone. The T_m values help to establish optimum annealing temperatures for the use of degenerate oligomers in PCR amplification.

4.4. Degenerate Oligomers as PCR Primers

Figures 5–7 show the results of PCR amplifications involving 17-mers containing P, K, and I. The template is an M13 ssDNA with an insert corresponding to the upper strand of the duplexes in the tables (e.g., entry 1), as shown in the figure legends. Figure 5 shows that P residues support amplification when paired with either A or G in the template and also when present at the 3'-terminus, under annealing conditions where a triple mismatched 17-mer (lane 7) is ineffective.

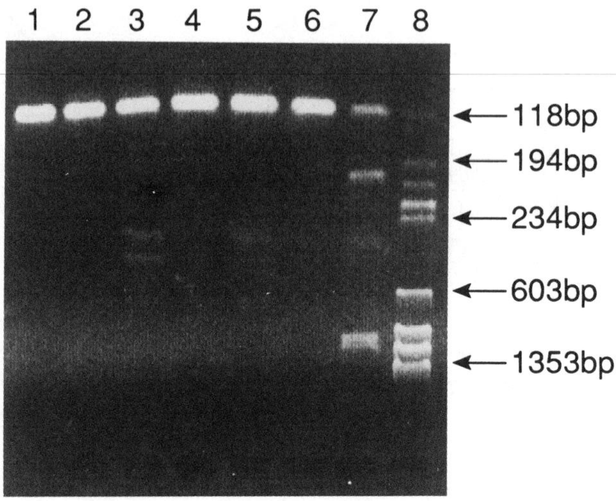


```
Forward Primer                 5'-GACGGATGAAGACGGGT-3'

5'---Template DNA-------GACGGATGAAGACGGGT-------78 bases-
--------------------------------------------------------------
----------------ACTTGGCCACCATTTTG--------------3'
```

		Lane	Base pairs
Reverse Primer	3'-TGAACCGGTGGTAAAAC-5'	1	Perfect
	3'-P————————————-5'	2	P/A
	3'-————P————-5'	3	P/A
	3'-——P————P————-5'	4	P/G, P/A
	3'-————P—P————-5'	5	P/A, P/A
	3'-——P——P—P————-5'	6	P/G, P/A, P/A
	3'-——T——C—C————-5'	7	T/G, C/A, C/A
	PhiX174 RF DNA Hae III Digest	8	

Fig. 5. Thermal cycle was: denaturing at 92°C for 1 min, annealing at 44°C for 1 min, chain extension at 70°C for 1 min, final extension at 70°C for 5 min, and the number of cycles = 30.

Figure 6 shows comparable experiments with K containing oligomers, but using a somewhat lower annealing temperature. Figure 7 gives examples where oligomers containing both P and K (lanes 2 and 3) lead to good amplification; comparable results are obtained with those containing P and I (lanes 4 and 5). Under the same conditions an

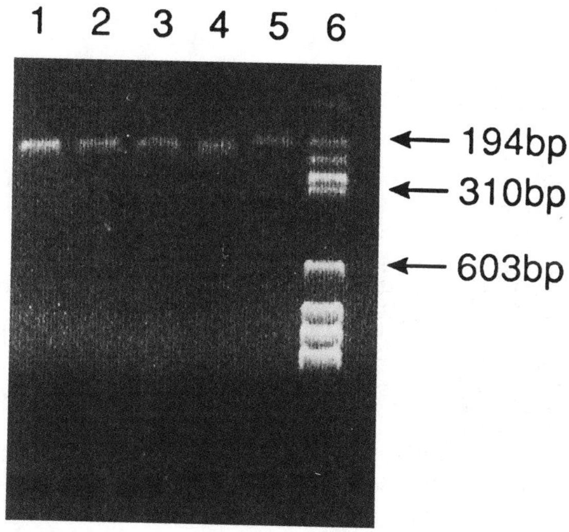

Figure 6.

		Lane	Base pairs
Forward Primer	5'-ACTTGGCCACCATTTTG-3'	1	Perfect
	5'-————————K————————-3'	2	K/T
	5'-————————K——K—————-3'	3	K/T, K/T
	5'-————K——K—————————-3'	4	K/C, K/T
	5'-————K——K——K——————-3'	5	K/C, K/T, K/T
	PhiX174 RF DNA Hae III Digest	6	

5'-----------ACTTGGCCACCATTTTG------155 bases----------

------Template DNA--CGGCCGCTTTTAGATT----------3'

Reverse Primer 3'-GCCGGCGAAAAATCTAA-5'

Fig. 6. Thermal cycle was: denaturing at 92°C for 1 min, annealing at 36°C for 1 min, chain extension at 62°C for 1 min, final extension at 62°C for 5 min, and the number of cycles = 30.

oligomer with four I residues gives incorrect priming (lane 7) although another gave a product of the correct mass.

The above evidence shows that primers containing the degenerate bases P and K (and P and I) can be used in PCR. Thus such primers can be extended by the polymerase and are recognized by it in the second strand synthesis; the bases inserted opposite P and K are not at present defined. From the above it follows that the multiplicity of primers based on protein amino acid sequences can be largely reduced.

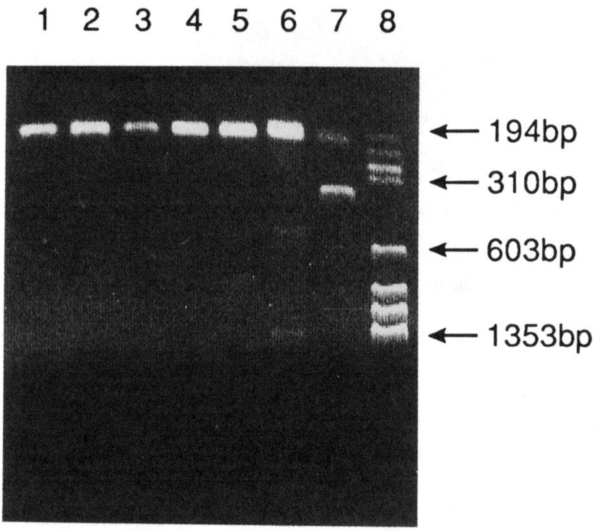

		Lane	Base pairs
Forward Primer	5 ' - ACTTGGCCACCATTTTG - 3 '	1	Perfect
	5 ' - —P—K————K—P—- 3 '	2	P/A, K/C, K/T, P/A
	5 ' - ————K—K–P——P—- 3 '	3	K/C, K/T, P/G, P/A
	5 ' - —P—I————I —P—- 3 '	4	P/A, I/C, I/T, P/A
	5 ' - ————I — I–P——P—- 3 '	5	I/C, I/T, P/G, P/A
	5 ' - —I —I———I ——I —- 3 '	6	I/A, I/C, I/T, I/A
	5 ' - ————I—I–I——I—- 3 '	7	I/C, I/T, I/G, I/A
	PhiX174 RF DNA Hae III Digest	8	

5 ' - - - - - - - - - - - - ACTTGGCCACCATTTTG - - - - - - 174 bases - - - - - - - - - - -
- - - - - - - Template DNA - - GAAGCGGACGGCAATCC - - - - - - - - - 3 '

Reverse Primer 3 ' - CTTCGCCTGCCGTTAGG - 5 '

Fig. 7. Thermal cycle was: denaturing at 92°C for 1 min, annealing at 36°C for 1 min, chain extension at 62°C for 1 min, final extension at 62°C for 5 min, and the number of cycles = 30.

References

1. Lathe, R. (1985) Synthetic oligonucleotide probes deduced from amino acid sequence data. Theoretical and practical considerations. *J. Mol. Biol.* **183,** 1–12.
2. Compton, T. (1990) Degenerate primers for DNA amplification, in *PCR Protocols, A Guide to Methods and Their Applications,* Chapter 5. (Innis, M. A., Gelfand, D. H., Sninsky, J. J., and White, T. J., eds.), Academic, New York, pp. 39–45.
3. Bej, A. K., Mahbubani, M. H., and Atlas, R. M. (1991) Amplification of nucleic acids by polymerase chain reaction (PCR) and other methods and their applications, in *Critical Reviews in Biochemistry and Molecular Biology* (Fasman, G. D., ed.), **26,** 301–344.
4. Ohtsuka, E., Matsuki, S., Ikehara, M., Takahashi, Y., and Matsubara, K. (1985) An alternative approach to deoxyoligonucleotides as hybridisation probes by insertion of deoxyinosine at ambiguous codon positions. *J. Biol. Chem.* **260,** 2605–2608.
5. Martin, F. H., Castro, M. M., Abou-Ela, F., and Tinoco, I. (1985) Basepairing involving deoxyinosine: implications for probe design. *Nucleic Acids Res.* **13,** 8927–8938.
6. Knoth, K., Roberds, S., Potect, C., and Tamkun, M. (1988) Highly degenerate inosine-containing primers specifically amplifying rare cDNA using polymerase chain reaction. *Nucleic Acids Res.* **16,** 10932.
7. Kong Thoo Lin, P. and Brown, D. M. (1989) Synthesis and duplex stability of oligonucleotides containing cytosine-thymine analogues. *Nucleic Acids Res.* **17,** 10373–10383.
8. Brown, D. M. and Kong Thoo Lin, P. (1991) Synthesis and duplex stability of oligonucleotides containing adenine-guanine analogues. *Carbohydrate Res.* **216,** 129–139.
9. Nederman, A. N. R., Stone, M. J., Kong Thoo Lin, P., Brown, D. M., and Williams, D. H. (1991) Basepairing of cytosine analogues with adenine and guanine in oligonucleotide duplexes: evidence for exchange between Watson-Crick and Wobble basepairs using ^{1}H NMR spectroscopy. *J. Chem. Soc., Chem. Commun.* **19,** 1357–1359.
10. Kong Thoo Lin, P. and Brown, D. M. (1991) Synthesis of oligodeoxyribonucleotides containing degenerate bases and their use as primers in the polymerase chain reaction. *Nucleic Acids Res.* **20,** 5149–5152.
11. Fissekis, J. D. and Sweet, F. (1973) The chemistry of some 5-(2-hydroxyalkyl) uracil derivatives and a synthesis of 5-vinyl uracil. *J. Org. Chem.* **28,** 264–269.
12. Hoffer, M. (1960) α-Thymidine. *Chem. Ber.* **93,** 2777–2781.
13. Jones, L. W. and Major, R. T. (1927) Substituted O-alkyl hydroxylamines chemically related to medicinally valuable amines. *J. Am. Chem. Soc.* **49,** 1527–1540.

14. Griengl, H., Bodenteich, M., Hayden, W., Wanek, E., Streicher, W., Stutz, P., Bachmayer, H., Ghazzouli, I., and Rosenwirth, B. (1985) 5-Haloalkyl-2'-deoxyuridines: a novel type of potent antiviral nucleoside analogue. *J. Med. Chem.* **28,** 1679–1684.
15. Reese, C. B. and Skone, P. A. (1984) The protection of thymine and guanine residues in oligodeoxynucleotide synthesis. *J. Chem. Soc., Perkin Trans. I* 1263–1271.
16. Seela, F., Westermann, B., and Bindig, U. (1988) Liquid-liquid and solid-liquid phase-transfer glycosylation of pyrrolo [2,3-*d*]-pyrimidines: stereospecific synthesis of 2-deoxy-β-D-ribofuranosides related to 2'-deoxy-7-carboguanosine. *J. Chem. Soc., Perkin Trans. I* 697–702.
17. Giner-Sorolla, A., O'Bryant, S. A., Nanos, C., Dollinger, M. R., Bendich, A., and Burchenal, J. H. (1968) The synthesis and biological properties of hydroxylaminopurines and related derivatives. *J. Med. Chem.* **11,** 521–523.
18. Zemlicka, J., Chladek, S., Holy, A., and Smrt, J. (1966) Oligonucleotidic compounds XIV. Synthesis of some diribonucleoside phosphates using the dimethylaminomethylene derivatives of 2', 3'-O-ethoxymethylene ribonucleosides. *Collect. Czech. Chem. Commun.* **31,** 3198–3211.

CHAPTER 8

Synthesis of [^{15}N]-Labeled DNA Fragments

Roger A. Jones

1. Introduction

DNA fragments labeled with ^{15}N have the potential to provide novel insight into base pairing, hydration, drug/nucleic acid, and protein/nucleic acid interactions. Synthetic routes to a variety of [^{15}N]-labeled pyrimidine nucleosides *(1–3)*, purines *(4–6)*, and purine nucleosides *(7–13)* have been reported. In some cases these labeled bases or monomers have been incorporated into nucleic acids, either by chemical synthesis *(9,10,14–17)* or by biosynthetic procedures *(18–22)*. The focus of this chapter will be on the preparation of [^{15}N]-labeled purine 2'-deoxynucleosides and their incorporation into DNA fragments by chemical synthesis.

2. Materials

2.1. Reagents

1. Lithium aluminum hydride ($LiAlH_4$)
2. *Tert*-Butyldimethylsilylchloride (BDMS-Cl)
3. Tetraethylammonium chloride (TEACl)
4. Tert-Butyl nitrite (TBN)
5. $NaIO_4$
6. $RuO_2 \cdot 2H_2O$
7. Benzyl bromide (BzBr)
8. 1*M* Tetra-*N*-butylammonium fluoride (TBAF) in THF
9. Linde 4 Å molecular sieves
10. [^{15}N]-KCN

From: *Methods in Molecular Biology, Vol. 26: Protocols for Oligonucleotide Conjugates*
Edited by: S. Agrawal Copyright ©1994 Humana Press Inc., Totowa, NJ

11. Cyanogen bromide (CNBr)
12. Na_2CO_3
13. $NaHCO_3$
14. Celite
15. CH_3I
16. Magnesium monoperoxyphthalate (MMPP)
17. KOH
18. Adenosine deaminase (Sigma [St Louis, MO] A-5773)
19. $[^{15}N]$-$NaNO_2$
20. Bio-Rad AG 1X2 hydroxide form resin
21. Phosphorus trichloride (PCl_3)
22. Phosphorus oxychloride ($POCl_3$)
23. 4,6-Diaminopyrimidine hemisulfate monohydrate
24. Aniline
25. 4-Bromoaniline
26. 2,4,6-Triaminopyrimidine
27. Raney nickel (50% slurry in water, Aldrich [Milwaukee, WI] 22,167-8)
28. *N*-Methylmorpholine
29. 5% Pd/C (50% water wet, Aldrich 27,670-7)
30. Diethoxymethyl acetate
31. Thymidine phosphorylase (dThd Pase, Sigma T-7006)
32. Purine nucleoside phosphorylase (PNPase, Sigma N-8264)
33. Imidazole-4,5-dicarboxylic acid
34. Ethyl chloroformate
35. 1,2,4-Triazole
36. *N,N*-Dimethylformamidine dimethyl acetal
37. Di-*N*-butylamine
38. Triethylamine (TEA)
39. 1,8-Diazabicyclo[5,4,0]undec-7-ene (DBU)
40. *N*-bromosuccinimide (NBS)
41. Pyridine
42. Tetrahydrofuran (THF)
43. Ethyl acetate
44. Dichloromethane
45. *N,N*-Dimethylformamidine (DMF)
46. Methanol
47. Acetonitrile
48. 4-Dimethylaminopyridine (DMAP)
49. $[^{15}N]$-Benzamide

2.2. HPLC Columns and Buffers

1. Reversed-phase analytical columns: Waters C-18 Novapak cartridges in an RCM 100 or a Beckman C-3 Ultrapore column (4.6 mm × 7.5 cm).
2. Reversed-phase preparative columns: Waters C-18 (15 × 190 mm steel column or a 25 × 100 mm Radial-Pak cartridge).
3. Beckman C-3 Ultrapore column (10 mm × 25 cm) or Dynamax C-18 column (21.4 mm × 25 cm or 41.4 mm × 25 cm).
4. Normal-phase preparative columns: Dynamax-60A silica column (21.4 mm × 25 cm or 41.4 mm × 25 cm).
5. Buffers: 0.1*M* triethylammonium acetate, pH 7 (TEAA), 0.1*M* ammonium acetate, pH 7 (AA), 0.1*M* ammonium bicarbonate, pH 7 (ABC).

3. Methods

3.1. Preparation of [6-^{15}N]-2'-Deoxyadenosine (6) and [1-^{15}N]-2'-Deoxyadenosine (10)

The transformation of 2'-deoxyadenosine to [6-^{15}N]-2'-deoxyadenosine (**6**) and from **6** to [1-^{15}N]-2'-deoxyadenosine (**10**) is shown in Figs. 1 and 2 *(8)*. After protection of the 3'- and 5'-hydroxyl groups by reaction with *tert*-butyldimethylsilyl chloride (BDMS-Cl), a nonaqueous diazotization reaction gives the corresponding 6-chloro derivative **3**. The ^{15}N is then introduced by reaction of **3** with [^{15}N]-benzylamine to give the [6-^{15}N]-benzyl derivative **4**. Although all attempts at reductive debenzylation of **4** have been unsatisfactory, oxidative conversion to the corresponding benzoyl compound proceeds readily. Ammonolysis then gives **5**, which can be desilylated using tetra-*N*-butylammonium fluoride (TBAF) in THF to yield [6-^{15}N]-2'-deoxyadenosine (**6**). Alternatively, alkylation of **5** with benzyl bromide *(23)*, followed by treatment with methanolic dimethylamine, affects Dimroth rearrangement to give the [1-^{15}N]-derivative **8**. Debenzylation and desilylation of **8** then are carried out exactly as with **4**.

3.1.1. [^{15}N]-Benzylamine/HCl

1. To 3.5 g (92 mmol) of $LiAlH_4$ suspended in 300 mL of anhydrous diethyl ether add 2 g (16 mmol) of [^{15}N]-benzamide over 3 d *via* continuous extraction by use of a Soxlet apparatus.

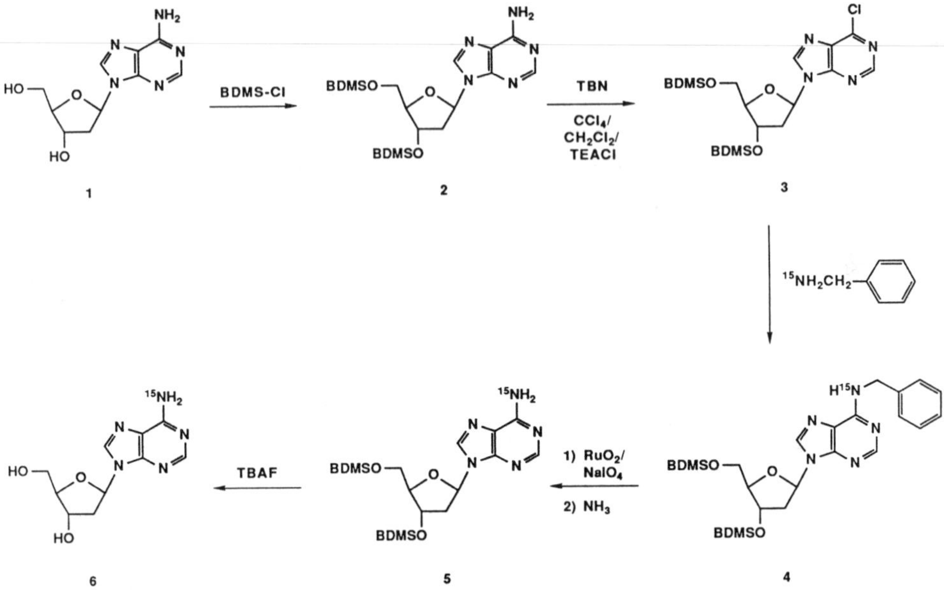

Fig. 1. Synthesis of [6-^{15}N]-2'-deoxyadenosine (**6**).

2. Stir the reaction mixture at room temperature one more day, cool in a dry ice/ethanol bath, and carefully hydrolyze by adding successively 3 mL water, 3 mL 5% aqueous sodium hydroxide, and 8 mL water (**Caution:** *See* Note 1).
3. Stir at room temperature for 10 min, filter, cool the filtrate in the dry ice/ethanol bath for 5 min, and add dropwise with vigorous stirring a 1.4-mL portion of concentrated HCl. Stir the resulting slurry at room temperature for 10 min and filter. You should get about 1.9 g (13 mmol, 81%) of [^{15}N]-benzylamine/HCl as white crystals, mp 258–260°C.

*3.1.2. 3',5'-O-Bis(*Tert*-Butyldimethylsilyl)-2'-Deoxyadenosine* (2)

1. To 10 g (40 mmol) of 2'-deoxyadenosine (**1**) and 7.5 g (110 mmol) imidazole, dried three times by evaporation of pyridine and dissolved in 80 mL of pyridine, add 20 mL of a solution of 13.6 g (90 mmol) BDMS-Cl in pyridine.
2. Stir the reaction mixture at room temperature for 20 h and then concentrate to dryness. Dissolve the residue in a 200-mL portion of a 1/1 (v/v) mixture of diethyl ether and petroleum ether and extract with 250-mL

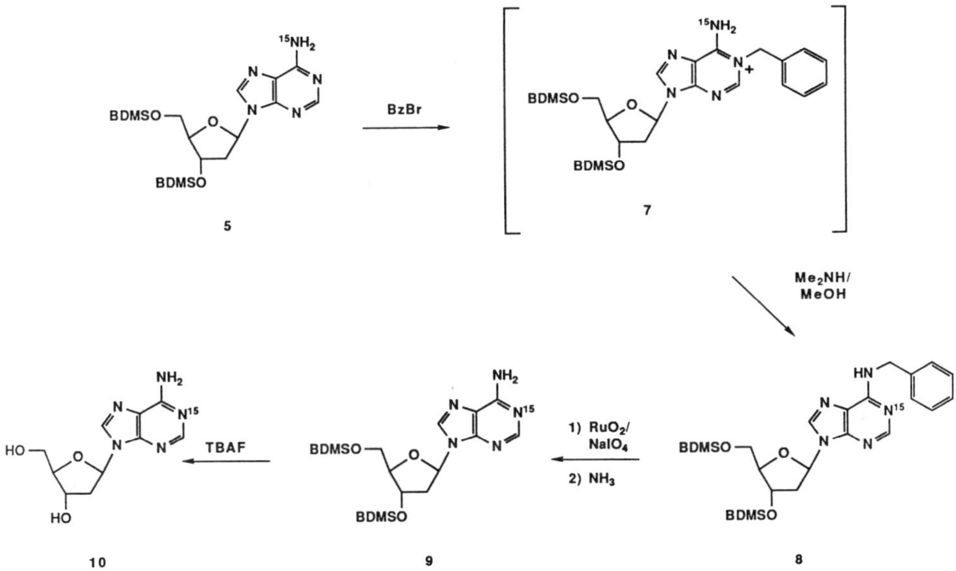

Fig. 2. Synthesis of [1-^{15}N]-2'-deoxyadenosine (**10**).

portions of 6% aqueous $NaHCO_3$. Back extract the aqueous layer with three 100-mL portions of the same organic solvent mixture. Combine the organic extracts and evaporate to dryness.

3. Dissolve the residue in 15 mL of toluene, precipitate the product by addition to 700 mL petroleum ether at 0°C, and filter. You should get about 15.5 g (32 mmol, 81%) of **2** as a white powder.

*3.1.3. 6-Chloro-9-{2-Deoxy-3,5-O-Bis (*Tert*-Butyldimethylsilyl)-β–D-Erythro-Pentafuranosyl}Purine* (3)

1. To 8.2 g (17 mmol) of (**2**) suspended in a mixture of 190 mL CCl_4 and 95 mL CH_2Cl_2 and heated in an oil bath at 55°C under a nitrogen atmosphere (*see* Note 2), add 28 g (170 mmol) tetraethylammonium chloride (TEACl) and 4.0 mL (34 mmol) tert-butyl nitrite (TBN).
2. Stir the reaction mixture at 55°C for 1 h, allow to cool to room temperature, and add to a 700-mL portion of cold, saturated aqueous $NaHCO_3$. Extract the mixture with three 80-mL portions of methylene chloride. Combine the extracts and concentrate to a gum.
3. Purify the residue by chromatography on silica gel using a gradient of 0.6 to 3% methanol in petroleum ether/methylene chloride (1/1, v/v) in 1 h at a flow rate of 40 mL/min. Concentrate the appropriate fractions to a gum. You should get about 5.6 g (11 mmol, 66%) of **3**.

3.1.4. [6-^{15}N]-6-N-Benzyl-2'-Deoxy-3',5'-O-Bis(Tert-Butyldimethylsilyl)Adenosine (4)

1. To 2.2 g (4.5 mmol) of (**3**) in 20 mL *tert*-butanol add 0.38 g (2.6 mmol) [^{15}N]-benzylamine hydrochloride and 1.2 mL (8 mmol) DBU (*see* Note 3). Stir the mixture at room temperature for 3 d, and concentrate to a gum.
2. Purify the residue by chromatography on silica gel using a gradient of 0–1.4% methanol in methylene chloride in 30 min at a flow rate of 10 mL/min. Concentrate the appropriate fractions to a solid foam. You should get about 1.3 g (2.3 mmol, 88% based on the [^{15}N]-benzylamine hydrochloride used) of **4**.

3.1.5. [6-^{15}N]-2'-Deoxy-3',5'-O-Bis(Tert-Butyldimethylsilyl)Adenosine (5) or [1-^{15}N]-3',5'-O-Bis(Tert-Butyldimethylsilyl)Adenosine (9)

1. To 1.3 g (2.3 mmol) of **4** or **8** dissolved in a mixture of 14 mL methylene chloride, 14 mL acetonitrile and 21 mL water add 2.1 g (9.6 mmol) $NaIO_4$ and 13 mg $RuO_2 \cdot xH_2O$.
2. Stir the mixture at room temperature for 3 h (*see* Note 4), filter, and wash the residue with methylene chloride. Add the filtrate to 80 mL of 6% aqueous $NaHCO_3$, and extract with three 50-mL portions of methylene chloride. Combine the organic layers and concentrate to a gum.
3. Treat the residue with a 10 mL portion of methanol saturated with NH_3 at room temperature overnight.
4. Concentrate the mixture and purify the residue by chromatography on silica gel using a gradient of 0–5% methanol in methylene chloride during 30 min at a flow rate of 10 mL/min. Concentrate the appropriate fractions to a solid foam. You should get about 0.88 g (1.83 mmol, 80%) of **5** or **9**.

3.1.6. [1-^{15}N]-6-N-Benzyl-2'-Deoxy-3',5'-O-Bis(Tert-Butyldimethylsilyl)Adenosine (8)

1. To 0.88 g (1.83 mmol) of (**5**) dissolved in 20 mL dried DMF add 0.43 mL (3.6 mmol) benzyl bromide and 2 g Linde 4 Å molecular sieves (*see* Note 5). Stir the reaction mixture in an oil bath at 40°C for 3 d under a nitrogen atmosphere.
2. Cool the reaction mixture to room temperature and treat with a 10 mL portion of a 1/1 mixture (v/v) of dimethylamine/methanol. After 1 h, filter the reaction mixture and concentrate to dryness.
3. Purify the residue by chromatography on silica gel using a gradient of 0.6–3% methanol in methylene chloride in 1 h at a flow rate of 10 mL/

min. Concentrate the appropriate fractions to a solid foam. You should get about 0.88 g (1.55 mmol, 85%) of **8**.

3.1.7. [6-^{15}N]-2'-Deoxyadenosine ***(6)*** *or [1-^{15}N]-2'-Deoxyadenosine* ***(10)***

1. To each mmol of **5** or **9** add 2.1 mL (2.1 mmol) 1*M* TBAF in THF (*see* Note 6).
2. Allow the mixture to stand at room temperature until the product has crystallized, usually 3–18 h. You should get a quantitative yield of **6** or **10**, slightly contaminated with TBAF. This material can be protected for oligonucleotide synthesis without further purification or can be recrystallized as described below.
3. **Optional:** The above products can be recrystallized in high yield by diffusion of diethyl ether into a methanol solution of the crude product (about 5 mL methanol per mmol of **5/9**). Place a beaker of the methanol solution (10 mL in a 50-mL beaker) in a desiccator containing several hundred mL of diethyl ether. Crystallization should be complete in 3–5 d.

3.2. Preparation of [1-^{15}N]-2'-Deoxyguanosine (17) and [2-^{15}N]-2'-Deoxyguanosine (21)

The syntheses of [1-^{15}N]-2'-deoxyguanosine (**17**) and [2-^{15}N]-2'-deoxyguanosine (**21**) each use the same set of reactions *(13)* that are based on Ueda's transformation of adenine nucleosides to the 6-thioguanine and 2,6-diaminopurine nucleosides *(24)*. The syntheses differ only in the ^{15}N source. In the former case (Fig. 3) the ^{15}N is derived from [6-^{15}N]-2'-deoxyadenosine (**6**), whereas for the latter (Fig. 4) it is from [^{15}N]-KCN. In the first step **6** or **1** is converted to the crystalline N^1-oxide (**11** or **18**) by oxidation with magnesium monoperoxyphthalate (MMPP). Alternatively, the oxidation also could be carried out using *m*-chloroperbenzoic acid *(25)*, which is available once again. Then a series of five reactions is carried out without intervening extraction or chromatography, beginning with reaction of **11** with cyanogen bromide or reaction of **18** with [^{15}N]-cyanogen bromide, generated *in situ* by reaction of [^{15}N]-KCN with bromine. The 6-methoxyamino derivative (**16** or **20**) then is purified by reversed-phase chromatography and the [^{15}N]-labeled deoxyguanosine (**17** or **21**) is obtained in quantitative yield (from **16** or **20**) by treatment with adenosine deaminase. These syntheses employ no protecting groups and require only one chromatographic purification.

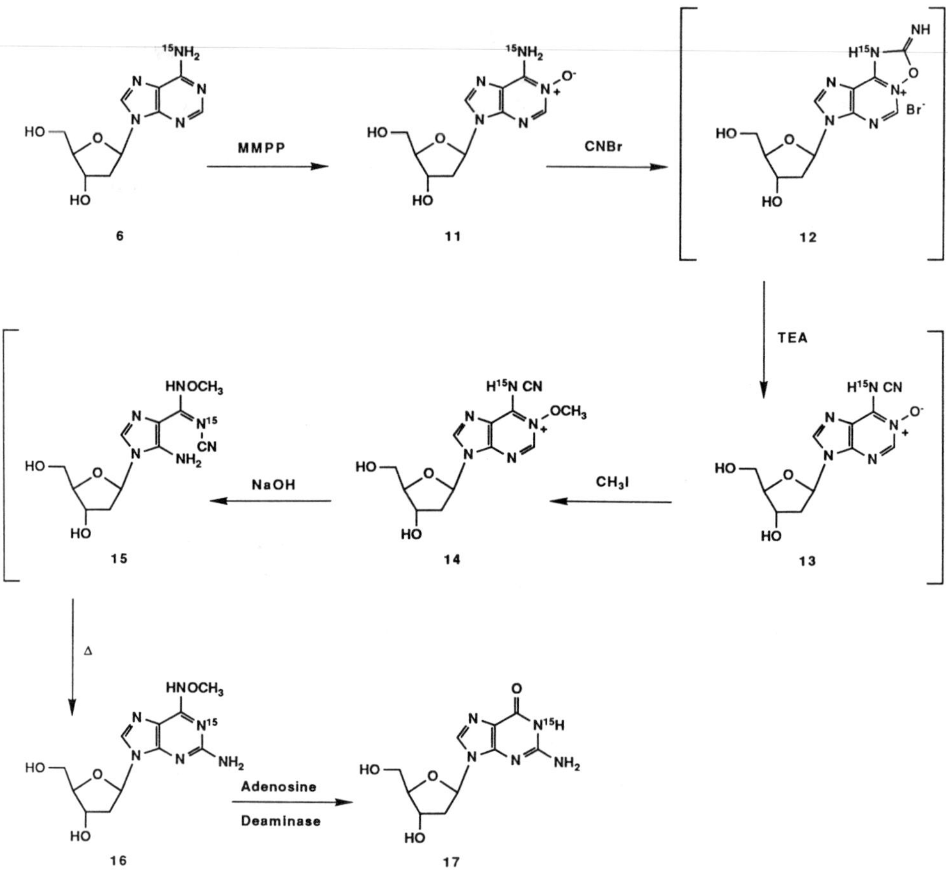

Fig. 3. Synthesis of [1-^{15}N]-2'-deoxyguanosine (**17**).

3.2.1. [6-^{15}N]-2'-Deoxyadenosine-N^{1}-Oxide (**11**) *or 2'-Deoxyadenosine-N^{1}-Oxide* (**18**)

1. To 1.4 g (5.6 mmol) of **6** or **1** dissolved in 140 mL 30% aqueous dioxane add 3.2 g (6.4 mmol) magnesium monoperoxyphthalate (MMPP).
2. Stir the mixture in the dark at room temperature for 48 h.
3. Concentrate the mixture to dryness, dissolve the residue in a minimum amount of water, and add methanol until cloudiness persists.
4. Cool the mixture at 0°C until crystallization is complete, and filter. You should get about 1.1 g (4.0 mmol, 70%) of **11/18**, mp 215°C dec.

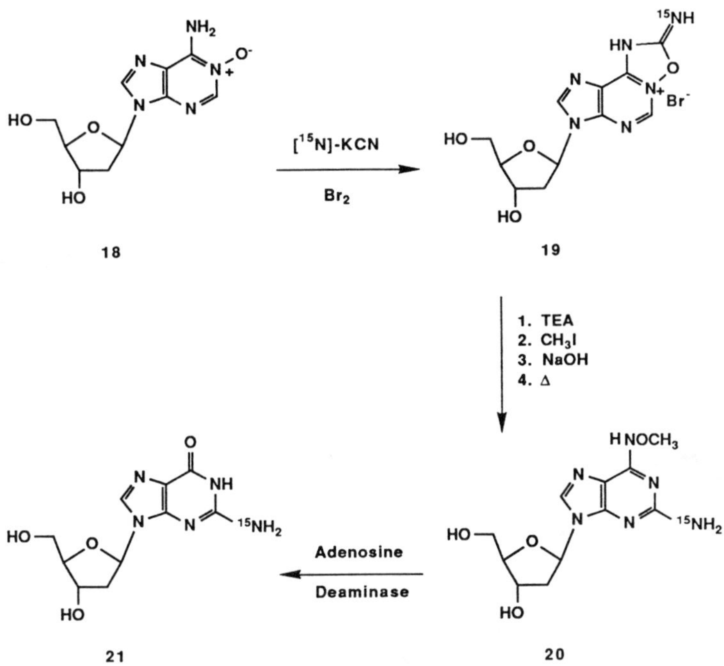

Fig. 4. Synthesis of [2-^{15}N]-2'-deoxyguanosine (**21**).

3.2.2. [1-^{15}N]-2-Amino-6-Methoxyamino-9-(2-Deoxy-β-D-Erythro-Pentofuranosyl)Purine (16)

1. To 0.88 g (3.3 mmol) of **11** add 80 mL anhydrous methanol and 0.4 g (3.8 mmol) CNBr.
2. Stir the mixture for 2 h and concentrate to dryness.
3. Dissolve the residue in a mixture of anhydrous DMF (6 mL) and triethylamine (1.1 mL, 7.9 mmol), under N_2.
4. Stir the mixture at room temperature for 40 min, and add 0.67 mL CH_3I (10.8 mmol).
5. Stir for a further 3.5 h and concentrate the mixture to dryness. Dissolve the residue in 75 mL 0.25*N* NaOH. After 30 min bring the pH to 7.4 by adding 1*N* HCl.
6. Add ethanol (65 mL), heat the mixture at 60°C for 4 h, and concentrate to dryness.

7. Purify the residue by chromatography on a Dynamax reversed-phase column (21.4 mm × 25 cm) using a gradient of 2–5% acetonitrile: 0.1*M* ammonium bicarbonate. Concentrate the appropriate fractions to a solid. You should get about 0.62 g (2.1 mmol, 64%), **16**.

3.2.3. [2-^{15}N]-2-Amino-6-Methoxyamino-9-(2-Deoxy-β–D-Erythro-Pentofuranosyl)Purine (**20**)

1. To 0.22 g (3.3 mmol) [^{15}N]-KCN dissolved in 75 mL anhydrous methanol and cooled to 0–10°C, add bromine (0.17 mL, 3.3 mmol), stir for 3 h, and add **18** (0.80 g, 3.0 mmol).
2. Stir the mixture an additional 2–3 h and concentrate to dryness.
3. Dissolve the residue in a mixture of anhydrous DMF (11 mL) and triethylamine (1.1 mL, 7.9 mmol), under N_2.
4. Stir the mixture at room temperature for 40 min, and add 0.67 mL CH_3I (10.8 mmol).
5. Stir for a further 3.5 h and concentrate the mixture to dryness. Dissolve the residue in 75 mL 0.25*N* NaOH. After 30 min bring the pH to 7.4 by adding 1*N* HCl.
6. Add ethanol (80 mL), heat the mixture at 60°C for 4 h, and concentrate to dryness.
7. Purify the residue by chromatography on a Dynamax reversed-phase column (21.4 mm × 25 cm) using a gradient of 2–5% acetonitrile: 0.1*M* ammonium bicarbonate. Concentrate the appropriate fractions to a solid. You should get about 0.44 g (1.5 mmol, 50%), of **20**.

3.2.4. [1-^{15}N]-2'-Deoxyguanosine (**17**) *and [2-^{15}N]-2'-Deoxyguanosine* (**21**)

1. To 0.59 g of **16** or **20** (2 mmol) dissolved in 40 mL 0.1*M* TEAA buffer (pH 6.8) (*see* Note 7), add adenosine deaminase (917 U). The product (**17** or **21**) will crystallize out as it is formed.
2. After 2 d at room temperature cool the mixture to 4°C and filter. You should get a first crop of about 0.40 g (1.5 mmol, 75%) of **17** or **21**, mp > 250°C.

3.3. Preparation of [7-^{15}N]-Adenine (30), [7-^{15}N]-2,6-Diaminopurine (31), [7-^{15}N]-2'-Deoxyadenosine (32), [7-^{15}N]-2,6-Diamino-9-(2-Deoxy-β–D-*Erythro*-Pentofuranosyl)Purine (33), [7-^{15}N]-2'-Deoxyinosine (34), and [7-^{15}N]-2'-Deoxyguanosine (35)

Diazotization of arylamines using [^{15}N]-$NaNO_2$ gives the corresponding β–[^{15}N] diazonium ion, which ordinarily does not undergo rearrangement of the ^{15}N to the a position. An azo coupling reaction of such a β–[^{15}N] diazonium ion is used for introduction of the ^{15}N atom in syntheses of both the [7-^{15}N]-labeled molecules described in this section *(11)*, and the [3-^{15}N]-labeled compounds described in Section 3.4. below *(12)*. For the [7-^{15}N] labels the azo coupling with either diaminopyrimidine **(22)** or triaminopyrimidine **(23)** gives crystalline azo derivatives **26** or **27** in high yield, without further purification (Fig. 5). Hydrogenolysis of the azo linkage then gives the corresponding [^{15}N]-labeled tri- or tetraaminopyrimidines **28** or **29**. Both **28** and **29** will discolor readily and are best immediately converted to, respectively, [7-^{15}N]-adenine **(30)** or [7-^{15}N]-2,6-diaminopurine **(31)** by treatment with diethoxymethyl acetate. The corresponding 2'-deoxynucleosides, **32** and **33**, are then obtained by an enzymatic transglycosylation procedure *(26–28)*. Finally, the hypoxanthine or guanine nucleosides, **34** or **35**, are obtained by deamination of **32** or **33** using adenosine deaminase. These reactions do not employ protecting groups, and the only chromatography is a low-resolution ion-exchange column after the transglycosylation step to remove the excess thymidine, thymine, and 2-deoxy-α–D-ribose-1-phosphate.

3.3.1. [5-^{15}N]-5-(4-Bromophenyl)Azo-4,6-Diaminopyrimidine **(26)**

1. To a cold suspension of 4-bromoaniline (0.86 g, 5.0 mmol) in 8.5 mL 7*M* HCl add dropwise a cold solution of [^{15}N]-$NaNO_2$ (0.39 g, 5.5 mmol) in 750 µL water.
2. After 2 h, add this solution of **24** dropwise to a 25-mL aqueous solution of 4,6-diaminopyrimidine hemisulfate monohydrate (**22**, 1.35 g, 7.5 mmol), Na_2CO_3 (0.63 g, 7.5 mmol) and sodium acetate/3 H_2O (12.6 g, 93 mmol).
3. After 48 h at 45 °C cool the mixture, collect the product by filtration, and wash it well with petroleum ether. You should get about 1.0 g (3.4 mmol, 70%) of **26**, mp 250°C dec. This material should be pure by tlc and can be used below without further purification.

3.3.2. [5-^{15}N]-5-Phenylazo-2,4,6-Triaminopyrimidine **(27)**

1. To a cold solution of aniline (0.65 mL, 7.1 mmol) in 50 mL dilute HCl (28 mmol) add dropwise a cold solution of [^{15}N]-$NaNO_2$ (0.5 g, 7.1 mmol) in 20 mL water.
2. After 15 min, add this solution of **25** dropwise to a 100-mL cold aqueous solution of 2,4,6-triaminopyrimidine (**23**, 0.89 g, 7.1 mmol) and sodium carbonate (0.76 g, 7.1 mmol).

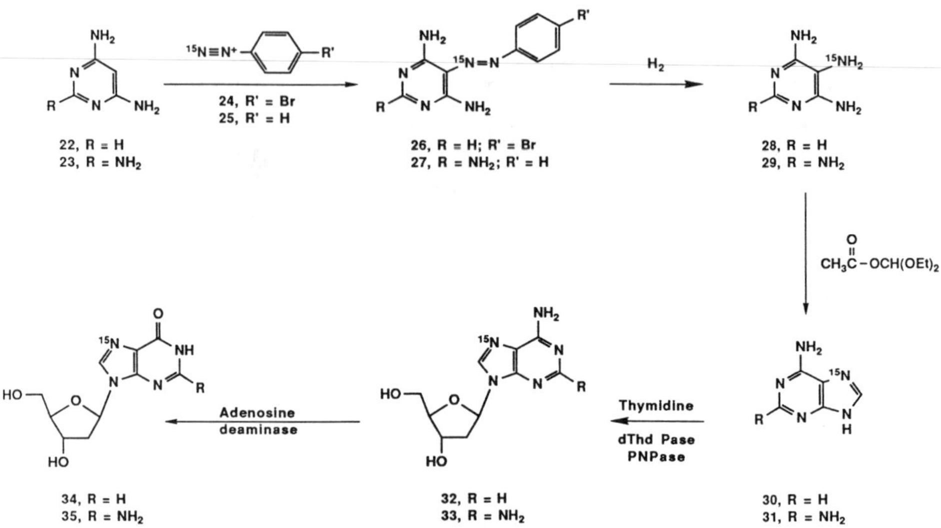

Fig. 5. Synthesis of [7-^{15}N]-adenine (**30**), [7-^{15}N]-2,6-diaminopurine (**31**), [7-^{15}N]-2'-deoxyadenosine (**32**), [7-^{15}N]-2,6-diamino-9-(2-deoxy-β-D-*erythro*-pentofuranosyl)purine (**33**), [7-^{15}N]-2-deoxyinosine (**34**), and [7-^{15}N]-2-deoxyguanosine (**35**).

3. After 2 h collect the product by filtration. You should get about 1.54 g (5.1 mmol, 72%) of the dihydrochloride of **27**, mp 264°C. This material should be pure by tlc and can be used below without further purification.

3.3.3. [5-^{15}N]-Triaminopyrimidine (28)

1. Hydrogenate a mixture of **26** (1.5 g, 5.0 mmol), 150 mL ethanol, and Raney nickel (6 mL of a 50% slurry in water) under 10 psi of H_2 for 90 min.
2. Filter the mixture (celite), concentrate the filtrate, and dissolve the residue in 50 mL water. Extract this aqueous solution with several 50-mL portions of diethyl ether to remove the aniline byproduct.
3. Concentrate the aqueous layer to a solid. You should get about 0.46 g (3.7 mmol, 74%) of **28** (*see* Note 8). This material should be pure by HPLC and can be used below without further purification.

3.3.4. [5-^{15}N]-Tetraaminopyrimidine (29)

1. Hydrogenate a mixture of **27** (1.5 g, 4.9 mmol), 100 mL methanol, and 5% Pd/C (1.0 g of 50% water wet) under 8 psi of H_2 for 4 h.

2. Filter the mixture (celite), concentrate the filtrate and dissolve the residue in 80 mL water. Extract this aqueous solution with several 50 mL portions of diethyl ether to remove the aniline byproduct.
3. Concentrate the aqueous layer to a solid. You should get about 0.88 g (4.5 mmol, 92%) of **29** (*see* Note 9) as the hydrochloride monohydrate. This material should be pure by HPLC and can be used below without further purification.

3.3.5. *[7-^{15}N]-Adenine* (30)

1. To **28** (0.45 g, 3.6 mmol) dissolved in 25 mL of anhydrous DMF add diethoxymethyl acetate (4.1 mL, 25 mmol).
2. After 6 h pour the reaction mixture into a 100-mL portion of cold methanolic HCl (0.1*M*) and allow the mixture to stand at room temperature for 7 h (*see* Note 10).
3. Concentrate the mixture, dissolve the residue in water, treat with decolorizing carbon, and filter.
4. Concentrate the solution and crystallize the residue from methanol by diffusion of ether (Section 3.1.7.). You should get about 0.43 g (2.5 mmol, 69%), of the hydrochloride of **30**, mp > 260°C.

3.3.6. *[7-^{15}N]-2,6-Diaminopurine* (31)

1. To **29** (0.88 g, 4.5 mmol) dissolved in 80 mL anhydrous DMF add diethoxymethyl acetate (5.1 mL, 31 mmol).
2. After 1 h pour the reaction mixture into a 100-mL portion of cold methanolic HCl (0.06*M*) and allow the mixture to stand at 4 °C for 18 h (*see* Note 11).
3. Concentrate the mixture, add a 50 mL portion of water, and concentrate the mixture again. Repeat this four times to remove acetic acid.
4. Crystallize the residue from water. You should get about 0.75 g (3.9 mmol, 86%), of the hydrochloride of **31**, mp > 280°C.

3.3.7. *[7-^{15}N]-2'-Deoxyadenosine* (32)

1. To 70 mL 20 m*M* potassium phosphate add **30** (6.0 mmol), and deoxythymidine (21 mmol). Adjust the pH to 7.2 using KOH, and add thymidine phosphorylase (67 U), and purine nucleoside phosphorylase (120 U).
2. Follow the reaction by HPLC using a gradient of 2–15% acetonitrile: 0.1*M* TEAA. After 3 d at 40°C the reaction should be more than 95% complete (*see* Note 12).
3. Filter the mixture and apply the filtrate to a column of Bio-Rad AG 1X2 hydroxide form resin (2.5 × 26 cm). Elute with a gradient of water to 35% methanol in water.

4. Concentrate the appropriate fractions and crystallize the residue from methanol by diffusion of ether (Section 3.1.7.). You should get about 5.0 mmol (83%) of **32**, mp 185–188°C.

3.3.8. *[7-^{15}N]-2'-Deoxyinosine* (**34**)

1. To 252 mg (1 mmol) of **32** in 10 mL water add 10 U of adenosine deaminase (*see* Note 13).
2. Maintain the reaction at 35–38°C for 30 min, at which time it should be complete (HPLC; 2–15% acetonitrile: 0.1*M* TEAA in 5 min at 4 mL/min).
3. You should get a quantitative yield of **34**.

3.3.9. *[7-^{15}N]-2,6-Diamino-9-(2-Deoxy-β-D-Erythro-Pentofuranosyl)Purine* (**33**)

1. To 100 mL 10 m*M* potassium phosphate add **31** (1.8 mmol), and deoxythymidine (19 mmol). Adjust the pH to 7.4 with KOH and add thymidine phosphorylase (125 units) and purine nucleoside phosphorylase (42 U).
2. Follow the reaction by HPLC using a gradient of 2–15% acetonitrile: 0.1*M* TEAA. After 6 d at 40°C the reaction should be more than 98% complete (*see* Note 14).
3. Filter the mixture and apply the filtrate to a column of Bio-Rad AG 1X2 hydroxide form resin (2.5 × 18 cm). Elute with a gradient of water to 40% methanol in water.
4. Concentrate the appropriate fractions. You should get a near quantitative yield of **33** that may contain trace amounts of a longer retention (HPLC) impurity. If desired, crystallize the product from methanol by ether diffusion (Section 3.1.7.) to get pure **33**, mp 145–148°C.

3.3.10. *[7-^{15}N]-2'-Deoxyguanosine* (**35**)

1. To the product fractions containing **33** from above (1.8 mmol) dissolved in 25 mL water add adenosine deaminase (28 U) (*see* Note 15).
2. Conversion to **35** should be quantitative within 4 d. If desired, crystallize **35** from 1/1 (v/v) methanol/water by ether diffusion (Section 3.1.7.). You should get about 0.41 g of **35** (1.5 mmol, 82% from **31**), mp > 280°C dec.

3.4. Preparation of [5-^{15}N]-5-Amino-4-imidazolecarboxamide (43), [3-^{15}N]-Adenine (46), and [3-^{15}N]-2'-Deoxyadenosine (47)

These syntheses (Fig. 6) use an azo coupling reaction of the β–[^{15}N] diazonium ion **40** with 2-bromoimidazole-4,5-dicarboxylate **(39)**

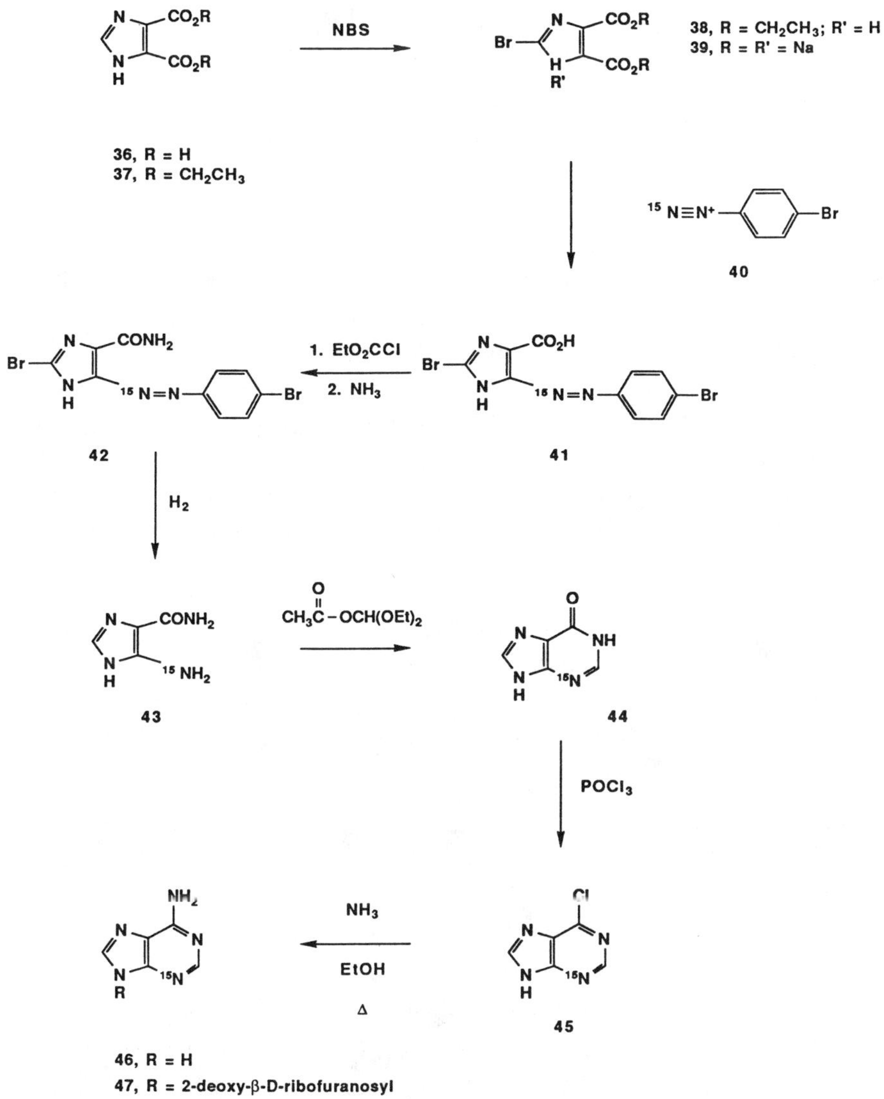

Fig. 6. Synthesis of [3-^{15}N]-6-chloropurine (**45**), [3-^{15}N]-adenine (**46**), and [3-^{15}N]-2'-deoxyadenosine (**47**).

for introduction of the ^{15}N atom *(12)*. The azo coupling occurs with concomitant decarboxylation to give **41**. In this scheme the bromine is used as a protecting group to direct the coupling reaction to the 4/5 position (in the absence of a substituent at the 2 position the coupling occurs there). The bromine is eventually cleaved during hydrogenolysis of the azo linkage, so that no additional steps are required for its removal. Before hydrogenolysis, **41** is converted to the corresponding carboxamide **42** by treatment with cold ethyl chloroformate followed by ammonia. Hydrogenolysis then gives [5-^{15}N]-5-amino-4-imidazolecarboxamide (**43**), which is converted to [3-^{15}N]-hypoxanthine (**44**) by reaction with diethoxymethyl acetate in DMF at reflux. Reaction of **44** with $POCl_3$ gives [3-^{15}N]-6-chloropurine (**45**) which is converted to [3-^{15}N]-adenine by treatment with ethanolic ammonia at 120°C. Conversion to the 2'-deoxynucleoside is affected by the enzymatic transglycosylation reaction described above in Section 3.3. These reactions employ only one protecting group, which is introduced before the ^{15}N atom and which does not require a separate deprotection step, and the only chromatography is the low-resolution ion-exchange column used after the transglycosylation step. In addition, the [5-^{15}N]-5-amino-4-imidazolecarboxamide (**43**) is a versatile intermediate that can be used to prepare other [3-^{15}N]-labeled bases and nucleosides *(7,29–31)*.

3.4.1. Ethyl Imidazole-4,5-Dicarboxylate (**37**)

1. Reflux a mixture of imidazole-4,5-dicarboxylic acid (**36**, 5 g, 32 mmol), 250 mL ethanol, and 30 mL H_2SO_4 for 11 h, until a homogeneous solution is attained.
2. Concentrate the mixture to remove most of the ethanol and add 300 mL H_2O. Add $NaHCO_3$ to neutralize the solution and extract using four 200-mL portions of ethyl acetate.
3. Dry the combined organic layers over anhydrous $MgSO_4$, filter, and concentrate the filtrate to a white solid.
4. Dissolve the solid in ethyl acetate and precipitate by dropwise addition to petroleum ether. Filter the precipitate and wash it with petroleum ether. You should get about 5.5 g (26 mmol, 82%) of **37**.

3.4.2. Ethyl 2-Bromoimidazole-4,5-Dicarboxylate (**38**)

1. To a solution of **37** (3.1 g, 15 mmol) in DMF (50 mL) add in one portion a solution of NBS (2.9 g, 16 mmol) in DMF (50 mL). Stir the mixture at 20°C for 1 h, and concentrate to dryness.

2. Dissolve the residue in ethyl acetate (300 mL) and wash with brine, saturated aqueous Na_2SO_3, and brine. Dry the organic layer over $MgSO_4$ and concentrate to dryness.
3. Dissolve the residue in ethyl acetate containing a small amount of methanol and precipitate by dropwise addition to n-pentane. Filter the precipitate and wash it with n-pentane. You should get about 3.7 g (12.7 mmol, 86%) of **38** (*see* Note 16).

*3.4.3. Sodium 2-Bromoimidazole-4,5-Dicarboxylate **(39)***

1. To a solution of **38** (2.3 g, 8.0 mmol) in 300 mL ethanol add 25 mL of a 10% solution of NaOH (2.6 g, 64 mmol) and reflux for 2 h with vigorous stirring.
2. Cool the mixture and filter. You should get a quantitative yield of **39**. This material should be pure by HPLC and can be used for the azo-coupling reaction without further purification.

*3.4.4. 5-(4-Bromophenyl)Azo-2-Bromoimidazole-4-Carboxylic Acid **(41)***

1. To a cold suspension of 4-bromoaniline (1.25 g, 7.3 mmol) in 29 mL 10% HCl add dropwise a cold solution of [^{15}N]-$NaNO_2$ (0.5 g, 7.1 mmol) in 10 mL of water.
2. After 20 min, add this solution of **40** dropwise to a 150-mL cold aqueous solution of **39** (2.4 g, 8 mmol) and sodium carbonate (8.4 g, 79 mmol).
3. After 24 h, filter the mixture and acidify the solution with concentrated HCl to precipitate **41**. Filter the precipitate and wash it with cold water. You should get about 2.3 g (6.1 mmol, 85% relative to the [^{15}N]-$NaNO_2$) of **41**, mp 198°C dec.

*3.4.5. 5-(4-Bromophenyl)Azo-2-Bromo-4-Imidazolecarboxamide **(42)***

1. To a mixture of **41** (2.3 g, 6.1 mmol) and TEA (1.7 mL, 12 mmol) in THF (100 mL) cooled to –10°C add dropwise ice-cooled ethyl chloroformate (1.2 mL, 12 mmol).
2. Stir for 20 min at –10°C and add excess anhydrous ammonia in THF (100 mL).
3. Stir the mixture for 3 h at 0°C and concentrate to dryness.
4. Dissolve the residue in 100 mL water, allow the solution to stand overnight at 0°C to crystallize, and filter. You should get about 2.2 g (5.9 mmol, 97%) of **42**, mp 212°C dec.

3.4.6. [5-^{15}N]-5-Amino-4-Imidazolecarboxamide (**43**)

1. Hydrogenate a mixture of **42** (1.9 g, 5.0 mmol), 100 mL of methanol, 1.4 g (25 mmol) of KOH and 5% Pd/C (0.94 g of 50% water wet) under 8 psi of H_2 for 5 h.
2. Filter the mixture (celite), concentrate the filtrate and dissolve the residue in 100 mL water. Extract this aqueous solution with several 50 mL portions of diethyl ether to remove the aniline byproduct.
3. Concentrate the aqueous layer and extract the residue with absolute ethanol. Concentrate the extract to a solid. You should get a quantitative yield of **43**, contaminated with small amounts of KOH and KBr. This product can be used below without further purification.

3.4.7. [3-^{15}N]-Hypoxanthine (**44**)

1. To a solution of crude **43** in 80 mL anhydrous DMF add diethoxymethyl acetate (4.1 mL, 25 mmol), reflux for 3 h, and concentrate.
2. Suspend the residue in 100 mL methanol, acidify with HCl, and stir overnight at room temperature.
3. Concentrate the mixture to a solid. This crude **44** can be used in the next step without further purification.

3.4.8. [3-^{15}N]-6-Chloropurine (**45**)

1. Reflux a mixture of crude **44**, dimethylaniline (1.7 mL, 13 mmol) and freshly distilled phosphorus oxychloride (50 mL, 540 mmol) for 1/2 h and allow to stand overnight at room temperature.
2. Concentrate the mixture to a viscous syrup and add 250 g ice. Adjust the pH to 12–14 with 10*N* NaOH.
3. Extract the alkaline solution in the cold with diethyl ether until the ether extract is colorless. Acidify the chilled aqueous solution with concentrated HCl and continuously extract this solution with diethyl ether (250 mL) for 48 h.
4. Concentrate the ether to dryness. You should get about 0.56 g (3.6 mmol, 72% from **42**) of **45**.

3.4.9. [3-^{15}N]-Adenine (**46**)

1. Saturate with ammonia a solution of **45** (0.56 g, 3.6 mmol) in 40 mL ethanol at 0°C in a steel bomb and then heat overnight at 120°C.
2. Cool the solution, concentrate to dryness, and crystallize the residue from methanol by diffusion of ether (Section 3.1.7.). You should get about 0.5 g of **46**, contaminated with a small amount of NH_4Cl. This can be used below without further purification.

3.4.10. [3-^{15}N]-2'-Deoxyadenosine (47)

1. To 72 mL of 10 m*M* potassium phosphate add the crude **46** from above and deoxythymidine (5.2 g, 22 mmol). Adjust the pH to 7.4 with KOH, and add thymidine phosphorylase (180 U) and purine nucleoside phosphorylase (122 U).
2. Follow the reaction by HPLC using a gradient of 2–15% acetonitrile: 0.1*M* TEAA. After 3 d the reaction should be more than 98% complete (*see* Note 17).
3. Filter the mixture and apply the filtrate to a column of Bio-Rad AG 1X2 hydroxide form resin (2.5 × 25 cm). Elute with a gradient of water to 30% methanol in water.
4. Concentrate the appropriate fractions to dryness. You should get about 0.6 g (2.3 mmol, 64% from **45**, 46% from **42**) of pure **47**.

3.5. Oligonucleotide Synthesis

The incorporation into DNA fragments of the above [^{15}N]-labeled 2'-deoxynucleosides can be carried out by the standard methods of oligonucleotide synthesis, which are described in detail in Chapters 2–4 of vol. 20 in this series. Both the phosphoramidite *(32)* and H-phosphonate *(33–35)* methods have been used to prepare [^{15}N]-labeled DNA fragments *(9,10,14–17)*. The H-phosphonate method, however, allows recovery and reuse of monomers, which is a distinct advantage for these valuable [^{15}N]-labeled monomers *(36)*. Further, the H-phosphonate method can be used successfully with a tentagel polystyrene/polyethylene glycol (PEG-PS) support *(36)*. The PEG-PS support allows a 2–3-fold increase in scale for a given size reactor, relative to the more commonly used controlled-pore-glass (CPG) supports. Using this support it is possible to do a 30–35 µmole scale synthesis in a standard 10–15 µmole cartridge, so that sufficient material for most ^{15}N NMR experiments can be obtained from one or two syntheses.

A capping step should be employed in these syntheses, rather than simply to rely on self-capping by the pivaloyl or adamantoyl chlorides used as the condensing agent. Inexpensive H-phosphonates, such as isopropyl- or cyanoethyl-H-phosphonates, can be used for capping *(37,38)*. In addition, devices to automate the trityl assay and to regulate better the amount of monomer used in the coupling steps are especially useful for these relatively expensive syntheses *(36)*. Such devices can be constructed using glass columns with optical liquid-level sensors that can control the appropriate solenoid valves in the synthesizer, which must be programmable. However, because the synthesizers vary so much in configuration, it is not possible to give a specific design that would be generally applicable. Any good electronics shop should be able to construct such a device, although installation in the synthesizer may affect the warranty!

3.6. Oligonucleotide Deprotection and Purification

Analytical HPLC using either Waters C-18 Novapak cartridges in an RCM 100 or a Beckman C-3 Ultrapore column (4.6 mm × 7.5 cm) works well. In both cases 5 min gradients at 4 mL/min (C-18) or 2 mL/min (C-3) can be used to affect rapid analysis. The [^{15}N]-labeled oligonucleotides should be purified by HPLC both before and after detritylation. The first purification cleanly separates tritylated product from untritylated failure sequences and also should be used to fractionate carefully the tritylated product. There are always a few impurities that are actually better resolved at this stage than they are after detritylation. Further, since any product that is inadvertently detritylated before this purification will be lost, it is important to use extra care in handling the crude material. For this reason the ammonia deprotection step is best carried out at room temperature for 2–3 d, rather than overnight at 65°C. The ammonia solution is then concentrated to remove most of the ammonia, so that it can be lyophilized. Simple evaporation of the ammonia solution to dryness is likely to result in detritylation.

After lyophilization, the residue should be dissolved in a minimal amount of 0.1*M* triethylammonium acetate (TEAA) buffer and filtered. Use of water rather than buffer is likely to result in detritylation. For the first HPLC purification a Waters C-18 reversed-phase column (15 × 190 mm steel column or a 25 × 100 mm Radial-Pak cartridge) with gradients of 2–40% or 2–50% acetonitrile: 0.1*M* TEAA in 45 min at a flow rate of 4 mL/min will provide clean separation of trityl vs nontrityl and allow fractionation of the tritylated material. The fractions should then be diluted with an equal volume of water, lyophilized, and detritylated using 0.1*M* acetic acid, pH ~3.2, for 20–40 min *(39)*. If the pH is >3.5, simply add more 0.1*M* acetic acid.

Alternatively, detritylation can be affected simply by evaporation to dryness. However, use of 0.1*M* acetic acid allows the detritylation to be carried out under controlled conditions where depurination can be minimized. The detritylation should be followed by HPLC, stopped by addition of a few drops of aqueous ammonia when complete, and the solution then lyophilized. There is noticeably less depurination with 0.1*M* acetic acid than there is with 80% acetic acid. During the detritylation the 0.1*M* acetic acid solution will become cloudy because of precipitation of the 4,4'-dimethoxytritanol, which is soluble in 80% acetic acid. If there is any doubt about the homogeneity of the fractions from the first purification, they are best detritylated separately since, for most impurities, the resolution will be better after detritylation.

After lyophilization, the residue is again dissolved in buffer and filtered to remove the 4,4'-dimethoxytritanol. For this purification a Beckman Ultrapore C-3 reversed-phase column (10 mm × 25 cm) with a gradient of 2–20% acetonitrile: 0.1*M* TEAA in 45 min at a flow rate of 2 mL/min is usually preferable, although the above C-18 column also can be used. In some cases, impurities not resolved using TEAA buffer will be resolved by additional chromatography on either column using 0.1*M* ammonium acetate (AA) buffer. The combined product fractions are then concentrated and lyophilized.

To remove all traces of triethylamine and acetic acid the lyophilized product fractions should be chromatographed using a C-18 column with a gradient of 0–20% acetonitrile: 0.1*M* ammonium bicarbonate (ABC), which is far more volatile than is TEAA or AA. In some cases additional minor impurities may be observed and removed during this chromatography as well, although in general ABC is a poor buffer for HPLC. After lyophilization the product is converted to the sodium form by ion exchange using sodium form Bio-Rad AG50W-X4 resin, and again lyophilized. The homogeneity of the product should be checked on both the C3 and C18 analytical columns. Finally, the product should be characterized by analysis of the ratio of 2'-deoxynucleosides produced upon enzymatic degradation of a 1 A_{260} U sample using a combination of a phosphodiesterase and a phosphatase. Venom phosphodiesterase and alkaline phosphatase are commonly used *(39)*, although other combinations may be equally effective. The relative amounts of the 2'-deoxynucleosides obtained can be determined by integration of the corresponding peaks separated by HPLC on the C-18 Novapak column using a gradient of 2–15% acetonitrile: 0.1*M* TEAA in 5 min at 4 mL/min.

4. Discussion

The syntheses of the [^{15}N]-labeled 2'-deoxynucleosides described above were designed to minimize use of protecting groups and to avoid tedious chromatographic separations in order to maximize overall yields. Further, the ^{15}N sources are relatively inexpensive. Thus, these [^{15}N]-labeled monomers can be obtained in sufficient quantity, and at reasonable cost, for use in oligonucleotide synthesis. They are, however, significantly more expensive than unlabeled monomers. The ability to recover and reuse monomers in the H-phosphonate method is therefore an important consideration for synthesis of [^{15}N]-labeled oligonucleotides. Finally, it is worth noting that these procedures, with some modifications, are all applicable to synthesis of the corresponding [^{15}N]-labeled ribonucleosides.

5. Notes

1. $LiAlH_4$ should be handled with care, particularly during the hydrolysis in Section 3.1.1., step 2.
2. The diazotization reaction will fail if oxygen is present.
3. Note that in this reaction the nucleoside **3** is used in excess.
4. In this reaction $NaIO_4$ reacts with a catalytic amount of RuO_2 to generate the RuO_4 that oxidizes the benzyl methylene group. There is always significant loss of the resultant benzoyl group during the oxidation, so that after step (ii) there will be two main product spots on TLC.
5. The 4 Å sieves are used to trap HBr to minimize depurination.
6. If excess TBAF is used the product will not crystallize and chromatography will be required.
7. The use of 0.1*M* TEAA buffer is not essential, even plain water can be used, so long as the pH is not too far from neutrality.
8. Solutions of the product (**28**) will darken rapidly. It should be converted to **30** without delay.
9. Solutions of the product (**29**) will darken rapidly. It should be converted to **31** without delay.
10. The methanolic HCl treatment is to hydrolyze putative ethoxyethylidene derivatives of **30**.
11. The methanolic HCl treatment is to hydrolyze putative ethoxyethylidene derivatives of **31**.

12. The yield will not necessarily increase with time indefinitely. Therefore the reaction should be carefully monitored by HPLC.
13. Ensure that the pH is not too far from neutrality.
14. The yield will not necessarily increase with time indefinitely. Therefore the reaction should be carefully monitored by HPLC.
15. Ensure that the pH is not too far from neutrality.
16. The diester, rather than the diacid, is used in the bromination because the latter gives a complex mixture of products.
17. The yield will not necessarily increase with time indefinitely. Therefore the reaction should be carefully monitored by HPLC.

References

1. DeGraw, J. I. and Lawson, J. A. (1978) Thymidine-6-^{13}C-a,a,a,-d_3-1,3-$^{15}N_2$, in *Nucleic Acid Chemistry* (Townsend, L. B. and Tipson, R. S., eds.), Wiley Interscience, New York, pp. 921–926.
2. Poulter, C. D. and Livingston, C. L. (1979) [3-^{15}N]-2',3',5'-Tri-O-Benzoyluridine. Detection of hydrogen bonding in A-U base pairs by ^{15}N NMR. *Tetrahedron Lett.* **9,** 755–758.
3. Niu, C.-H. (1984) Synthesis of [4-$^{15}NH_2$]- and [1,3-$^{15}N_2$]-cytidine derivatives for use in NMR-monitored binding tests. *Anal. Biochem.* **139,** 404–407.
4. Leonard, N. J. and Henderson, T. R. (1975) Purine ring rearrangements leading to the development of cytokinin activity. Mechanism of the rearrangement of 3-benzyladenine to N^6-benzyladenine. *J. Am. Chem. Soc.* **97,** 4990–4999.
5. Barrio, M. D. C. G., Scopes, D. I. C., Holtwick, J. B., and Leonard, N. J. (1981) Syntheses of all singly labeled [^{15}N] adenines: Mass spectral fragmentation of adenine. *Proc. Natl. Acad. Sci. USA* **78,** 3986–3988.
6. Sethi, S. K., Gupta, S. P., Jenkins, E. E., Whitehead, C. W., Townsend, L. B., and McCloskey, J. A. (1982) Mass spectrometry of nucleic acid constituents. Electron ionization spectra of selectively labeled adenines. *J. Am. Chem. Soc.* **104,** 3349–3353.
7. Golding, B. T., Slaich, P. K., and Watson, W. P. (1986) Conversion of 'AICA-riboside' into [^{15}N]-guanosines. *J. Chem. Soc., Chem. Commun.* 901–902.
8. Gao, X. and Jones, R. A. (1987) Nitrogen-15-labeled deoxynucleosides. Synthesis of [6-^{15}N]- and [1-^{15}N] deoxyadenosines from deoxyadenosine. *J. Am. Chem. Soc.* **109,** 1275–1278.
9. Kupferschmitt, G., Schmidt, J., Schmidt, T., Fera, B., Buck, R., and Rüterjans, H. (1987) ^{15}N labeling of oligodeoxynucleotides for NMR studies of DNA-ligand interactions. *Nucleic Acids Res.* **15,** 6225–6241.
10. Massefski, W., Jr., Redfield, A., Sarma, U. D., Bannerji, A., and Roy, S. (1990) [7-^{15}N]Guanosine-labeled oligonucleotides as nuclear magnetic resonance probes for protein-nucleic acid interaction in the major groove. *J. Am. Chem. Soc.* **112,** 5350–5351.
11. Gaffney, B. L., Kung, P.-P., and Jones, R. A. (1990) Nitrogen-15-labeled deoxynucleosides. 2. Synthesis of [7-^{15}N] labeled deoxyadenosine, deoxyguanosine, and related deoxynucleosides. *J. Am. Chem. Soc.* **112,** 6748–6749.

12. Rhee, Y. S. and Jones, R. A. (1990) Nitrogen-15-labeled deoxynucleosides. 3. Synthesis of [3-^{15}N] labeled 2'-deoxyadenosine. *J. Am. Chem. Soc.* **112,** 8174–8175.
13. Goswami, B. and Jones, R. A. (1991) Nitrogen-15-labeled deoxynucleosides. 4. synthesis of [1-^{15}N]- and [2-^{15}N]-labeled 2'-deoxyguanosines. *J. Am. Chem. Soc.* **113,** 644–647.
14. Gao, X. and Jones, R. A. (1987) Nitrogen-15-labeled oligodeoxynucleotides. Characterization by ^{15}N NMR of d[CGTACG] containing ^{15}N6- or ^{15}N1-labeled deoxyadenosine. *J. Am. Chem. Soc.* **109,** 3169–3171.
15. Wang, C., Gao, X., and Jones, R. A. (1991) Nitrogen-15-labeled oligodeoxynucleotides. 2. Solvent isotope effects on the chemical shift of the adenine N1 in an A/T base pair. *J. Am. Chem. Soc.* **113,** 1448–1450.
16. Wang, C., Gao, H., Gaffney, B. L., and Jones, R. A. (1991) Nitrogen-15-labeled oligodeoxynucleotides. 3. Protonation of the adenine N1 in the A/C and A/G mispairs of the duplexes {d[CG(^{15}N^{1})AGAATTCCCG]}$_2$ and {d[CGGGAATTC(^{15}N^{1})ACG]}$_2$. *J. Am. Chem. Soc.* **113,** 5486–5488.
17. Gaffney, B. L., Wang, C., and Jones, R. A. (1992) Nitrogen-15-labeled oligodeoxynucleotides. 4. Tetraplex formation of d[G(^{15}N^{7})GTTTTGG] and d[T(^{15}N^{7}GGGT] monitored by ^{1}H detected ^{15}N NMR. *J. Am. Chem. Soc.* **114,** 4041–4050.
18. Griffey, R. H., Poulter, C. D., Yamaizumi, Z., Nishimura, S., and Hurd, R. E. (1982) ^{1}H NMR studies of ^{15}N-labeled *Escherichia coli* tRNA^{fMet}. Use of $^{1}J_{1H\text{-}15N}$ couplings to identify imino resonances of uridine-related bases. *J. Am. Chem. Soc.* **104,** 5810–5811.
19. Roy, S., Papastavros, M. Z., Sanchez, V., and Redfield, A. G. (1984) Nitrogen-15-labeled yeast tRNA^{Phe}: Double and two-dimensional heteronuclear NMR of guanosine and uracil ring NH groups. *Biochemistry* **23,** 4395–4400.
20. Gewirth, D. T., Arbo, S. R., Leontis, N. B., and Moore, P. B. (1987) Secondary structure of 5S RNA: NMR experiments on RNA molecules partially labeled with nitrogen-15. *Biochemistry* **26,** 5213–5220.
21. Davis, D. R., Yamaizumi, Z., Nishimura, S., and Poulter, C. D. (1989) ^{15}N-labeled 5S RNA. Identification of uridine base pairs in *Escherichia coli* 5S RNA by ^{1}H-^{15}N multiple quantum NMR. *Biochemistry* **28,** 4105–4108.
22. Davis, D. R. and Poulter, C. D. (1991) ^{1}H-^{15}N NMR studies of *Escherichia coli* tRNA^{Phe} from *hisT* mutants: A structural role for pseudouridine. *Biochemistry* **30,** 4223–4231.
23. Robins, M. J. and Trip, E. M. (1973) Sugar-modified N^6-(3-methyl-2-butenyl)adenosine derivatives, N^6-benzyl analogs, and cytokinin-related nucleosides containing sulfur or formycin. *Biochemistry* **12,** 2179–2187.
24. Ueda, T., Miura, K., and Kasai, T. (1978) Synthesis of 6-thioguanine and 2,6-diaminopurine nucleosides and nucleotides from adenine counterparts *via* a facile rearrangement in the base portion. *Chem. Pharm. Bull.* **26,** 2122–2127.

25. MacCoss, M., Ryu, E. K., White, R. S., and Last, R. L. (1980) A new synthetic use of nucleoside N^1-oxides. *J. Org. Chem.* **45,** 788–794.
26. Krenitsky, T. A., Koszalka, G. W., and Tuttle, J. V. (1981) Purine nucleoside synthesis, an efficient method employing nucleoside phosphorylases. *Biochemistry* **20,** 3615–3621.
27. Krenitsky, T. A., Rideout, J. L., Chao, E. Y., Koszalka, G. W., Gurney, F., Crouch, R. C., Cohn, N. K., Wolberg, G., and Vinegar, R. (1986) Imidazo[4,5-*c*]pyridines (3-deazapurines) and their nucleosides as immunosuppressive and antiinflammatory agents. *J. Med. Chem.* **29,** 138–143.
28. Krenitsky, T. A., Hall, W. W., Selph, J. L., Truax, J. F., and Vinegar, R. (1989) Nucleosides of azathioprine and thiamiprine as antiarthritics. *J. Med. Chem.* **32,** 1471–1475.
29. Chern, J.-W. and Townsend, L. B. (1985) A novel and efficient synthesis of the naturally occurring nucleoside doridosine. *Tetrahedron Lett.* **26,** 6419–6422.
30. Groziak, M. P. and Townsend, L. B. (1986) A new and efficient synthesis of guanosine. *J. Org. Chem.* **51,** 1277–1282.
31. Chern, J.-W., Lin, G.-S., Chen, C.-S., and Townsend, L. B. (1991) Nucleosides. 3. Reactions of AICA-riboside with isothiocyanates. A convenient synthesis of isoguanosine and xanthosine derivatives. *J. Org. Chem.* **56,** 4213–4218.
32. Matteucci, M. D. and Caruthers, M. H. (1981) Synthesis of deoxyoligonucleotides on a polymer support. *J. Am. Chem. Soc.* **103,** 3185–3191.
33. Froehler, B. C., Ng, P. G., and Matteucci, M. D. (1986) Synthesis of DNA via deoxynucleoside H-phosphonate intermediates. *Nucleic Acids Res.* **14,** 5399–5407.
34. Froehler, B. C. and Mateucci, M. D. (1986) Nucleoside H-phosphonates: Valuable intermediates in the synthesis of deoxyoligonucleotides. *Tetrahedron Lett.* **27,** 469–472.
35. Garegg, P. J., Lindh, I., Regberg, T., Stawínski, J., and Strömberg, R. (1986) Nucleoside H-phosphonates. III. Chemical synthesis of oligodeoxyribonucleotides by the hydrogen phosphonate approach. *Tetrahedron Lett.* **27,** 4051–4054.
36. Gao, H., Gaffney, B. L., and Jones, R. A. (1991) H-phosphonate oligonucleotide synthesis on a poylethylene glycol/polystyrene copolymer. *Tetrahedron Lett.* **32,** 5477–5480.
37. Andrus, A., Efcavitch, J. W., McBride, L. J., and Giusti, B. (1988) Novel activating and capping reagents for improved hydrogen-phosphonate DNA synthesis. *Tetrahedron Lett.* **29,** 861–864.
38. Gaffney, B. L. and Jones, R. A. (1988) Large-scale oligonucleotide synthesis by the H-phosphonate method. *Tetrahedron Lett.* **29,** 2619–2622.
39. Gaffney, B. L. and Jones, R. A. (1989) Thermodynamic comparison of the base pairs formed by the carcinogenic lesion O^6-methylguanine with reference both to Watson-Crick pairs and to mismatched pairs. *Biochemistry* **28,** 5881–5889.

CHAPTER 9

Analysis and Purification of Synthetic Oligonucleotides by High-Performance Liquid Chromatography

William J. Warren and George Vella

1. Introduction

Synthetic oligonucleotides are important in a wide variety of applications ranging from use as hybridization probes *(1)*, primers for DNA sequencing and the polymerase chain reaction *(2,3)*, to utilization as potential therapeutics in antisense and related technology investigations *(4)*. In brief, their synthesis is most frequently performed via a series of systematic reactions that result in the stepwise addition of specific nucleotides [(protected at their 5' hydroxy end with a dimethoxytrityl group (DMT)] to the solid phase support containing the growing chain. Variation in choice of incoming nucleotide monomers as well as modifications during or after the standard phosphodiester synthesis protocol can result in a final product containing atypical base, sugar, and/or phosphate backbone structures. Examples of such modifications may include substitution of a sulfur for an oxygen atom on the phosphate backbone creating phosphorothioate DNA or substitution of a methyl or borane group in lieu of oxygen to create methyl-phosphonate or boranophosphonate DNA. These and other modifications from conventional phosphodiester DNA are currently being investigated in antisense-related therapeutic studies as ways to increase cellular uptake, biological activity, and survival in both in vitro and in vivo experiments *(4)*. A more complete discussion con-

From: *Methods in Molecular Biology, Vol. 26: Protocols for Oligonucleotide Conjugates*
Edited by: S. Agrawal

cerning the synthesis of phosphodiester as well as modified oligonucleotide species are addressed in volume 20 of this series.

To keep up with the ever increasing demand for product in light of the ubiquitous use of PCR and interest in these compounds as pharmaceutical agents, automated solid-phase DNA synthesizers capable of producing from 0.05 to hundreds of micromoles of product in one synthesis have been developed. These instruments possess high coupling-efficiency capability with, in some instances, simultaneous, multi-strand synthesis flexibility. Yet, despite constant technological improvements, coupling efficiency in each synthesis cycle still remains greater than 98% but below 100%. In addition, depurination and strand cleavage result in "failure sequences" of shorter total length that contaminate the desired full-length synthetic product and reduce overall yield *(5)*. The amount of these shorter length product contaminants contained in crude synthesis reaction mixtures is proportional to the length of the desired product synthesized and related to overall yield as depicted in Fig. 1.

For some applications, such as site-directed mutagenesis or X-ray crystallographic experiments, the analysis of crude reaction mixtures and subsequent purification of the full-length oligonucleotide product from the contaminants contained in the synthesis mixture are mandatory. Traditionally, polyacrylamide slab gel electrophoresis has been employed for both analysis and purification needs *(6)*. Although high resolution separations of multiple samples can be achieved with this methodology, this technique has limitations. Post-electrophoresis visualization of the DNA often requires the use of radiolabeling or staining techniques if enhanced detection sensitivity is required. As a result, quantitation of the amount of full-length product from the smaller failure sequences is only semi-quantitative making direct determination of the average coupling efficiency impossible without the use of off-line gel scanning devices. Furthermore, product isolation involves post-electrophoresis extraction from the gel and often results in poor product recovery and contamination of the oligonucleotide product with gel-matrix constituents that can be inhibitory to subsequent in vitro applications *(7)*. Of increased concern is that conventional slab-gel electrophoresis cannot readily purify the multimilligram to gram quantities of material required for therapeutic investigations.

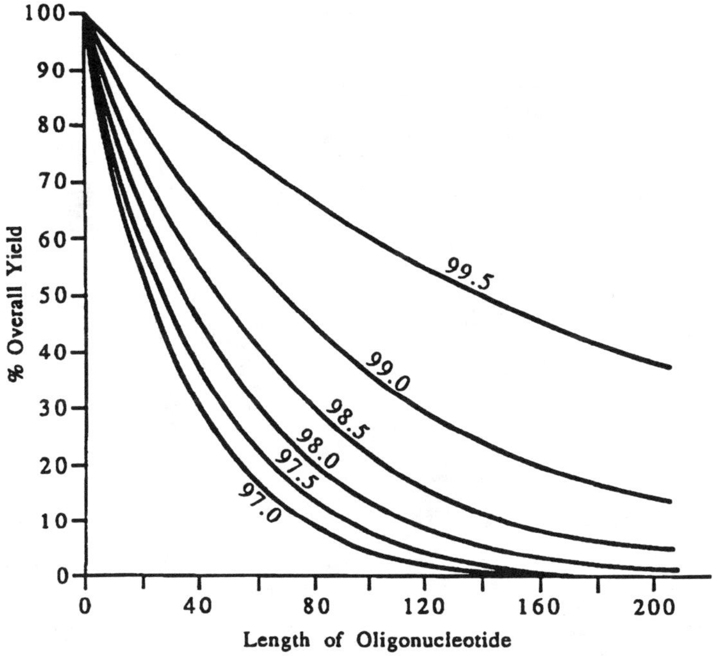

Fig. 1. Synthetic oligonucleotide length compared to theoretical yield at various coupling efficiencies.

The use of high-performance liquid chromatography (HPLC) to separate, quantitate, and isolate synthetic oligonucleotides is becoming commonplace and addresses many of the shortcomings associated with traditional gel electrophoresis, especially those of large scale isolation. More recently, the technique of capillary electrophoresis (CE) has been commercialized and is currently being optimized for the separation and quantitation of single- and double-stranded DNA. Combined, HPLC, capillary electrophoresis, and gel electrophoresis provide a formidable array of analytical and preparative techniques for synthetic oligonucleotide as well as other nucleic acid species. As depicted in Fig. 2, some or all of these techniques may be used at various stages of post-DNA synthesis work-up.

This chapter primarily details both the principles and practice of HPLC for the characterization and isolation of synthetic DNA. In addition, several post synthesis sample manipulation protocols (e.g.,

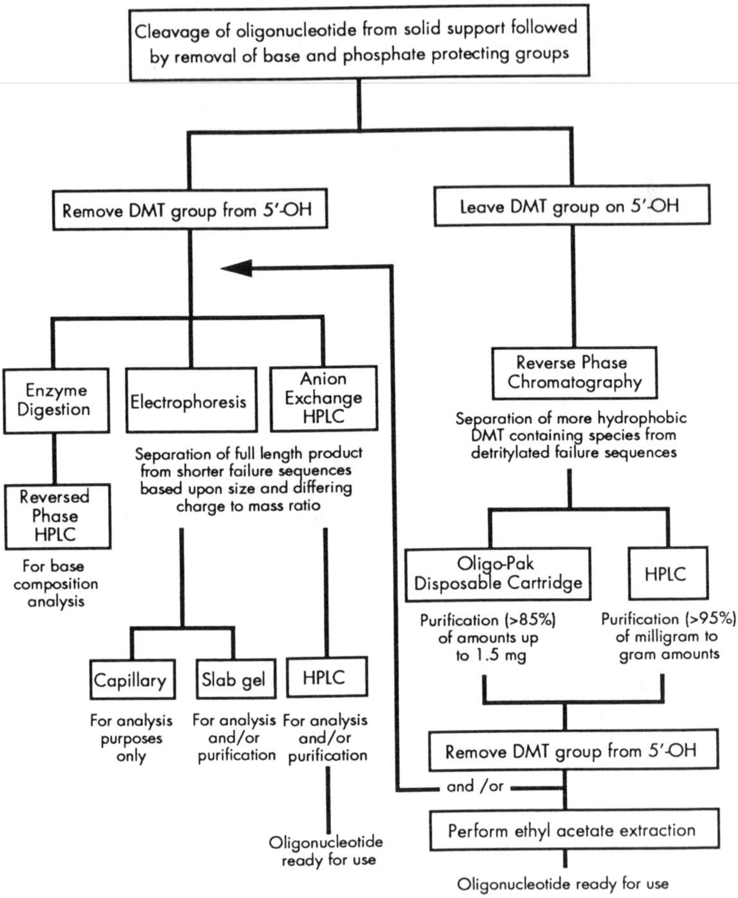

Fig. 2. Analysis and purification scheme for synthetic oligonucleotides.

removal and deprotection of oligonucleotide from solid-phase support) routinely used prior to or after chromatography are included.

2. Post Synthesis Sample Treatment for HPLC

2.1. Cleavage and Deprotection Protocol

Once a synthesis is complete, the oligomer must be cleaved from the solid-phase support (e.g., controlled pore glass). The base and phosphate groups must then be fully deprotected prior to the utilization of any subsequent HPLC or capillary electrophoresis technique. The principles behind oligonucleotide cleavage and deprotection are detailed in Chapter 1 of this volume and also in Chapters 2–4 of vol.

20. In brief, cleavage and deprotection can be performed in one step as described below. The same general protocol can also be used with larger scale syntheses provided that proportionally greater amounts of ammonium hydroxide and triethylamine are used.

2.1.1. Materials and Methods

1. Transfer the solid phase support (e.g., controlled pore glass) from the synthesis column to an appropriately sized screw cap tube.
2. Add 1–2 mL of fresh concentrated ammonium hydroxide for <1.0 µmol scale synthesis or 6 mL for 15-µmol scale synthesis. Ensure that the screw-cap tube is well sealed. Use of concentrated ammonium hydroxide (i.e., 30%) is *essential* to ensure complete cleavage and deprotection. To ensure a concentrated solution, refrigerate the ammonium hydroxide after initial use and keep tightly capped. Purchase the concentrated ammonium hydroxide in small quantities only.
3. Deprotection of oligonucleotides synthesized using standard beta-cyanoethyl phosphoramidites is accomplished by either heating the tube at 55°C for 5 h or by letting the reaction proceed at room temperature for 24 h. If the oligomer will be purified by reverse-phase chromatography, the highest possible yield will result by using the 24 h room temperature treatment since this minimizes any detritylation caused by heat. Use of novel base protected nucleotides (e.g., Expedite™ Amidites, Millipore, Bedford, MA) substantially reduces the time required for deprotection because of the increased lability of the exogenous amine protecting groups to ammonium hydroxide as described in Chapter 1.
4. Record the volume of solution before and after cleavage/deprotection. If there is a significant loss (>20%) owing to volatile ammonium hydroxide leakage from the tube, the reactions may not be complete. In such cases, additional 30% ammonium hydroxide should be added and the incubation continued.
5. Decant the DNA-containing supernatant into a test tube or flask suited for sample concentration. If the DMT group was left on the 5'-OH end of the oligomer following synthesis completion, as required for product purification by reverse-phase chromatography *(see below),* add 30 µL of triethylamine to the supernatant (or enough to produce a 1% solution). This will help maintain the basic pH of the solution during evaporation and minimize detritylation of the oligomer that would result in a reduced product yield or inaccurate product quantitation using reverse-phase HPLC *(see below).*
6. Cool the sample on ice and then dry the solution at room temperature using a rotary evaporator or a Savant (Farmingdale, NY) SpeedVac™. A pellet should be visible that can be stored at –20°C until needed.

2.2. Estimation of Total Yield from Synthesis Support

An estimation of the total crude oligonucleotide yield (i.e., material containing both full-length as well as shorter "failure sequences") can be determined spectrophotometrically for syntheses up to 15 µmol as described below.

2.2.1. Materials and Methods

1. Dissolve the pellet of crude oligomer in 1 mL of Milli-Q® water (Millipore) or distilled water.
2. Dilute an aliquot of this stock solution as described. This will bring the DNA absorbance readings into the linear range of the spectrophotometer:
   0.2 µmol synthesis: 10 µL to 1 mL of water
   1.0 µmol synthesis: 5 µL to 1 mL of water
   15.0 µmol synthesis: 1 µL to 2 mL of water
3. Read the absorbance at 260 nm against a water blank.
4. Calculate the total number of A_{260} U contained as follows:
   0.2 µmol synthesis: Total A_{260} U = Absorbance read X100
   1.0 µmol synthesis: Total A_{260} U = Absorbance read X200
   15.0 µmol synthesis: Total A_{260} U = Absorbance read X2000
   1 A_{260} U = Approx 33 µg of oligonucleotide

2.3. Product Detritylation and Extraction After Removal from Solid-Phase Support

Removal of the DMT group from the 5'-OH of the deoxyribose sugar can be performed in an automated fashion while the DNA is still on the solid-phase support. This is usually done when only electrophoresis or anion-exchange HPLC is to be used for purification and/or analysis purposes. Alternatively, if oligonucleotide analysis or purification by reverse-phase HPLC is desired, detritylation of the desired DNA product is performed following the chromatography. The following detritylation procedure is recommended for this application or prior to electrophoresis or anion-exchange HPLC of samples that have not been previously detritylated on the solid-phase support.

2.3.1. Materials and Methods

1. The fraction(s) containing the DMT-protected product should be brought to dryness using a rotary evaporator or a Savant SpeedVac.

2. Resuspend the residue in 100 µL of 80% acetic acid in Milli-Q water or distilled water. After a 20-min incubation at room temperature, the sample should be again brought to dryness.
3. Resuspend pellet in 0.5 mL of 1% triethylamine (TEA) in Milli-Q water or distilled water, then concentrate to dryness. The TEA addition neutralizes the acetic acid thus minimizing the exposure of the oligomer to acid.
4. If required, the oligonucleotide-free DMT species can be removed from the sample using the following liquid–liquid extraction protocol. The dried detritylated sample, from step 3, is first resuspended in 300 µL of Milli-Q water or distilled water.
5. 300 µL of ethyl acetate is then added and the sample is vortexed. The upper layer, which contains the free DMT group, is then discarded. Repeat this procedure two additional times.
6. Concentrate sample to dryness. Solid should be stored at –20°C until needed.

3. Analytical High-Performance Liquid Chromatography

3.1. Anion-Exchange Chromatography

3.1.1. Principles

Anion-exchange HPLC has been shown to be effective for the analysis of synthetic oligonucleotides *(8,9)*. Separations of fully detritylated samples rely primarily on the interaction of the negatively-charged phosphate groups on the DNA backbone with the positively-charged cations contained on the anion-exchanger. Elution of the detritylated oligonucleotides, in order of increasing chain length, is accomplished using a simple gradient of increasing ionic strength. Component resolution depends on HPLC column choice, gradient profile, and the overall length of the oligonucleotide mixture analyzed. Use of a high-performance anion-exchange column can provide *N* from *N*-1 chain-length resolution particularly for samples < 30 bases in length. Compared to slab gel electrophoretic techniques, HPLC instrument design enables real-time, on-line monitoring of the DNA species as they elute from the column and pass through the UV detector. Staining or post-run visualization techniques are not required. In general, detection of as little as 300 ng of nucleic acid/injection is possible, which translates into a 50 µL injection from a sample at 0.0002 A_{260} U/µL.

3.1.2. Chromatographic System and Columns

The following Waters HPLC instrument can be used for the analysis of synthetic oligonucleotides by anion-exchange chromatography. Any similar set of modules should provide equivalent results.

- 600 Gradient Solvent Delivery System.
- U6K Manual Injector or 717 Autosampler.
- Temperature Control System.
- 486 UV/VIS Absorbance Detector.
- Millennium™ 2010 Chromatography Manager.

A wide variety of anion-exchange packing materials suitable for the fractionation of oligonucleotides according to chain length are available from several manufacturers. In general, they consist either of silica particles or rigid organic polymers in a variety of particle and pore sizes. These materials are then permanently functionalized with positively-charged alkylamine or polyethyleneimine groups. For most analytical separations, the use of the high efficiency, small particle size material is recommended (e.g., 5 µm). Although silica-based materials can normally withstand higher operating flow rates, polymer based packing materials can be used at higher pH values because of their increased stability at alkaline pHs. An additional advantage of using polymer-based packings is that these materials can be cleaned with weak acid or base solutions thereby eliminating the possibility of column contamination by samples previously chromatographed.

In addition to base particle composition, both "weak" and "strong" anion-exchange fuctionalities are also available. Whereas strong exchangers remain ionized (i.e., positively charged) at pH values up to 12, weak exchangers possess lower pK values and are therefore normally used at lower pH values (e.g., <8.0) *(10)*. Because anion-exchange HPLC column choice can significantly affect the quality of synthetic oligonucleotide separations obtained, application literature frequently available from the manufacturers can be helpful in choosing an appropriate column for the task at hand.

3.1.3. Chromatographic Buffers

1. Milli-Q water or distilled water.
2. Anion-exchange buffer A: 25 m*M* Tris-HCl, 1 m*M* EDTA, pH 8.0/ Acetonitrile (90/10), (v/v).

3. Anion-exchange Buffer B: 25 m*M* Tris-HCl, 1 m*M* EDTA, 1.0*M* NaCl (or 2.0*M* NaCl), pH 8.0/Acetonitrile (90/10), (v/v).
4. 0.1*N* Phosphoric acid.

3.1.4. Sample Preparation and Chromatography

Use of a high-resolution anion-exchange column, (e.g., Waters Gen-Pak™ FAX; 4.6 × 100 mm), is ideally suited for the rapid analysis of synthetic oligonucleotide mixtures. This particular polymer-based packing material is composed of 2.5 µm, nonporous particles functionalized with diethylaminoethyl (DEAE) groups. Compared to anion-exchangers with larger particle sizes (e.g., 10 through 40 µm), higher resolution separations are obtained with this material. Samples should be detritylated prior to analysis since DMT-containing species elute at slightly longer retention times during the gradient separation compared to the same sequence without the DMT group. This appears to be owing to nonionic interactions of the DMT group with the polymer-based backbone of the packing material. Thus, analysis of samples containing partially detritylated products makes data interpretation difficult.

Prior to chromatography, the sample, diluted in Milli-Q water, should be microfuged or filtered through a 0.22 µm membrane. An $A_{260\ nm}$ reading (*see* Section 2.2.) should be used to determine the sample volume required for a minimum 0.12 $A_{260\ nm}$ unit injection (i.e., 4 µg). This injection mass will give ample signal generation over background noise while maintaining linearity of the HPLC detector. Figures 3 and 4 indicate the resolving power of this chromatographic technique for the analysis of the detritylated, pd(A) samples. As noted in Fig. 4, improved component resolution is easily obtained using appropriate gradient adjustments. As with any HPLC column, proper use and storage will ensure maximum column life, making anion-exchange HPLC analysis of synthetic oligonucleotides a cost-effective technique. In addition, the incorporation of a 0.1*N* phosphoric acid column wash procedure on a pH tolerant, polymer-based anion-exchanger minimizes potential sample carryover between injections *(11)*.

The analysis of detritylated heteropolymers yield more complex chromatographic profiles, as shown in Fig. 5. It is believed that although these synthetic oligonucleotides contain nearly identical

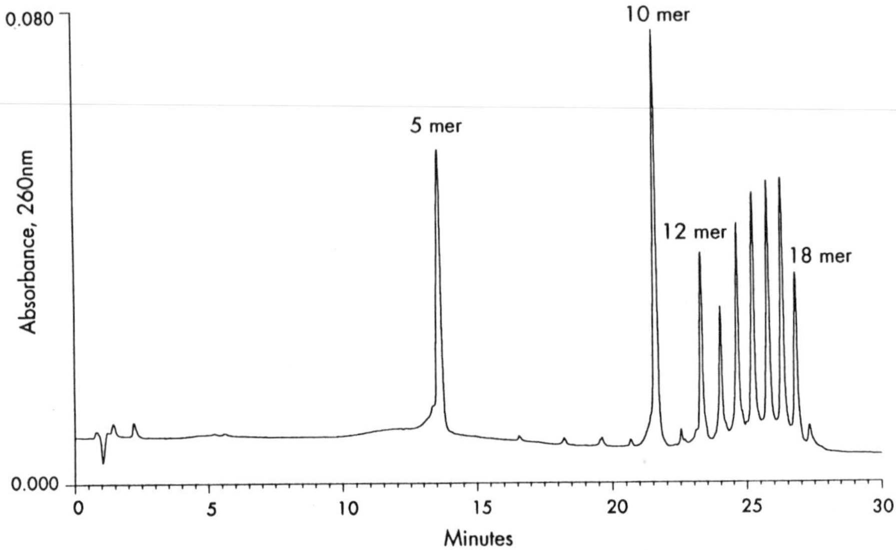

Fig. 3. Analysis of a commercially available phosphodeoxyadenine preparation (Pharmacia LKB Biotechnology, Piscataway, NJ) by anion-exchange HPLC.
Sample: pd(A) 5, 10, and 12–18mer.
Column: Waters Gen-Pak™ FAX (4.6 × 100 mm).
Buffer A = 25 m*M* Tris-HCl, 1 m*M* EDTA, pH 8.0 with acetonitrile (90/10).
Buffer B = A containing 1.0*M* NaCl.
Injection: 1 µg.
Gradient: 10% B to 60% B in 30 min.
Flow: 0.75 mL/min.
Temperature: 30°C.

charge-to-mass ratios, differing secondary structure conformations result in differing degrees of interaction between the DNA and the anion-exchanger. Incorporation of 10% acetonitrile in the HPLC buffers as well as chromatography performed at an elevated temperature (e.g., up to 80°C) minimize these undesirable secondary interactions. Therefore, absolute oligonucleotide length cannot be ascertained by this analytical technique. Although gradient adjustment (as described in Fig. 4) can be used in improved component resolution, baseline resolution is not always achieved. Yet, the ability to identify full-length detritylated product from smaller failure sequences (i.e., *N* from *N*-1) is still possible, thus permitting the quantitation of the product.

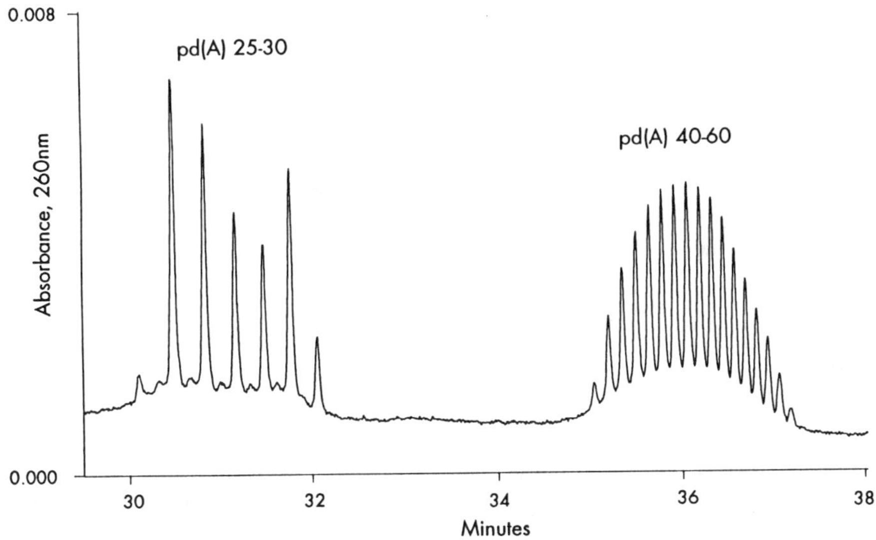

Fig. 4. Analysis of a commercially available phosphodeoxyadenine preparation (Pharmacia LKB Biotechnology, Piscataway, NJ) by anion-exchange HPLC.

Sample: pd(A) 25–30 and 40–60 at 0.001 A_{260} U/μL.

Column: Waters Gen-Pak FAX (4.6 × 100 mm).

Buffer A = 25 m*M* Tris-HCl, 1 m*M* EDTA, pH 8.0 with acetonitrile (90/10).

Buffer B = A containing 1.0*M* NaCl.

Injection: 100 ng.

Gradient: 10% B to 70% B in 30 min. Then hold for 10 min at 70% B.

Flow: 0.75 mL/min.

Temperature: 80°C.

Note: Baseline resolution of the pd(A) 40–60 standard can be obtained using a more shallow gradient (e.g., Load at 10% B to 50% B in 0.01 min. Then 50% B to 70% B in 30 min. Flow: 0.75 mL/min).

3.1.5. Analysis of Chemically-Modified Synthetic Oligonucleotides

As previously described, phosphodiester DNA bears a negative charge at every phosphate residue at neutral pH, making strong or weak anion-exchange HPLC an excellent analytical tool. Modifications to the phosphate backbone for purposes of increasing bioactivity of putative antisense or third-strand therapeutics can significantly effect the charge characteristics of the molecules. For example, the phosphodiester bonds of DNA oligonucleotides can be esterified with

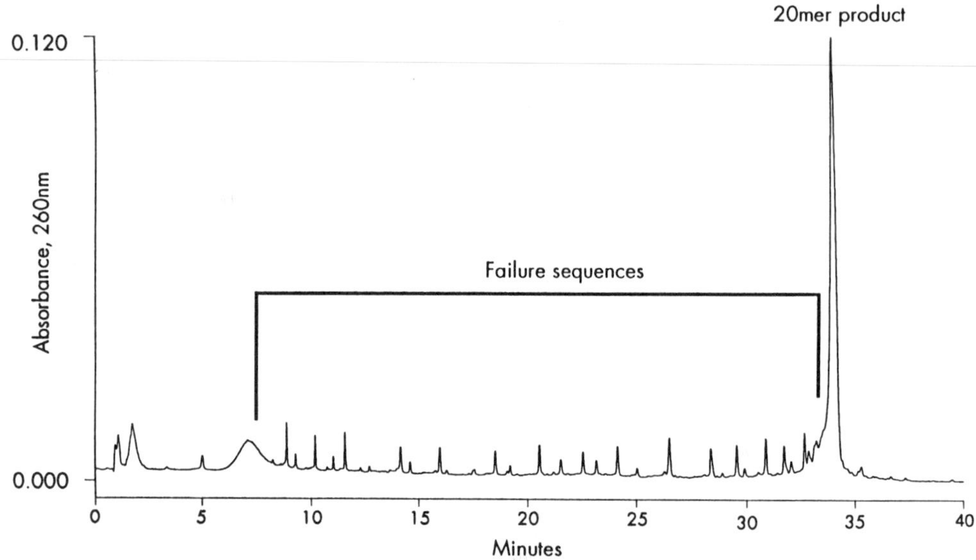

Fig. 5. Analysis of a synthetic oligonucleotide by anion-exchange HPLC.
Sample: 20-mer, phosphodiester (DMT-off)
Column: Waters Gen-Pak FAX (4.6 × 100 mm).
Buffer A = 25 m*M* Tris-HCl, 1 m*M* EDTA, pH 8.0 with acetonitrile (90/10).
Buffer B = A containing 1.0*M* NaCl.
Injection: 45 µg.
Gradient: Load at 15% B to 35% B in 2 min. Then, 35% B to 55% B in 30 min. Then hold for 10 min at 55% B.
Flow: 0.75 mL/min.
Temperature: 80°C.

methyl alcohol to produce O-methylated oligonucleotides that possess a neutral charge on the phosphate backbone. Anion-exchange HPLC or gel electrophoresis cannot be used for the analysis of this or other classes of neutral oligonucleotides. Other separation techniques, such as reverse-phase HPLC *(see below),* are useful alternatives.

Compared to noncharged DNA, sulfurization of the phosphodiester bond generates highly charged phosphorothioated oligonucleotides that bind very tightly to anion-exchange columns and thus require significantly greater concentrations of salt for product elution *(12).* Comparison of anion-exchange HPLC separations of detritylated phosphorothiate and phosphodiester 21mer samples possessing the

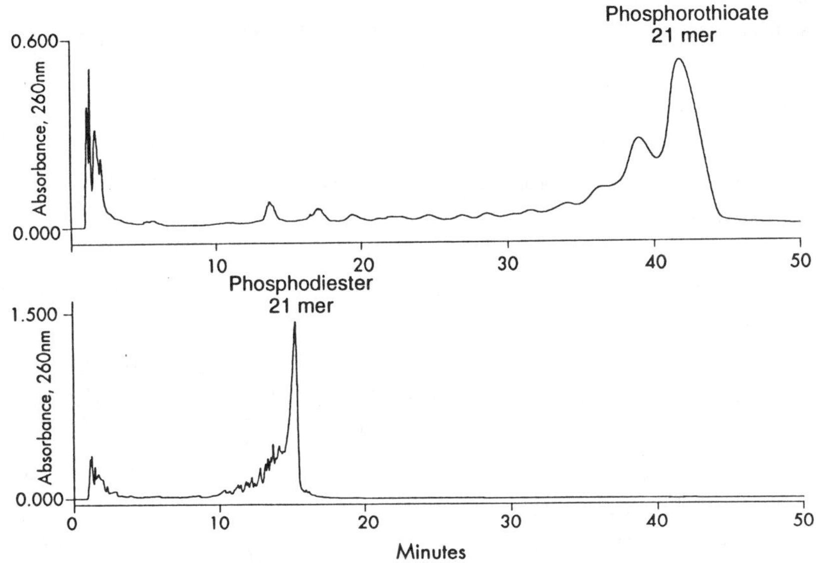

Fig. 6. Analysis of a detritylated phosphorothioate and a phosphorodiester synthetic oligonucleotide of same sequence by anion-exchange HPLC.
Samples: Top: 21-mer, phosphorothioate
Bottom: 21-mer, phosphodiester
Column: Waters Gen-Pak FAX (4.6 × 100 mm).
Buffer A = 25 m*M* Tris-HCl, 1 m*M* EDTA, pH 8.0 with acetonitrile (90/10).
Buffer B = A containing 2.0*M* NaCl.
Injection: 80 µg from each sample.
Gradient: 10% B to 100% B in 60 min.
Flow: 0.75 mL/min
Temperature: 30°C.

same sequence is seen in Fig. 6. Thus, care must be taken in choosing an appropriate chromatographic or electrophoretic technique based on the physicochemical characteristics of the oligonucleotide species studied.

3.1.6. Anion-Exchange HPLC vs Capillary Electrophoresis

Electrophoresis using gel-filled capillaries is an efficient, recently introduced, analytical technique for the evaluation of synthetic oligonucleotide mixtures *(13–15)* and has been discussed in detail in Chapter 11 of this volume. In brief, the basis of the separation involves the differential migration of the various-sized DNA species as they are electrophoretically driven through the gel matrix in a manner similar to that employed with conventional slab-gel techniques. During the separation, the smaller length failure sequences contained in the reaction mixture move more rapidly through the gel-filled capillary, because of their relatively smaller Stokes radii, than the larger full-length oligonucleotide product. Because gel-filled capillaries are of small internal diameter (e.g., 0.75 µm), excellent heat dissipation from the gel matrix occurs. Therefore, CE separations can be performed at higher field strengths (250 V/cm) than are possible using conventional slab gels. This results in very efficient separations in relatively short time intervals.

Although the reagents for this application can be prepared in the laboratory with a moderate degree of effort *(16)*, gel-filled capillaries containing crosslinked polyacrylamide together with the necessary electrolyte (e.g., Tris/Borate/Urea buffer), and oligonucleotide standards can be conveniently purchased from several manufacturers (e.g., Millipore). The utilization of these gel-filled devices is straightforward and fully described in the care and use manuals provided with these kits. As with anion-exchange HPLC separations, the analysis of fully detritylated samples is recommended. This is because the presence of the DMT group on a full-length product will increase its migration time through the capillary, compared to the same detritylated sequence, making electropherogram interpretation difficult *(14)*.

A representative gel-filled capillary analysis of a detritylated synthetic oligonucleotide reaction mixture is shown in Fig. 7. Compared to the analysis of the same sample using anion-exchange HPLC (Fig. 5), a similar degree of component resolution is obtained. As with HPLC techniques, direct UV detection of the DNA as it elutes from the capillary permits quantitation of the product and eliminates the need for post-run staining and visualization techniques inherent with slab-gel methodologies. However, owing to the limited sample mass that can be introduced into the capillary separation, for all practical purposes, CE techniques are primarily analytical. Furthermore, compared to HPLC columns, the life of cross-linked polyacrylamide-filled capillaries is shorter and is dramatically influenced by the care exercised during the use and storage of the device. In general, use of filtered samples and electrolyte as well as avoidance of rapid temperature fluctuations imposed on the gel-filled device are recommended. In

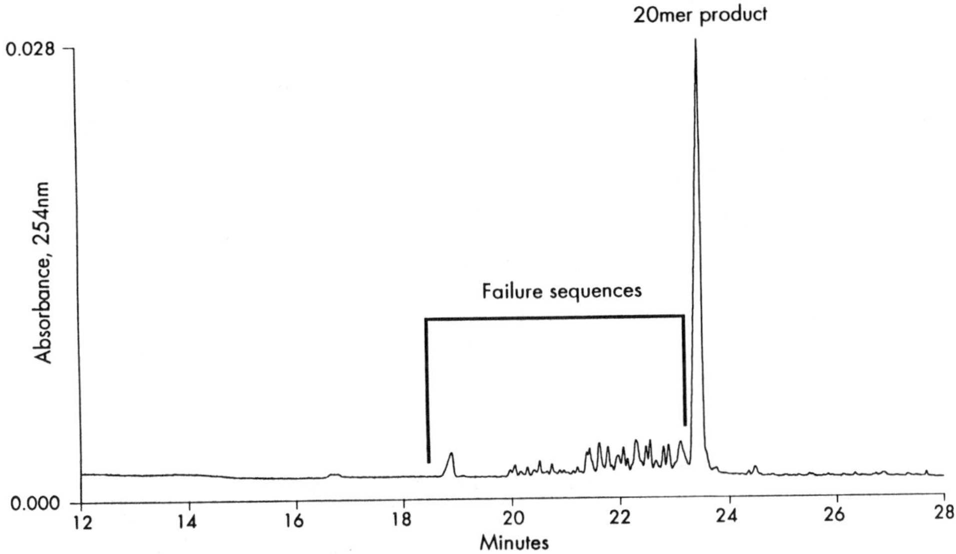

Fig. 7. Analysis of a synthetic oligonucleotide by gel-filled capillary electrophoresis on Waters Quanta™ 4000 Capillary Electrophoresis System.
Sample: 20-mer, phosphodiester (DMT-off)
Capillary: Waters μPAGE-5 gel-filled (5% T and 5% C) 75 μ*M* id, 60 cm (total) 51 cm (effective).
Electrolyte: 100 m*M* Tris-Borate, pH 8.3 with 7*M* urea.
Injection: –5 kV for 5 s.
Run: –13 kV.
Detection: 254 nm.
Temperature: Ambient.

spite of these limitations, the automated analysis of synthetic oligonucleotides using gel-filled capillary electrophoresis is gaining increased popularity.

3.1.7. Calculation of Average Coupling Efficiency from Anion-Exchange HPLC Analysis

The precise quantitation of full-length product from shorter length failure sequences can be obtained using anion-exchange HPLC. These data can then be applied in the following formula to approximate the average coupling efficiency for the DNA synthesizer used in the preparation of the oligonucleotide.

$$\text{Average coupling efficiency} = \text{Overall yield}^{1/(N-1)}$$

Overall yield = X/Y
X = the area of the product peak
Y = the combined areas of all peaks including product peak
N = the length of the oligonucleotide

Example for 20-mer:

X = 2,208,160 area counts
Y = 2,208,160 + 216,877 area counts

Overall yield = 0.91
Average coupling efficiency = $0.91^{1/19} = 0.995 \times 100 = 99.5\%$

3.2. Reverse-Phase Chromatography

3.2.1. Principles

Reverse-phase HPLC chromatography can be used for the analysis of synthetic oligonucleotide reaction mixtures or in the determination of oligonucleotide base composition. For crude reaction mixture analysis, the DMT group should be left on the sample prior to chromatography. Separation of the failure sequences from the more hydrophobic, DMT-containing species is based on the differential elution from the nonpolar column packing material of these various molecules as the concentration of organic solvent (e.g., acetonitrile or methanol) is slowly increased within the column. For accurate quantitation of the total amount of full-length product generated during the synthesis, it is essential that detritylation of the full-length oligonucleotide be avoided. As with anion-exchange HPLC techniques, instrument design enables direct, on-line detection of low masses of synthetic oligomers. Staining or postrun visualization protocols are not required.

Reverse-phase HPLC is also useful in obtaining base composition data of purified oligonucleotide products as detailed in Section 3.2.6. Similar to the analysis of synthesis reaction mixtures, analysis of the enzymatically generated nucleoside components can be used to either confirm performance of the DNA synthesizer or determine whether base modifications have occurred during the synthesis and/or cleavage and deprotection protocols *(17,18)*.

3.2.2. Chromatographic System and Columns

The same instrument components used for anion-exchange chromatography can be used for reverse-phase HPLC of synthetic oligonucleotides. As with anion-exchange HPLC columns, a sizable array of commercially available reverse-phase packing materials are manufactured for analytical purposes. Again, both silica particle and rigid organic polymers of differing particle and pore sizes are available. The hydrophobic characteristic of these base materials is often obtained by bonding C_4, C_8, or C_{18} carbon chains to the rigid particles. Application literature available from HPLC column manufacturers is helpful in choosing an appropriate column although in all cases, use of a small particle size, reverse-phase column packing material (e.g., 5 μm) is recommended for high resolution separations at the analytical scale.

3.2.3. Chromatographic Buffers

1. Milli-Q water or distilled water.
2. Reverse-phase buffer A: 100 m*M* Triethylammonium acetate, pH 6.5/ Acetonitrile (95/5), (v/v).
3. Reverse-phase buffer B: Acetonitrile/Milli-Q water (95/5), (v/v).

3.2.4. Sample Preparation and Chromatography of Phosphodiester Synthetic Oligonucleotides

Use of a high resolution, silica-based, reverse-phase HPLC column, (e.g., Waters Delta-Pak™, C_{18}, 300Å, 5 μm) can be successfully used for the rapid analysis of DMT-containing samples. Prior to chromatography, samples should be microfuged or filtered through a 0.22 μm membrane to remove particulates. Based on the A_{260} U reading, an appropriate volume should be injected onto the column to yield a minimum injected mass of approx 4 μg (i.e., 0.12 A_{260} U). Figure 8 shows a typical reverse-phase HPLC chromatogram for the analysis of a 20-mer phosphodiester synthetic reaction mixture. As previously described, the DMT group must be left on the sample following synthesis to more accurately quantitate the amount of full-length product generated during the synthesis. In addition, DMT groups are found on shorter length DNA products because of depurination or incomplete end-capping during the synthesis procedure *(5)*. For these reasons, reverse-phase chromatography may be

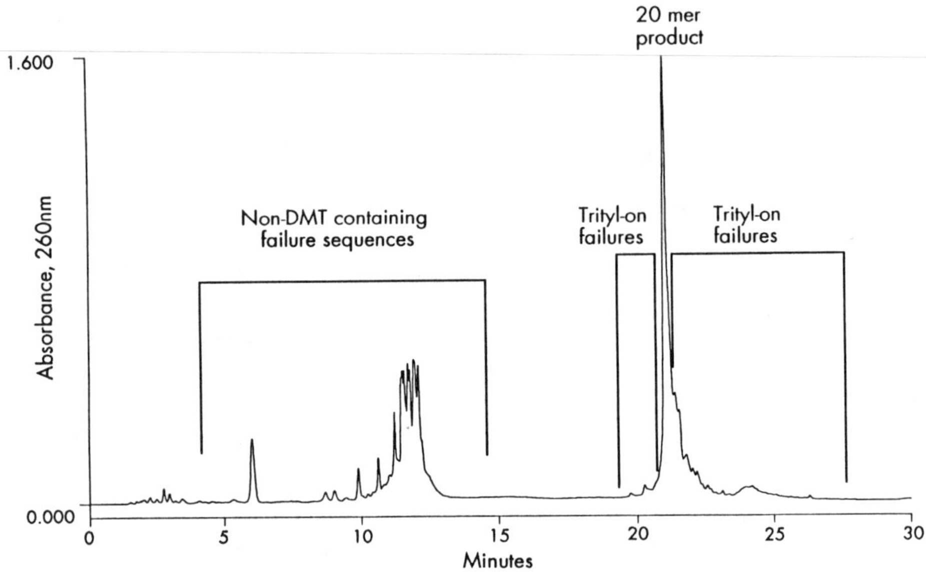

Fig. 8. Reverse-phase HPLC analysis of synthetic oligonucleotide.
Sample: 20-mer, phosphodiester (DMT-on).
Column: Waters Delta-Pak, C_{18}, 300Å, 5μ (3.9 × 150 mm).
Eluent A = 0.1*M* Triethylammonium acetate, pH 6.5 with acetonitrile (95/5).
Eluent B = Acetonitrile with Milli-Q™ water (95/5).
Gradient: 0% B to 40% B in 25 min. Then hold for 5 min at 40% B.
Flow: 1.0 mL/min.
Temperature: 30°C.

less accurate than anion-exchange HPLC or capillary electrophoresis in the determination of full-length phosphodiester or phosphorothioate oligonucleotide product yields.

3.2.5. Reverse-Phase Analysis of Modified Synthetic Oligonucleotides

Compared to anion-exchange HPLC or gel electrophoresis, reverse-phase HPLC is better suited for the study of modified synthetic oligonucleotide reaction mixtures. As indicated in Fig. 9, the analysis of a 21-mer phosphorothioate crude reaction mixture on a 5 μm, high-performance reverse-phase HPLC column indicates the presence of several major DMT-containing products. Anion-exchange HPLC does

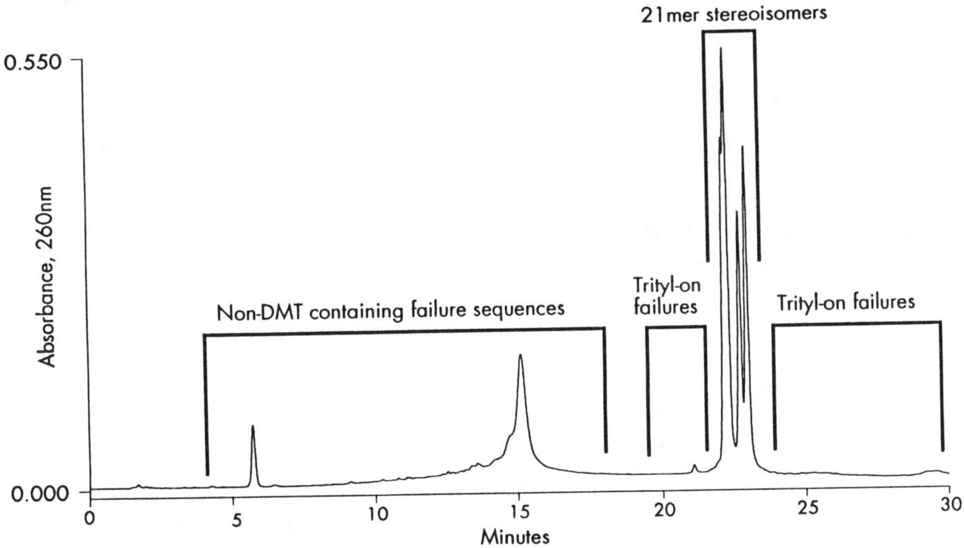

Fig. 9. Reverse-phase HPLC analysis of synthetic oligonucleotide.
Sample: 21mer, phosphorothioate (DMT-on).
Column: Waters Delta-Pak, C_{18}, 300Å, 5 μm (3.9 × 150 mm).
Eluent A = 0.1*M* Triethylammonium acetate, pH 6.5 with acetonitrile (95/5).
Eluent B = Acetonitrile with Milli-Q™ water (95/5).
Gradient: 0% B to 40% B in 25 min. Then hold for 5 min at 40% B.
Flow: 1.0 mL/min.
Temperature: 30°C.

not adequately resolve the various sterioisomeric species present in the sample (*see* Fig. 6). In addition to the analysis of phosphorothioate samples, reverse-phase HPLC is also essential for the analysis of neutral oligonucleotides (e.g., methyl-phosphonates) that cannot be analyzed by either anion-exchange HPLC or gel electrophoresis.

3.2.6. Sample Preparation and Chromatography for Base Composition Analysis

Base composition determination of nucleosides generated by the enzyme digestion of purified synthetic oligonucleotides has been well documented using reverse-phase HPLC *(17,18)*. The following sample processing procedure can be used prior to the HPLC determinations.

3.2.6.1. Materials and Methods

1. Concentrate to dryness approx 0.5 A_{260} U of HPLC purified, detritylated oligonucleotide sample.
2. Dissolve the DNA pellet in 50 µL of 50 m*M* Tris-HCl, pH 8.0 containing 10 m*M* $MgCl_2$.
3. Add 3 U of snake-venom phosphodiesterase I (e.g., Part # 20240; United States Biochemical, Cleveland, OH) and 3 U of bacterial alkaline phosphatase (e.g., Part # 70035; USB) to the dissolved sample.
4. Incubate mixture for a minimum of 6 h at 37°C to ensure that the oligonucleotides are completely digested to their corresponding nucleosides.
5. The processed sample should be microfuged then ultrafiltered (5000 dalton cutoff membrane) prior to the analysis of an aliquot (e.g., 5 µL) by reverse-phase HPLC.

As indicated in the base composition analysis of a 16-mer oligonucleotide sample (Fig. 10), good nucleoside resolution is obtained using this mode of chromatography. As controls, nucleotides and nucleoside standards should also be processed and chromatographed in an identical manner. The data generated by this technique can be used to ascertain the quality of DNA obtained following its synthesis, cleavage, deprotection, and purification.

4. Preparative Liquid Chromatography

4.1. Purification Using Anion-Exchange HPLC Columns

4.1.1. Chromatographic System

The same instrument components used for reverse-phase HPLC purification of synthetic oligonucleotides can also be used with this mode of preparative column chromatography.

4.1.2. Chromatographic Buffers

1. Milli-Q water/distilled water.
2. Anion-exchange buffer A: 25 m*M* Tris-HCl, 1 m*M* EDTA, pH 8.0/Acetonitrile (90/10), (v/v).
3. Anion-exchange buffer B: 25 m*M* Tris-HCl, 1 m*M* EDTA, 1.0*M* NaCl, pH 8.0/Acetonitrile (90/10), (v/v).
4. 0.1*N* Phosphoric acid.

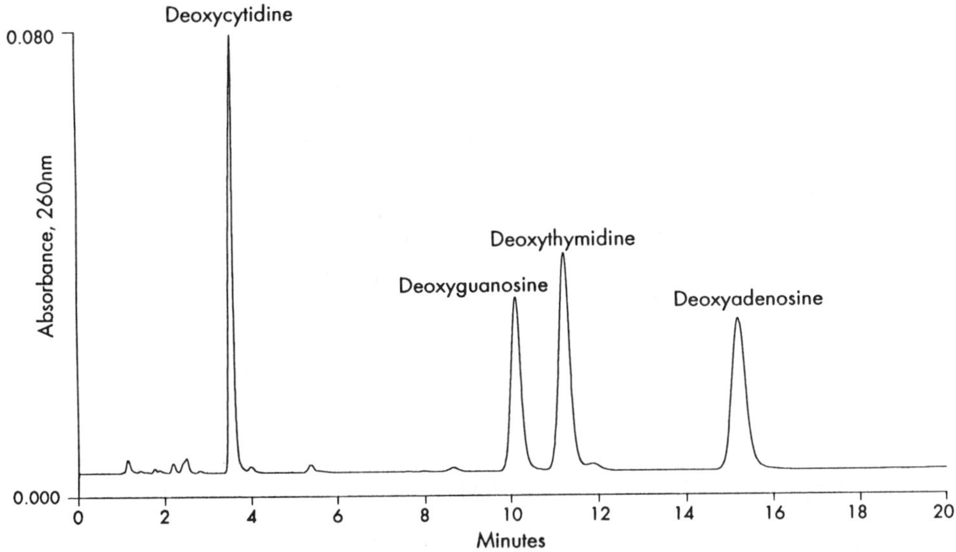

Fig. 10. Base-composition analysis by reverse-phase HPLC
Sample: Snake-venom phosphodiesterase and bacterial alkaline phosphatase digested 16mer, phosphodiester (DMT-off).
Column: Waters Nova-Pak, C_{18}, 60Å, 4 µm (3.9 × 150 mm).
Eluent A = 50 m*M* Sodium phosphate, pH 5.7.
Eluent B = Methanol with Milli-Q™ water (95/5).
Gradient: 0% B hold for 10 min. Then 0% B to 30% B in 20 min.
Flow: 1.0 mL/min.
Temperature: 30°C.

4.1.3. Column Selection, Sample Preparation, and Chromatography

Although nonporous, small-particle anion-exchangers (e.g., Gen-Pak FAX) are well-suited for the analysis of detritlyated oligonucleotides samples, limited mass capacity (i.e., <100 µg) on this class of packing material hinders their effective use for large-scale preparative applications. For this reason, anion-exchangers of larger particle size and of greater porosity are used for these preparative isolations. Although the ionic interaction of the analyte is the same on both types of packing material, the diffusion kinetics are slower on porous beads

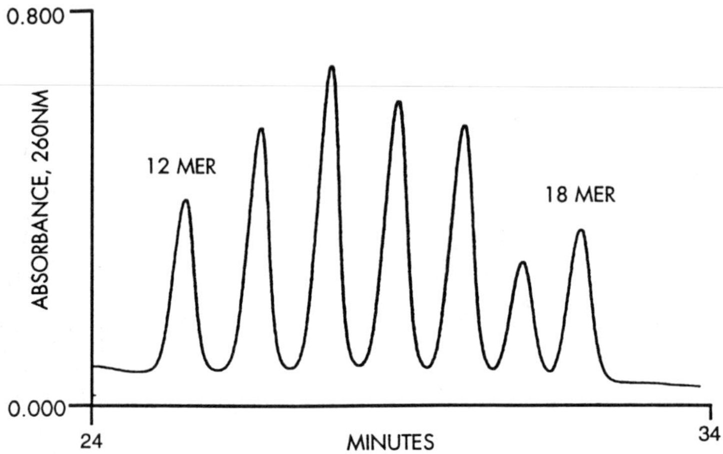

Fig. 11. Anion-exchange HPLC of a commercially available phosphodeoxyadenine preparation (Pharmacia LKB Biotechnology, Piscataway, NJ).
Sample: pd(A) 12–18-mer.
Column: Waters Protein-Pak™ Q 8HR, 1000Å, 8 µm (10 × 100 mm).
Buffer A = 25 m*M* Tris-HCl, 1 m*M* EDTA, pH 8.0 with acetonitrile (90/10).
Buffer B = A containing 1.0*M* NaCl.
Injection: 6 µg
Gradient: Load at 0% B to 30% B in 0.01 min. Then, 30% B to 50% B in 30 min. Then hold for 5 min at 50% B.
Flow: 1.0 mL/min.
Temperature: Ambient.

which result in slightly larger peak volumes. However, the surface area is significantly larger, thereby providing greater load capacity for detritylated samples. Figure 11 shows the separation of an oligonucleotide sample mixture on a porous anion-exchange packing, Protein-Pak™ Q 8HR (1000Å, 8 µm pore size). Excellent resolution of the phosphodeoxyadenine (pd(A) 12–18) sample is achieved on this packing material and, due to its high mass capacity, it is ideally suited for preparative separations. In addition, the incorporation of a 0.1*N* phosphoric acid column wash procedure to this pH-tolerant, polymer-based anion-exchanger minimizes potential sample carryover between purifications.

The same sample preparation, methods development, and scaling strategy employed for preparative reverse-phase HPLC purifications

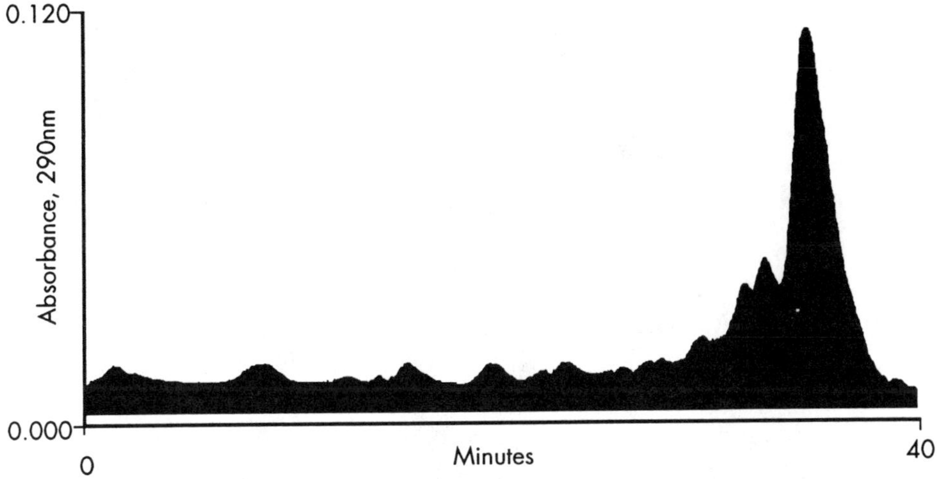

Fig. 12. Anion-exchange HPLC analysis of a synthetic oligonucleotide.
Sample: 20-mer, phosphodiester (DMT-off).
Column: Waters Protein-Pak Q 8HR (10 × 100 mm).
Buffer A = 10 m*M* NaOH.
Buffer B = A + 1.5*M* NaCl.
Gradient: 5% B to 40% B in 40 min.
Flow: 0.8 mL/min.
Temperature: Ambient.
Note: Chromatography was performed at an elevated pH (e.g., pH 10.0) to minimize strong secondary structural characteristics of this G-rich 20-mer sequence.

of synthetic oligonucleotides can also be successfully employed with this mode of chromatography. An important difference between both techniques, however, is that only fully detritylated samples should be chromatographed by anion-exchange HPLC as previously detailed. Figure 12 is representative of the separation of 40 A_{260} U of a detritylated synthetic oligonucleotide mixture on a strong anion-exchanger contained within a 10 mm × 100 mm column. Using an appropriate gradient of increasing sodium chloride concentration, good component resolution of sequences up to 30 bases in length is possible. As shown, chromatography can be performed at an elevated pH (e.g., pH 10.0) on this class of polymer-based, strong anion-exchangers, thereby minimizing oligonucleotide secondary structure interactions that can compromise component resolution. In addition, gradient separations using volatile buffer systems (i.e., 3.0*M* ammonium acetate, pH 4.6) can also be employed if desired to simplify product desalting following chromatography. Thus, anion-exchange chromatography following reverse-phase HPLC of crude oligonucleotide reaction mixtures can produce product possessing greater than 99% homogeneity.

4.2. Purification Using Reverse-Phase Devices

4.2.1. Chromatographic System

Single use reverse-phase cartridge devices are available from a variety of manufacturers for the purification of DMT-protected oligonucleotides. Use of the Oligo-Pak™ cartridge (Millipore) is described for demonstrative purposes although similar results can be obtained using other devices. Compared to reverse-phase HPLC separations, a conventional disposable syringe is used for sample and solvent delivery, thus eliminating the need for HPLC instrumentation.

4.2.2. Chromatographic Buffers

1. Acetonitrile (HPLC grade).
2. 1.0*M* Triethylammonium acetate (TEAA), pH 7.0.
3. Milli-Q water or distilled water.
4. 3% Ammonium hydroxide/Milli-Q water (v/v).
5. 20% Acetonitrile/Milli-Q water (v/v).
6. 2% Trifluoroacetic acid (TFA)/Milli-Q water (v/v).
7. 40% Acetonitrile/Mill-Q water (v/v).

4.2.3. Sample Preparation and Chromatography

Oligo-Pak reverse-phase devices are designed for the rapid purification of DMT-protected oligonucleotides using a standard Luer® lock syringe for sample introduction, washes, and product elution. Contained within this polypropylene device is approx 1.0 g of a 1000Å, 40 µm styrene divinylbenzene polymeric packing material. Prior to sample introduction, the packing is prepared for use by "wetting" the hydrophobic material with 10 mL of 100% acetonitrile followed by 15 mL of 1.0*M* triethylammonium acetate, pH 7.0. Up to 50 A_{260} U of base deprotected, DMT-containing oligonucleotide mixture in 15% ammonium hydroxide (i.e., 1:1 dilution of protected oligonucleotide in Milli-Q H_2O) is then slowly applied to the cartridge. Depending on the sequence length, this can be the entire solution from a synthesis at the 0.2-µmol scale or 20% from a 1.0 µmol scale synthesis. The failure sequences and protecting groups are then washed from the column with 15 mL of 3% ammonium hydroxide solution. The DMT-protected product can then be eluted from the cartridge with 3 mL of 40% acetonitrile.

If detritylated DNA is desired for subsequent experimentation, the DMT group can be cleaved from the full-length product directly on the cartridge with 10 mL of 2% TFA prior to elution of the desired oligonucleotide with 20% acetonitrile. Because the styrene divinylbenzene material is stable from pH 2–13, on-column detritylation with TFA as well as sample loading in ammonium hydroxide are possible. Following purification, the oligonucleotide product can be concentrated to dryness. The solid should be stored at –20°C until needed. Figure 13 compares a reverse-phase HPLC analysis of a 20-mer phosphodiester oligonucleotide (DMT-on) before and after purification on the Oligo-Pak cartridge. The effectiveness of this technique is clearly demonstrated by the absence of failure sequences (peaks detected between 2 and 15 min) in the Oligo-Pak purified sample compared to the HPLC analysis of the initial synthesis reaction mixture. A limitation of this technique is that it cannot resolve the desired DMT-protected full-length product from DMT-protected "failure sequences" that can arise from product depurination or incomplete endcapping during the synthesis reaction *(5)*. When final product purity must exceed 85% or when multimilligram to gram amounts of purified material are required, reverse-phase HPLC with precise gradient elution control is recommended.

4.3. Purification Using Reverse-Phase HPLC Columns

4.3.1. Chromatographic System

The following Waters HPLC instrumentation, or equivalent, is well suited for the purification of microgram to gram amounts of synthetic oligonucleotides.

- Delta-Prep 4000 Preparative Chromatography System.
- 486 UV/VIS Tunable Wavelength Detector with semi-prep flow cell.
- Millennium 2010 Chromatography Manager.

4.3.2. Chromatographic Buffers

1. Milli-Q water/distilled water
2. Reverse-phase buffer A: 0.2*M* sodium acetate, pH 7.2 with 20% methanol.

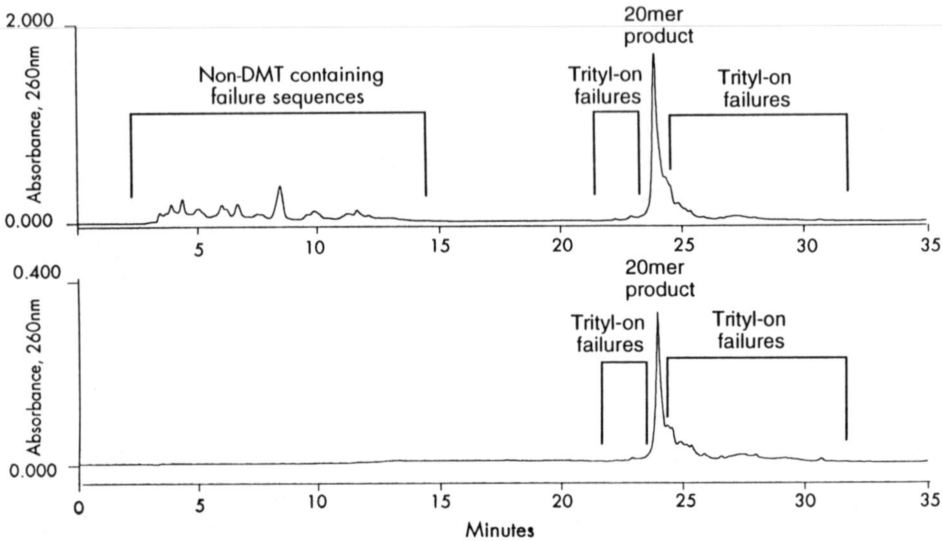

Fig. 13. Reverse-phase HPLC analysis of synthetic oligonucleotide before (Top) and after (Bottom) Oligo-Pak purification.
Sample: 20-mer, phosphodiester (DMT-on).
Column: Waters Delta-Pak™, C18, 300Å, 5 µm (3.9 × 150 mm).
Eluent A = 0.1*M* Triethylammonium acetate, pH 6.5 with acetonitrile (95/5).
Eluent B = Acetonitrile with Milli-Q™ water (95/5).
Gradient: 0% B to 40% B in 25 min. Then hold for 10 min at 40% B.
Flow: 1.0 mL/ min.
Temperature: 30°C.

3. Reverse-phase buffer B: 0.2*M* sodium acetate, pH 7.2 with 40% methanol.
4. Reverse-phase buffer C: 100% methanol.

4.3.3. Column Selection, Sample Preparation, and Chromatography

A large variety of packing materials, including silica- and polymer-based, are available for the reverse-phase HPLC purification of tritylated synthetic oligonucleotides. Although polymer-based, pH-tolerant materials offer the convenience of separating material immediately following cleavage and deprotection from the controlled pore glass support (i.e., in 30% ammonium hydroxide), silica based products are available in a wider range of column configurations and can withstand greater solvent flow rates often desired for large-scale purifications. The ability to select a reverse-phase packing material that not only has high oligonucleotide binding capacity but is also available in a variety of column sizes is especially important when multimilligram or gram scale purifications are required. As a consequence, silica-based reverse-phase packing materials appear to be favored at the present time for these applications. Regardless of the choice of reverse-phase material for preparative applications, good chromatographic practices must be followed to ensure satisfactory performance while providing reasonable column longevity.

Purification of DMT-protected oligonucleotides by reverse-phase HPLC not only offers the ability to purify large amounts of material in a single run, but can also deliver product of greater overall homogeneity. Compared to the use of a simple syringe with Oligo-Pak cartridges, eluent delivery to an HPLC column is precise and reproducible. As a consequence, HPLC gradient conditions can be developed for the separation of the DMT-protected full-length product from branched chain synthesis products or DMT-containing, shorter length sequences. As indicated in Figs. 8 and 9, elution of these "failure" species during the gradient program normally occurs immediately before and after the elution of the tritylated full-length product. Thus, HPLC gradient elution protocols coupled to proper fraction collection techniques can be used to yield product in excess of 95% homogeneity.

In preparative scale applications, the total amount of synthetic oligonucleotide that can be chromatographed in a single run depends in part on the total DNA binding capacity of the selected reverse-phase packing as well as on the column size used for the purification. However, the availability of reverse-phase packing materials in various column dimensions and the cost associated with the preparative columns are also factors that must be considered. One economical approach for preparative scale, reverse-phase oligonucleotide chromatography, employs Radial Compression Technology *(19)*. Compared to the use of relatively expensive stainless steel columns filled with preparative scale reverse-phase packing material, this technique uses less expensive, polyethylene cartridge columns that are radially compressed to provide optimum column efficiency at a large range of flow rates. A practical approach that can be used in developing a method for the purification of milligram to gram amounts of synthetic oligonucleotide is outlined below:

1. First, select an appropriate reverse-phase packing material that is available in a variety of column configurations. For the initial methods development work, separations on a small scaling column are recommended. The advantage of using a small column (2–5 mL column volume) initially is that the separation run times are relatively short and the appropriate choice of mobile phase, flow rate, and gradient profile may be determined quickly without consuming much solvent or sample.
2. Once conditions are found that adequately resolve the failure sequences from the desired DNA product, a loading study should be performed by injecting increasing amounts of the crude product onto the scaling column. This information is critical in determining the maximum amount of sample that will be able to be chromatographed on the preparative column without overloading or seriously compromising the quality of the separation. Having determined the maximum allowable load on the scaling column under the optimized chromatographic conditions, the separation can be scaled-up to an appropriate column size to meet the purification needs.
3. Following the initial method development and sample loading studies, the separation can be scaled to a larger column containing the identical type of reverse-phase packing material. To ensure consistent and predictable transfer of the separation, flow rate, gradient duration, and sample load must be adjusted appropriately. The equations shown in Table 1 are used for this purpose.

An example adopting this scale-up strategy is demonstrated using a phosphorothioated synthetic oligonucleotide of 20 bases in length (Figs. 14 and 15). The packing material chosen for the separation was Waters Bondapak™ $HC_{18}HA$ (125Å, 37–55 µm, 14% carbon load) contained in an 8 × 100 mm Radial-Pak™ scaling column and a 47 × 300 mm PrepPak® large-scale purification cartridge column. The separation was first optimized on the 8 × 100 mm cartridge column so that the maximum amount of DMT-containing species would bind to the column. Sample loading at an experimentally determined concentration of organic solvent (e.g., 20% methanol for the particular example shown) was used to determine the maximum capacity for the desired 20-mer product by preventing the 5'-OH failure sequences from binding to reverse-phase sites. Separations performed at pH 7.2 enhance product recovery by minimizing product detritylation dur-

Table 1
Scaling Factors for Flow Rate, Sample Load, and Gradient Duration Using the Identical Packing Material Contained in Columns of Increased Internal Diameter and Volume

Equation 1:		
Flow Rate (prep)/Flow Rate (scaling)	=	r^2 (prep)/ r^2 (scaling)
Equation 2:		
Mass Load (prep)/Mass Load (scaling)	=	Length (prep)/Length (scaling) × r^2 (prep)/r^2 (scaling)
Equation 3*:		
Gradient Duration (prep)/Gradient Duration (scaling)	=	Length (prep)/ Length (scaling)

* Adjustment in gradient duration using Eq. 3 assumes that both separations are performed at the same linear velocity as determined in Eq. 1.

Note: When transferring a purification from one HPLC system to another, the gradient may need additional adjustment because of differences in system delay volumes between the point of mixing of the mobile phases and the top of the column. For this reason, the selection of an HPLC instrument capable of precise eluent delivery at both methods development (1.0 mL/min.) as well as large-scale preparative flow rates (i.e., hundreds of mL/min.) simplifies methods development and scale up and is thus recommended.

ing sample loading and gradient elution. The gradient program was then adjusted so that an acceptable level of product purity would result at the completion of the separation. For the particular phosphorothioated 20-mer sequence and separation conditions used in this example (*see* Fig. 14), up to 63 mg of crude material could be loaded on the 8 × 100 mm scaling column without column overload. Recovered product from these separations was found to possess greater than 95% homogeneity as determined by off-line gel electrophoretic techniques (data not shown). Higher mass loadings for this particular DNA sequence exceeded the column's capacity as indicated by the presence of full-length, trityl-on product in the void volume of the column.

Based on calculations obtained from the formula in Table 1, it was predicted that 6.3 g of material could be successfully chromatographed on the same Bondapak $HC_{18}HA$ material packed in a 47 × 100 mm PrepPak cartridge column. The large scale isolation and conditions used are shown in Fig. 15. Greater than 95% pure, full-length prod-

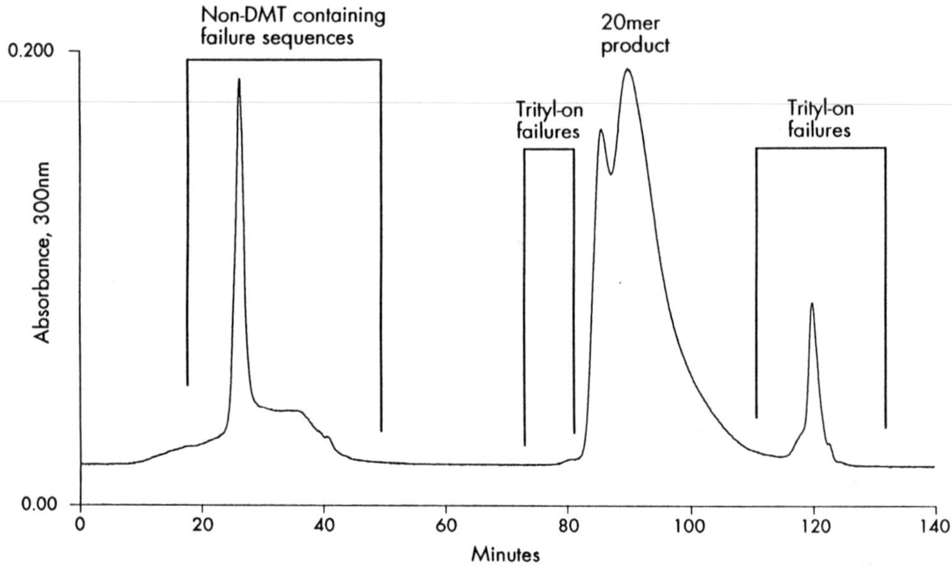

Fig. 14. Reverse-phase HPLC loading study of synthetic oligonucleotide.
Sample: 20-mer, phosphorothioate (DMT-on).
Column: Waters Bondapak™ $HC_{18}HA$, 125Å, 37–55 µm (8 × 100 mm Radial-Pak cartridge column).
Eluent A = 0.2*M* Sodium acetate, pH 7.2 with 20% methanol.
Eluent B = 0.2*M* Sodium acetate, pH 7.2 with 40% methanol.
Eluent C = 100% methanol.
Injection: 63 milligrams in Eluent A.
Gradient: Load sample through pump and hold at 100% A for 70 min. Then from 100% A to 100% B in 10 min. At 110 min, go from 100% B to 100% C in 10 min. Then hold for 20 min at 100% C.
Flow: 3.0 mL/min.
Temperature: Ambient.
Note: Sodium acetate and methanol have been substituted for the reverse-phase eluents triethylammonium acetate and acetonitrile to reduce potential biotoxicity of the purified final product. In addition, the separation was monitored at 300 nm to prevent detector overload.

uct, without column overload was obtained, thus confirming the effectiveness of this scaling technique. Although this discussion demonstrated a scale-up strategy using a reverse-phase mode of separation, this approach can be applied to any mode of chromatography, including anion-exchange.

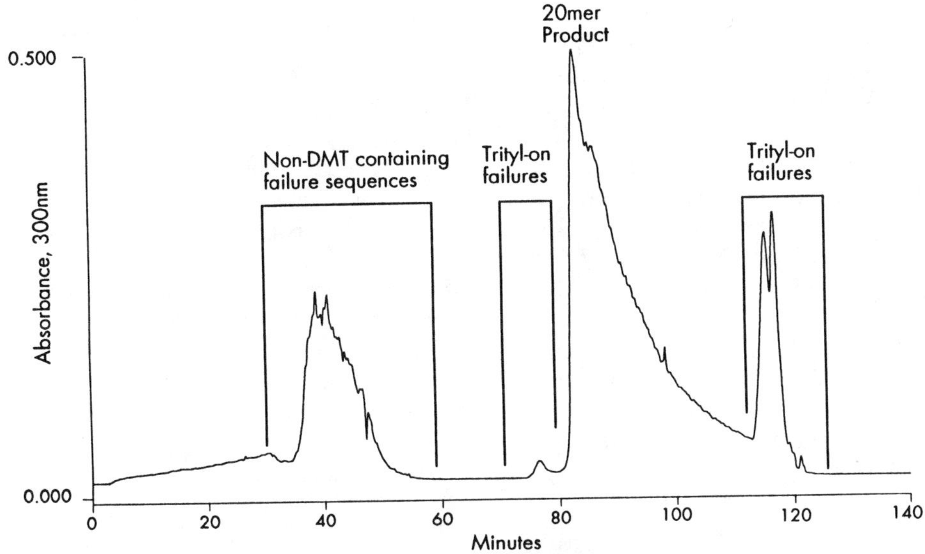

Fig. 15. Large-scale reverse-phase HPLC purification of synthetic oligonucleotide.
Sample: 20-mer, phosphorothioate (DMT-on).
Column: Waters Bondapak $HC_{18}HA$, 125Å, 37–55 μ*M* (47 × 300 mm PrepPak cartridge column).
Eluent A = 0.2*M* Sodium acetate, pH 7.2 containing 20% methanol.
Eluent B = 0.2*M* Sodium acetate, pH 7.2 containing 40% methanol.
Eluent C = 100% methanol.
Injection: 6.3 grams in Eluent A.
Gradient: Load sample through pump and hold at 100% A for 70 min. Then from 100% A to 100% B in 10 min. At 110 min, go from 100% B to 100% C in 10 min. Then hold for 20 min at 100% C.
Flow: 100 mL/min.
Temperature: Ambient.

References

1. Joudrier, P. E., Foard, D. E., Floener, L. A., and Larkins, B. A. (1987) Isolation and sequences of cDNA encoding the soybean protease inhibitors PI IV and C-II. *Plant Mol. Biol.* **10,** 35–42.
2. Sanger, F., Nicklen, S., and Coulson, A. F. (1977) DNA sequencing with chain-terminating inhibitors. *Proc. Natl. Acad. Sci. USA* **74,** 5463–5467.
3. Mullis, K. B. and Faloona, F. A. (1987) Specific synthesis of DNA *in vitro* via a polymerase-catalyzed chain reaction. *Meth. Enzym.* **155,** 335–350.

4. Bielinska, A., Shivdasani, R. A., Zhang, L., and Nabel, G. J. (1990) Regulation of gene expression with double-stranded phosphorothioate oligonucleotides. *Science* **16,** 997–999.
5. Paivinen, A., Aguiar, H., Reiss, P., and Bonner, A. (1986) New challenges in automated DNA synthesis. *J. Anal. Purif.* **Oct.,** 28–37.
6. Rickwood, D. and Hames, B. B., eds. (1990) *Gel Electrophoresis of Nucleic Acids: A Practical Approach,* 2nd ed. IRL, Oxford, England, pp. 125–149.
7. Vornham, A. V. and Kerschner, J. (1986) Purification of small oligonucleotides by polyacrylamide gel electrophoresis and transfer to diethylaminoethyl paper. *Anal. Biochem.* **152,** 221–225.
8. Drager, R. R. and Regnier, F. E. (1985) High-performance anion-exchange chromatography of oligonucleotides. *Anal. Biochem.* **145,** 47–56.
9. Kato, Y., Kitamura, T., Mitsui, A., Yamasaki, Y., Hashimoto, T., Murotsu, T., Fukushige, S., and Matsubara, K. (1988) Separation of oligonucleotides by high-performance ion-exchange chromatography on a non-porous ion exchanger. *J. Chrom.* **447,** 212–220.
10. Harris, E. L. V. and Angal, S., eds. (1989) *Protein Purification Methods: A Practical Approach.* IRL, Oxford, England, pp. 202–216.
11. Warren, W., Wheat, T., and Knudsen, P. (1991) Rapid analysis and quantitation of PCR products by high-performance liquid chromatography. *BioTech.* **11,** 250–255.
12. Metelev, V. and Agrawal, S. (1992) Ion-exchange high-performance liquid chromatography analysis of oligodeoxyribonucleotide phosphorothioates. *Anal. Biochem.* **200,** 342–346.
13. Warren, W. J. and Vella, G. (1993) Analysis of synthetic oligodeoxyribonucleotides by capillary gel electrophoresis and anion exchange HPLC. *Biotech.,* in press.
14. Guttman, A. and Cooke, N. (1991) Capillary gel affinity electrophoresis of DNA fragments. *Anal. Chem.* **63,** 2038–2042.
15. Turner, K. (1991) New dimensions in capillary electrophoresis columns. *LC/GC* **9,** 350–353.
16. Heiger, D. N., Cohen, A. S., and Karger, B. L. (1990) Separation of DNA restriction fragments by high performance capillary electrophoresis with low and zero crosslinked polyacrylamide using continuous and pulsed electric fields. *J. Chrom.* **516,** 33–48.
17. Eadie, J. S., McBride, L. J., Efcavitch, J. W., Hoff, L. B., and Cathcart, R. (1987) High-performance liquid chromatographic analysis of oligodeoxyribonucleotide base composition. *Anal. Biochem.* **165,** 442–447.
18. Crowther, J. B., Caronia, J. P., and Hartwick, R. A. (1982) An improved method for the determination of oligonucleotide chain length using phosphodiesterase hydrolysis and ion-pair high-performance liquid chromatography. *Anal. Biochem.* **124,** 65–73.
19. Little, J. N., Cotter, R. L., Prendergast, J. A., and McDonald, P. D. (1976) Preparative liquid chromatography using radially compressed columns. *J. Chromat.* **126,** 439–445.

CHAPTER 10

Sequence Analysis of Oligodeoxyribonucleotides

Diane M. Black and Peter T. Gilham

1. Introduction

There are three main analytical methods for determining nucleotide sequences in oligodeoxyribonucleotides. In two of the methods, Mobility Shift Analysis *(1,2)* and Chemical Cleavage Analysis *(4,5)*, sequences are inferred by visualizing patterns of fragments derived from the oligomer by enzymatic or chemical cleavage, whereas the third, Nested Set Analysis *(6)*, involves the direct determination of the terminal nucleotide for each member of a nested set of oligomer subfragments. For analytical work on chemically-synthesized oligomers, the direct nested set method has some advantage over the other two, especially in cases involving oligomers whose purity might be in question or in analyses requiring assignments of identities and locations of modified or unusual bases. Brief summaries of the experimental procedures used in the three methods, together with a detailed description of all the steps involved in the Nested Set Analysis, are given below.

1.1. Mobility Shift Analysis

This approach uses a two-dimensional fractionation of a nested set of fragments derived from the oligonucleotide by partial exonuclease action. The set members, which contain radioactive labels at their common ends, are separated by electrophoresis on cellulose acetate at low pH in the first dimension, after which the fragments are transferred to a DEAE-cellulose thin-layer sheet for homochromatography

From: *Methods in Molecular Biology, Vol. 26: Protocols for Oligonucleotide Conjugates*
Edited by: S. Agrawal Humana Press Inc., Totowa, NJ

in the second dimension. Since the mobility of any set member depends on the fragment's size and base composition, and since the nucleotide sequence of the member (of chain length n) is fully contained within its larger neighbor (chain length $n + 1$), the difference in relative positions of the two members on the chromatogram must be caused solely by the extra terminal nucleotide within the $n + 1$ member. The relative positioning in the final pattern allows the identification of this terminal nucleotide in each case, thus yielding the sequence of the original oligomer. There are a few difficulties with this technique: The sequence cannot always be determined reliably by visual inspection alone; corrections relative to standards must be made for the position of each fragment in the pattern *(2)* and, even when the analysis is performed carefully, some deviations and ambiguities may occur *(3)*. Also, the method would be adaptable to the analysis of oligomers containing a modified nucleotide only if the nucleotide possessed an electrophoretic mobility markedly different from the mobilities of the four common ones.

1.2. Chemical Cleavage Analysis

In this technique, the terminally labeled oligomer is cut at specific sites in four different base-excising reactions, and the products are separated according to chain length, using polyacrylamide gel electrophoresis. The sequence may then be inferred from the resulting gel pattern by observing which base-excising reaction is responsible for cleavage at each position of the chain. Difficulties with this method include the possibility that extraneous bands may appear in the final autoradiogram, thereby making interpretation difficult *(5)*. Again, adaptation of the procedure for positioning of modified bases within a sequence would not be easy; it would require the development of new chemical reactions that could specifically excise nucleotides containing such bases.

1.3. Nested Set Analysis

This procedure produces sequence information by allowing direct visualization of the nucleotide at each position of the chain; the method is in contrast with the above two approaches in which the identity of each nucleotide is inferred from mobility shift or chemical cleavage patterns. The ability to directly "see" each nucleotide arises from the fact that, subsequent to the formation of the nested set, the set members are labeled at their noncommon ends instead of the common ends as in the above two analytical procedures. As a consequence, the position and identity of modified nucleotides may be easily established, along with the positioning of the regular nucleotides. The method, described in detail below, requires the initial enzymatic conversion of the molecule in question to a nested set of fragments that possess common 3' ends and ^{32}P labels at their noncommon 5' ends. The important steps in the analysis are based on the ability to separate the set members on a chromatographic sheet according to chain length and to degrade the separated members *in situ* with a second nuclease. The labeled, terminal nucleotides released by this treatment are then identified by chromatography in the second dimension, to yield the sequence of the original molecule. The strategy of analyzing nested fragments labeled at their noncommon ends has also been employed in another sequence analysis method published somewhat earlier *(7)*. However, the method is a rather lengthy one, requiring separation of the set members by two-dimensional homochromatography, followed by elution of each fragment from the thin layer plate, and subsequent enzymatic digestion of each fragment in solution to convert it to mononucleotides.

2. Materials

2.1. Enzymes and Substrates

1. Lyophilized spleen phosphodiesterase (Pharmacia Biotech, Inc., Piscataway, NJ).
2. Lyophilized calf intestinal phosphatase (Boehringer Mannheim, Indianapolis, IN).
3. Polynucleotide kinase (Boehringer Mannheim).
4. Nuclease Bal 31 (Bethesda Research Laboratories, Inc., Gaithersburg, MD).
5. Terminal deoxynucleotidyl transferase (Bethesda Research Laboratories).
6. Test oligomers synthesized chemically using techniques *(8–10)* discussed in accompanying chapters of this book.
7. [γ-^{32}P]ATP (3000 Ci/mmol, Amersham Corp., Arlington Heights, IL).
8. [α-^{32}P]ATP (300 Ci/mmol, ICN Radiochemicals, Irvine, CA).

9. N^6-Methyldeoxyadenosine 5'-phosphate prepared by a previously reported method *(11)*.
10. Nucleotide markers and yeast tRNA (type X, catalog no. R-9001, Sigma, St. Louis, MO).

2.2. Chromatographic Sheets and Solvents

1. PEI-cellulose MN 300, UV-254 precoated plastic sheets, 20 × 20 cm, (Brinkmann Instruments Co., Westbury, NY) are washed prior to use. Each sheet is soaked in 1 L 10% NaCl for 10 min, then in 1 L distilled water for 10 min, and again in water, with occasional agitation and with no intermediate drying. The same NaCl solution may be used for washing 4 sheets sequentially, but fresh distilled water is used for each sheet. Finally, the sheets are dried at room temperature for several hours and stored at –20°C wrapped in foil.
2. TLC Solvent A is deionized formamide—2.7*M* Tris-HCl buffer (1:1, v/v). The Tris-HCl buffer is prepared by mixing 206 g Tris base and 158 g Tris-HCl with water up to a final volume of 1 L. The formamide is deionized by adding 2 g mixed bed resin (AG 501-X8[D] from Bio-Rad Laboratories, Hercules, CA), to 100 mL formamide, and allowing it to stand for 30 min. The Tris-HCl and formamide are mixed just before use.
3. TLC Solvent B is 1.8*M* Tris-HCl (pH 8.0)—deionized formamide (1:1, v/v). The Tris-HCl is prepared by adjusting the pH of a 1.8*M* solution of Tris base with HCl at 25°C. The Tris-HCl and formamide are mixed just before use.
4. TLC Solvent C is 0.2*M* sodium acetate, adjusted to pH 4.35 with acetic acid.
5. TLC Solvent D is prepared by diluting 10 mL acetic acid with water, adjusting the pH to 3.65 with pyridine, and adding water to a final volume of 100 mL *(12)*. This solvent is freshly prepared each week.
6. TLC Solvent E is i-PrOH-AcOH—1% aqueous $(NH_4)_2SO_4$ (45:35:20, v/v), *(11)*. This solvent is freshly prepared each week.

3. Method

3.1. Partial Spleen Phosphodiesterase Digestion

The reaction mixtures contain 10/*x* nmol oligomer, where *x* is the chain length, in 50 µL 10 m*M* sodium 2-(*N*-morpholino)-ethanesulfonate (MES) (pH 6.5, 25°C). For molecules up to undecamers in size, the mixtures are treated with 5 mU spleen phosphodiesterase; for longer molecules, twice as much of the phosphodiesterase is added. Incubation is at 37°C, and 5 µL aliquots, withdrawn at 1, 3, 6, 10, 15, 20, 30, 40, 50, and 60 min, are added to a single Eppendorf tube kept on solid carbon dioxide. The combined aliquots are then heated at 100°C for 5 min to inactivate the enzyme.

3.2. Removal of 3' Phosphate Groups

To 5 µL of the spleen phosphodiesterase digest is added 5 µL 250 m*M* Tris-HCl (pH 7.6), 10 µL water, and 1 µL calf intestinal phosphatase solution (0.01 U/µL in 100 m*M* Tris-HCl [pH 8.0]—glycerol [1:1, v/v]). Incubation is at 37°C for 15 min, and the enzyme is inactivated by heating at 100°C for 5 min.

3.3. Labeling with Polynucleotide Kinase

To the phosphatase-treated reaction mixture are added 5 µL of a solution that contains 50 m*M* $MgCl_2$ and 25 m*M* dithiothreitol, and 1 µL polynucleotide kinase (5–6 U). The mixture is combined with 33 pmol dried [γ-^{32}P]ATP and incubated at 37°C for 15 min.

3.4. Separation by Chain Length by TLC in the First Dimension

One µL of the labeled oligomer mixture is applied to a point 1.5 cm from the bottom and 1.5 cm from one side of a thin layer sheet. Chromatography tanks are filled to a depth of 0.5 cm with developing solutions. For molecules up to undecamers in length, the sheet is developed with water to the origin, and then, without intermediate drying, with Solvent A to the top (about 6 h). For longer molecules, such as an icosamer, two thin layer sheets are used. The first sheet, used to separate the shorter fragments, is developed as for short molecules described above. The second sheet, used to separate the longer fragments, is prepared by stapling a wick of Whatman 3MM paper to the top. The wick, which is the width of the sheet and extends from the covered tank, permits chromatography to proceed for an extended period of time. The sheet is developed with water to the origin, and then, without intermediate drying, with Solvent B for 38 h. After development, each sheet is dried in a stream of cool air for 5 min, soaked in 500 mL methanol for 5 min to remove Tris and formamide, soaked in another 500 mL methanol for 5 min, and dried.

3.5. *In Situ Enzymatic Digestion to Mononucleotides*

A solution containing 3 U nuclease *Bal* 31 and 1 A_{260} U tRNA in 100 µL 20 m*M* Tris-HCl (pH 8.0, 25°C)–12 m*M* $CaCl_2$–12 m*M* $MgCl_2$–1 m*M* EDTA is streaked from a 20-µL capillary at about 4 µL/cm over the line of oligomers on the thin layer. The sheet is immediately covered with Saran Wrap™, clamped between two Lucite™ sheets, and incubated at 37°C for 1 h.

3.6. *Identification of 5' Terminals by TLC in the Second Dimension*

Identification of the labeled nucleotides is effected with one of the solvent systems listed (together with relevant mobilities) in Table 1. In order to prevent streaking of the spots, the sheet is soaked for 5 min in a mixture of 5 mL of the selected solvent and 500 mL water prior to chromatography. The sheet is dried, developed to the top in the second dimension with the selected solvent, dried, and subjected to autoradiography.

3.7. *Identification of 3' Terminals*

The oligomer (200 pmol) is incubated with 15 pmol [α-^{32}P]ATP and 17 U terminal transferase in 10 µL 10 m*M* potassium 4-(2-hydroxyethyl)-1-piperazine-ethanesulfonate (HEPES) (pH 7.2, 25°C), 5 m*M* $MgCl_2$, 0.2 m*M* dithiothreitol, at 37°C for 2 h. The oligomers are then degraded to 3' mononucleotides by adding 10 µL 100 m*M* MES (pH 6.5), and 2 µL of spleen phosphodiesterase solution (10 mU/µL in H_2O), and incubating at 37°C for 2 h. One microliter of the digest is spotted on a PEI-cellulose thin layer sheet next to a spot containing a set of 3' nucleotide standards, dried, developed to the top in Solvent D, and subjected to autoradiography.

4. Discussion

The analytical strategy depends on the conversion of the molecule in question to a nested set of oligomers that have common 3' ends and ^{32}P labels at their noncommon 5' ends. The set is created by treating the oligomer with spleen phosphodiesterase, freezing portions of the mixture at timed intervals, and then heating the combined portions to inactivate the enzyme. At this point, the solution contains nucleoside

Table 1
Mobilities of 5'-Nucleotides Relative to Deoxyguanosine 5'-Phosphate

TLC Solvent	dG	dA	dT	dC	m^6dA	m^5dC
C	1.0	2.9	4.7	5.9	2.9	6.7
D	1.0	2.6	2.4	3.2	2.9	3.3
E	1.0	11.5	6.5	19.0	16.0	23.5

3' phosphates as well as the desired set and, in order to prevent labeling of these monomers in the next step, their phosphate groups are removed by treatment with phosphatase. Following inactivation of the phosphatase, the set of fragments is labeled with polynucleotide kinase and [γ-^{32}P]ATP. For the analysis of small oligomers (with chain lengths of up to about 12), the members of the set are separated by length on a single PEI-cellulose thin layer sheet. For larger oligomers, however, it is preferable to use two sheets: one to separate the smaller members of the set and the other to separate the larger species by extending the chromatography over a longer time. Nuclease *Bal* 31 is used to digest the oligomers *in situ* to 5' mononucleotides, which are then separated by chromatography in the second dimension. Subsequent autoradiography gives a pattern of spots that allows the sequence to be read.

The technique was tested initially in our laboratory *(6)* in the analysis of a number of molecules, including one containing a 5-methyldeoxycytidine residue: GATCATCTTCTp, CCATGm^{5}CAT, ATATCATAT, CGAGTTTGACGp, CGCTAAACTCGp, AGTCAT, TGACGTGA, ATCTTATTTTGTGTTTGTA, and GGTCACAAGATCTGAAGCAG; the TLC patterns corresponding to the analyses of three of these are shown in Fig. 1A–C. The patterns obtained from the other test oligomers were equally unambiguous in interpretation except the one corresponding to a particular region in the sequence of the icosamer, the last member in the above list. The pattern of nucleotide spots deriving from the left half of this molecule is reproduced in Fig. 1D in order to demonstrate a potential limitation to the size and structure of large oligomers that can be completely analyzed by the method. The arrangement of these spots indicates a less than satisfactory first dimension resolution of the nested set members that have chain lengths of 14, 13, and 12, with 5' terminals of A, A, and G, respectively. The difficulty seems to be related to the chain lengths of the set members as well as

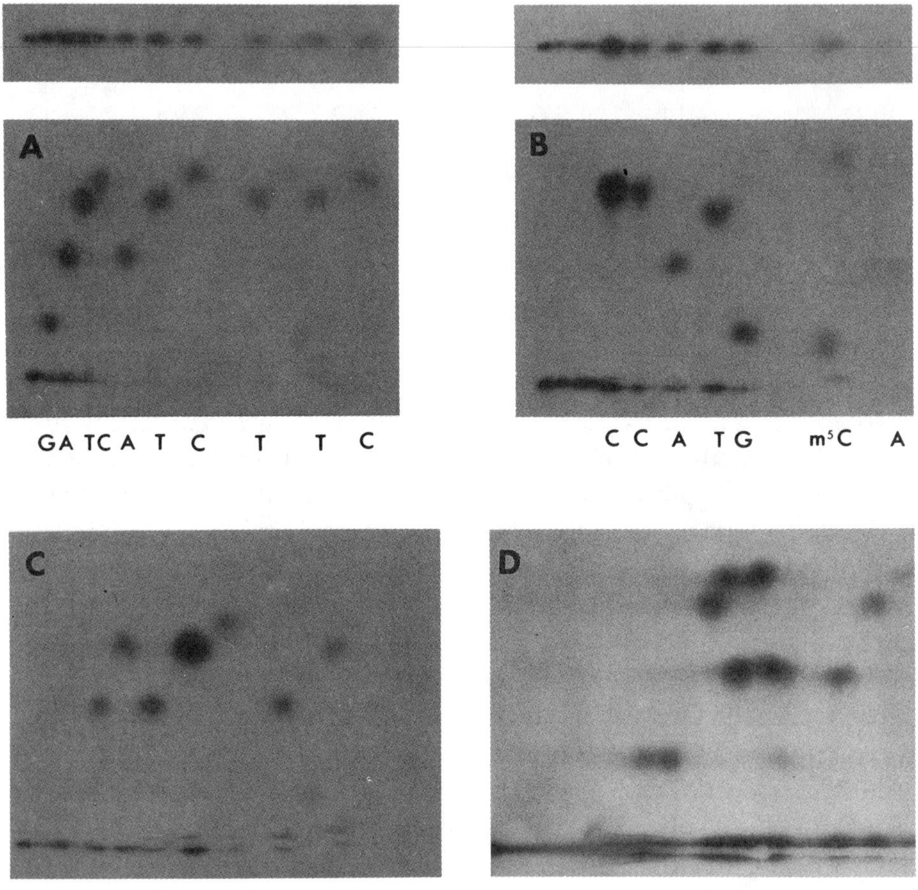

Fig. 1. TLC patterns deriving from the analyses of oligomers: **A:** GATCA-TCTTCTp; **B:** CCATGm^{5}CAT; **C:** ATATCATAT; **D:** GGTCACAAGATCTG-AAGCAG. In each case, the first separation proceeded from left to right with TLC Solvent A (for **A–C**) and TLC Solvent B (for **D**), and the second dimension from bottom to top using TLC Solvent C. At the top of the figure are the intermediate patterns obtained for oligomers A and B by autoradiography after the first dimension separation and prior to the second. In each of the patterns **A–C**, the spot indicated by an arrow arises from contaminating inorganic phosphate. The pattern of the icosamer **D** represents the analysis of the longer members of the nested set using extended chromatography in the first dimension. The sequence corresponding to the right half of this molecule was obtained by using the standard first dimension separation that is illustrated in the analyses of molecules **A–C**.

the nature of their 5' terminals because the shorter oligomers corresponding to the sequences to the right of this area are well resolved by chromatography with TLC Solvent A, even though they contain a similar arrangement (A, A, G) of 5' terminals deriving from the sequence near the 3' end of the icosamer. This is the only problem detected so far, although it is possible that there are other special sequence arrangements that may cause difficulty in oligomer resolution when they are located near the 5' ends of long oligonucleotides. Some examples of the use of the method in analyzing other oligonucleotides are given in refs. *13–17*.

5. Notes

1. The spleen phosphodiesterase used to form the nested sets seems to have little base specificity, and the digestion conditions described above can be used successfully for molecules ranging up to icosamers in size. However, it has been noticed that spots derived from dinucleotides in the nested sets are not as intense as the others; whether this is because of some cleavage specificity in the phosphodiesterase or a preferred labeling of larger molecules by the kinase has not been investigated.
2. The fact that there are no intermediate purification steps in the method is an advantage. However, the avoidance of intermediate purification requires the use of modified experimental protocols for the two labeling procedures so that extraneous spots do not appear in the final patterns. In the case of the kinase reaction, the set of oligomers is added in molar excess over the [γ-^{32}P]ATP, so that no ATP remains at the end of the incubation. Nevertheless, a radioactive contaminant that comigrates with inorganic phosphate always appears on the patterns; it runs near the trimer position in the first dimension and just below pdG with Solvent C in the second (Fig. 1). The inorganic phosphate is present in commercial preparations of [γ-^{32}P]ATP, and its radioactive spot becomes more intense with prolonged incubation of the kinase reaction, an increase that probably results from a side reaction of the kinase in which it catalyzes the hydrolysis of the γ phosphate group *(18)*.
3. The chromatographic separation in the first dimension is a modification of a previously reported step-gradient TLC system using high concentrations of Tris-HCl and urea *(19)*. Although oligomers of a nested set can be well resolved in this gradient system, the stepwise elution procedure was found to be somewhat tedious in the present application. However, the testing of a number of modifications of this system showed that equivalent resolution can be achieved by using a single solvent

containing a high concentration of formamide. The mobility of an oligomer in the formamide system is mainly dependent on its chain length, although there is some variation caused by differing base compositions, with dG residues causing extra retardation. It will be necessary for users to take this into account if mixed oligonucleotide probes are to be analyzed by the method.

4. Several enzymes have been tested for their ability to digest oligomers when they are adsorbed to PEI-cellulose sheets. Snake venom phosphodiesterase and the nucleases S1 and P1 tend to leave some dimers and larger oligomers undigested. Dimers migrate below pdG whereas the larger species remain at the origin during the second chromatographic step with Solvent C. In contrast to these enzymes, nuclease Bal 31 has the capacity to completely degrade oligonucleotides along the line where the capillary containing the enzyme solution contacts the thin layer sheet; undigested oligomers on each side of the line remain at the origin during chromatography in the second dimension. However, enzymatic digestion proceeds to completion only when the enzyme solution contains tRNA, which presumably competes with the labeled oligomers for binding sites on the PEI-cellulose, thus rendering them more accessible to enzyme attack. The products of the digestion are nucleoside 5'-phosphates, and these mononucleotides, including the 5'-phosphates of 5-methyldeoxycytidine and N^6-methyldeoxyadenosine, are well separated by one or other of the solvent systems listed in Table 1.
5. The 3' terminal residue of the oligomer is not detected in the two-dimensional separation technique because, at the end of the phosphodiesterase and phosphatase treatments (Sections 3.1. and 3.2.), this residue exists as a nucleoside, which is not labeled by the kinase. However, a 3'-terminal labeling procedure has been adapted for use in identifying the nucleotide at this position, when this information is required. It involves the use of terminal deoxynucleotidyl transferase to add a single ^{32}P-labeled adenosine 5'-phosphate to the 3' end of a separate sample of the oligomer, followed by degradation of the product with spleen phosphodiesterase. The terminal deoxyribonucleoside is thus released as a labeled nucleoside 3' phosphate, which is then identified by a one-dimensional separation on a PEI-cellulose sheet. The relative mobilities of the 3' nucleotides in Solvent D used for this separation are essentially the same as those listed for the 5' nucleotides in Table 1; inorganic phosphate migrates below dGp, and other contaminants in commercial ATP preparations remain at the origin. Under the conditions normally used for terminal transferase reactions, several ribonucleotide residues are added and, in the present application, this would cause difficulty in

interpreting the TLC patterns because two nucleotide spots would always be present. However, in testing this system, it was noted that, by using a large oligomer to ATP ratio (10:1 or greater) and by substituting Mg^{2+} for the commonly used Co^{++} ion, multiple additions are not observed.

6. The procedures described above constitute a simple and accurate method for determining the sequences of oligodeoxyribonucleotides of up to about 20 nucleotides in length. The analysis employs readily available equipment and is quite sensitive, using <0.1 A_{260} U of oligomer for the spleen phosphodiesterase reaction mixture with only a small portion of this mixture required for application to the PEI-cellulose thin layer sheet.

References

1. Brownlee, G. G. and Sanger, F. (1969) Chromatography of ^{32}P-labelled oligonucleotides on thin layers of DEAE-cellulose. *Eur. J. Biochem.* **11,** 395–399.
2. Tu, C. D., Jay, E., Bahl, C. P., and Wu, R. (1976) A reliable mapping method for sequence determination of oligodeoxyribonucleotides by mobility shift analysis. *Anal. Biochem.* **74,** 73–93.
3. Tu, C. D. and Wu, R. (1980) Sequence analysis of short DNA fragments, in *Methods in Enzymology,* vol. 65 (Grossman, L. and Moldave, K., eds.), Academic, New York, pp. 620–638.
4. Maxam, A. M. and Gilbert, W. (1980) Sequencing end-labeled DNA with base-specific chemical cleavages, in *Methods in Enzymology,* vol. 65 (Grossman, L. and Moldave, K., eds.), Academic, New York, pp. 499–560.
5. Banaszuk, A. M., Deugau, K. V., Sherwood, J., Michalak, M., and Glick, B. R. (1983) An efficient method for the sequence analysis of oligodeoxyribonucleotides. *Anal. Biochem.* **128,** 281–286.
6. Black, D. M. and Gilham, P. T. (1985) A new method for sequence analysis of oligodeoxyribonucleotides. *Nucleic Acids Res.* **13,** 2433–2442.
7. Wu, R., Tu, C. D., and Padmanabhan, R. (1973) Nucleotide sequence analysis of DNA. XII. The chemical synthesis and sequence analysis of a dodecadeoxynucleotide which binds to the endolysis gene of bacteriophage lambda. *Biochem. Biophys. Res. Comm.* **55,** 1092–1099.
8. Gough, G. R., Singleton, C. K., Weith, H. L., and Gilham, P. T. (1979) Protected deoxyribonucleoside-3' aryl phosphodiesters as key intermediates in polynucleotide synthesis. Construction of an icosanucleotide analogous to the sequence at the ends of Rous Sarcoma Virus 35S RNA. *Nucleic Acids Res.* **6,** 1557–1570.
9. Gough, G. R., Collier, K. J., Weith, H. L., and Gilham, P. T. (1979) The use of barium salts of protected deoxyribonucleoside-3' *p*-chlorophenyl phosphates for construction of oligonucleotides by the phosphotriester method. High yield synthesis of dinucleotide blocks. *Nucleic Acids Res.* **7,** 1955–1964.
10. Gough, G. R., Brunden, M. J., Nadeau, J. G., and Gilham, P. T. (1982) Rapid preparation of hexanucleotide triester blocks for use in polydeoxyribonucleotide synthesis. *Tetrahedron Lett.* **23,** 3439–3442.

11. Griffin, B. E. and Reese, C. B. (1963) The synthesis of N^1 and N^6-methyl-adenosin 5'-pyrophosphates. Possible substrates for polynucleotide phosphorylase. *Biochim. Biophys. Acta* **68,** 185–192.
12. Bernardi, A. (1974) A fast method of analysis of the 5' terminal nucleotides of deoxyribooligonucleotides. *Anal. Biochem.* **59,** 501–507.
13. Marconi, R. T. and Hill, W. E. (1988) Identification of defined sequences in domain V of *E. coli* 23S rRNA in the 59S subunit accessible for hybridization with complementary oligodeoxyribonucleotides. *Nucleic Acids Res.* **16,** 1603–1614.
14. Caron, P. R. and Grossman, L. (1988) Incision of damaged versus nondamaged DNA by the *Escherichia coli* UvrABC proteins. *Nucleic Acids Res.* **16,** 7855–7865.
15. Marconi, R. T. and Hill, W. E. (1989) Evidence for a tRNA/rRNA interaction site within the peptidyltransferase center of the *Escherichia coli* ribosome. *Biochemistry* **28,** 893–899.
16. Rabow, L. E., Stubbe, J., and Kozarich, J. W. (1990) Identification and quantitation of the lesion accompanying base release in bleomycin-mediated DNA degradation. *J. Am. Chem. Soc.* **112,** 3196–3203.
17. Marconi, R. T., Lodmell, J. S., and Hill, W. E. (1990) Identification of a rRNA/chloramphenicol interaction site within the peptidyltransferase center of the 50S subunit of the *Escherichia coli* ribosome. *J. Biol. Chem.* **265,** 7894–7899.
18. Van de Sande, J. H., Kleppe, K., and Khorana, H. G. (1973) Reversal of bacteriophage T4 induced polynucleotide kinase action. *Biochemistry* **12,** 5050–5055.
19. Randerath, K., Randerath, E., Chia, L. S. Y., Gupta, R. C., and Sivarajan, M. (1974) Sequence analysis of nonradioactive RNA fragments by periodate-phosphatase digestion and chemical tritium labeling: characterization of large oligonucleotides and oligonucleotides containing modified nucleosides. *Nucleic Acids Res.* **1,** 1121–1141.

CHAPTER 11

Gel-Capillary Electrophoresis Analysis of Oligonucleotides

Alex Andrus

1. Introduction

Gel-capillary electrophoresis is a new alternative to traditional electrophoretic techniques for the analysis of oligonucleotides *(1–5)*. The advantages are dramatically decreased analysis time, excellent resolution, in-capillary detection, reduced sample quantities, and automation. Capillary electrophoresis (CE) is already established as an important analytical tool for other biomolecules, such as proteins, peptides, and high-mol-wt, double-stranded nucleic acids *(6,7)*. Now the CE method has been extended to single-stranded oligonucleotides using special polymeric, gel-filled capillaries. In a gel-matrix capillary, as with polyacrylamide slab gel electrophoresis (PAGE), DNA separates primarily, and predictably, on the ratio of mass-to-charge under the influence of an electric field. The same elution pattern of small oligonucleotides followed by the largest, usually the product, is obtained. Single base resolution can often be attained beyond 100 bases. The analysis, called an electropherogram, is *quantitative*. It can be displayed, stored, integrated, and printed like HPLC chromatograms. Slab gels (PAGE), on the other hand, are visualized at a single time point in the analysis, do not easily yield to integrative quantitation, and are thus a *subjective* method of analysis. The resolution of Gel CE surpasses the current traditional techniques, HPLC and PAGE, for oligonucleotide analysis. The combination of the gel materials,

From: *Methods in Molecular Biology, Vol. 26: Protocols for Oligonucleotide Conjugates*
Edited by: S. Agrawal Copyright ©1994 Humana Press Inc., Totowa, NJ

acrylamide/urea and others, and heating of the capillary confers a significant denaturing effect for predictable elution patterns devoid of secondary structure artifacts. The gel capillaries sustain multiple injections depending on the gel matrix and storage and handling conditions. The parameters affecting oligonucleotide analysis; resolution limits, time, sample preparation and concentration, reproducibility, electrophoresis settings, and adaptability to automation; are examined in this chapter.

2. The Capillary Electrophoresis System

There are a variety of commercial electrophoresis instruments available, with a range of features, specifications, and degrees of automation. The generic instrument configuration, shown in Fig. 1, consists of a high-voltage power supply with electrodes immersed in static buffer reservoirs that are spanned by the fused silica, gel filled capillary. At a fixed point, the capillary is mounted at the detector site. A variable wavelength, UV/VIs detector collects light passed directly through the capillary, giving the typical HPLC type analog voltage signal to be processed by an A/D converter (*see* Note 1).

The system used to generate the data shown here is the Model 270A/HT Capillary Electrophoresis System equipped with MicroGel capillaries (Applied Biosystems, Foster City, CA) *(3–5)*.

2.1. Capillaries

The typical gel capillary is a long (50 cm), thin-fused silica tube, coated on the outside with polyimide. The small internal diameter of the capillary, 25–100 µm, allows fast and efficient heat dissipation, permitting the use of high field strengths to achieve rapid analysis times. At the detection point, the polyimide coating is removed to give a clear window location, which is about 5 mm long. The gel matrix can be loaded inside the capillary by a variety of methods under carefully controlled conditions. The gel concentration must be constant from one end to the other and identical for all capillaries upon which data is to be compared. In addition, a void free gel must be produced, since they can cause capillary failure, erratic current, and spurious peaks. Small bubbles may even travel through a capillary, causing a prolonged period of absorbance that obliterates all sample detection. Since several commercial sources of gel-filled capillaries now exist, it would appear that preparing one's own is not a relatively worthwhile endeavor.

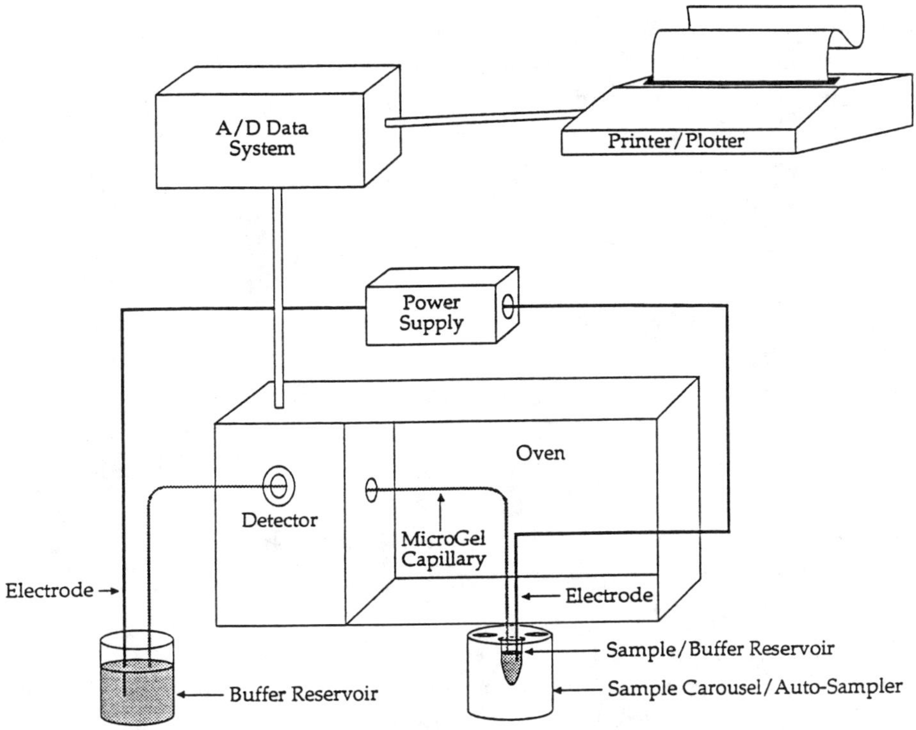

Fig. 1. Schematic diagram of a Capillary Electrophoresis instrument.

The gel materials that are useful for CE are crosslinked or chain-entangled elastic polymers with hydrophilic and denaturing properties. Polyacrylamide, the traditional polymer used for slab electrophoretic analysis of nucleic acids, has been extensively investigated. The optimum acrylamide, *bis*-acrylamide, buffer, and denaturant formula has been intensively sought *(1,2)*. The commercial development of polyacrylamide capillaries and their use even by experts has been hampered by instability problems *(8)*. It is clear that when urea is present, capillary stability is compromised and the capillary will have a short shelf life. Acrylamide/urea capillaries also must be refrigerated off the instrument when not in use. MicroGel capillaries do not contain acrylamide or urea, but are filled with a viscous-elastic, high-mol-weight polymer *(3–5)*. Although not containing traditional chemical denaturants, such as urea or formamide, hydrogen bonding between nucleobases is almost totally disrupted upon interaction with the MicroGel matrix and the elevated capillary temperature. MicroGel capillaries are stable enough to endure many sample injections. Given proper handling and analysis conditions, typically 50–100 analyses/MicroGel capillary are attained. The buffer chambers and the MicroGel capillaries are filled with 75 m*M* Tris-phosphate buffer/10% methanol, pH 7.5. It is important to change the buffers every few hours of electrophoresis time, especially on the sample side. On most systems, this is the smaller volume buffer chamber, which undergoes ion depletion during the course of electrophoresis. Changes in ionic strength or pH of the buffer or sample solutions have large effects on elution time, resolution, and peak size. Because there may be some variations in the absolute elution times between runs, exact identification may be facilitated by using relative migration times. When a reference marker is included in the same run as the unknown sample, a relative elution time may be calculated by dividing the elution time of the unknown oligonucleotide peaks by the time of the reference peak. This relative elution time is very reproducible and may be used to predict the elution time of the same oligonucleotide in subsequent runs.

Elution time variability of oligonucleotides on MicroGel capillaries is about 1.0% when an internal standard is included with the sample in a run. Accurate peak area quantitation also requires an internal standard because of possible differences in molar extinction coefficients. Injection anomalies can be caused by sample conductivity differences or changes in electrode alignment.

Installation of a MicroGel capillary takes only a few minutes on the Model 270A. Setup and maintenance is easy and fast compared to more labor-intensive analytical methods, such as HPLC or PAGE. The capillary oven chamber can be precisely temperature controlled between ambient and 60°C. MicroGel capillaries should be held at constant temperature between 30–50°C. Temperature cycling or temperatures higher than 50°C can slowly degrade the gel-capillary matrix. All examples shown here were conducted at 40°C. Higher temperatures disrupt secondary structures of hydrogen bonding that can lead to extra or broadened peaks.

2.2. Establishing a Method

The Model 270 allows the control of many variables that affect the analysis. Together with the data system, parameters such as resolution, speed, and sample effects can be optimized. Since both the injection and elution of the oligonucleotide occur under the influence of a precisely controlled electric field, the settings are critical. The capillary establishes an electrical circuit at very high voltage and very low current. The voltage is held constant and the current may change as a function of the resistance. The following are some recommended method parameters.

2.3. Gel-Capillary Electrophoresis Parameters

Detector:	Wavelength = 260 nm
	Rise time = 1 s
	Range = 0.5
	Zero = yes (auto zero at the beginning of each run)
Sample:	Loading time = 5 s
	Voltage = –5 kV
	Temperature = 40°C
Time:	Run Time = 22 min
	Voltage = –15 kV
	Temperature = 40°C

2.4. Data Acquisition/Integrator Parameters

Acquisition Time = 15–50 min
Sampling rate = 5 points/s
Signal = 0–1 V
Integration format = baseline-to-baseline or peak-to-peak
Area reject = 10,000

The data-acquisition system used to process the data shown here is the Model 1020S PE/Nelson Personal Integrator with a Panasonic KX-P1080i dot-matrix printer and a Hewlett Packard Model 7550A Graphics Plotter.

3. Sample Preparation

1. Quantitate sample; optional desalting procedure (*see* Note 3).
2. Dry aliquot in 500 µL tube; concentration: 0.1 (0.2 > 40-mer) A_{260} U.
3. Redissolve in 100 µL deionized water.

Single-stranded, synthetic oligonucleotides, DNA or RNA, from four to about 150 bases, can be effectively analyzed by Gel CE. Various oligonucleotide analogs can also be evaluated, including phosphorothioates. Analysis can be done directly from crude reaction mixtures containing ammonium hydroxide or from lyophilized (dried in a vacuum centrifuge) samples dissolved in water. No spurious peaks will be caused on the electropherogram owing to the ammonium hydroxide, but changes to the sample pH may inhibit sample introduction (*see* Note 2).

Introduction of an oligonucleotide onto the capillary occurs by *electrokinetic* injection, whereby the highly charged sample is induced to migrate into the capillary inlet under the influence of an electric field. This is different than the physical placement of a sample onto an HPLC column, via direct injection, or PAGE, by physical loading of the sample in a well at the top of the gel. The electrokinetic injection voltage is less than the running voltage and the duration is typically 3–20 s. The amount of oligonucleotide sample that enters the capillary increases, but not linearly, with higher concentration, longer injection duration, or higher voltage. It can be highly influenced however, by the presence of other salts that alter the conductivity of the sample solution.

The crude sample is usually sufficiently free of salts to allow the proper conductance for adequate loading. Samples containing inhibiting levels of salt and other contaminants should undergo purification (e.g., Oligonucleotide Purification Cartridge, OPC) or desalting operations, such as precipitation, prior to Gel CE analysis. The presence of small amounts of organic solvents does not affect injection. Samples that have been purified by HPLC and eluted in salt containing media can be analyzed directly as long as they do not contain excessive amounts of salts, which will limit the electrokinetic loading. If salt contamination is suspected, a desalting procedure may be necessary (*see* Note 3). Oligonucleotide concentrations of about 0.1–2.0 A_{260} U (3–66 µg)/mL in deionized water may be appropriate. There is a wide range in sample concentrations because of the nature of electrokinetic injection, which is sensitive to conductance and the presence of competing salts in the sample. Nonionic solutions containing, for example, urea or formamide can also be used at comparable concentrations. At least 20 µL of sample solution (about 0.004 A_{260} U of DNA) in a 500

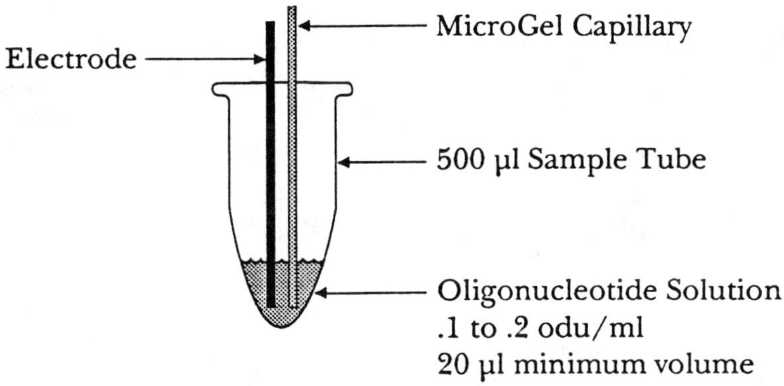

Fig. 2. Sample tube with capillary and electrode.

µL microcentrifuge (Eppendorf style) tube is required to ensure adequate immersion of the capillary tip and electrode during injection (Fig. 2). Several repetitive injections can be made from this amount of sample. However, after a few injections, the sample introduced onto the capillary will diminish, resulting in smaller peaks, owing to the carry-over of salts from the buffer by the capillary and electrode and depletion of the sample.

Electropherograms generated from gel-filled capillaries present graphic and highly resolved representations of oligonucleotide purity. With a data system, quantitative results can be obtained from peak areas. As with any chromatographic data, reliable conventions must be developed to obtain consistent interpretations. Variables such as attenuation, risetime, and peak width on a data acquisition system can dramatically affect purity analysis. Once data acquisition parameters are set, consistent sample concentration, injection duration, and injection voltage should also be set. Generally, synthetic yields are consistent enough that analysis conditions can be held constant. Long oligonucleotides often require higher sample concentrations or increased injection times because of reduced product yields, and therefore reduced purity. Run times also need to be increased for longer oligonucleotides.

3.1. Analysis of Crude 18-mers

Figure 3A shows an electropherogram of a typical crude oligonucleotide, an 18-mer made under standard synthesis conditions. In precise terms, the sample was prepared after measuring the absorbance of an aliquot at 260 nm. After chilling the ammonia solution on ice, an aliquot containing 0.1 A_{260} U was transferred to a 500 μL microcentrifuge tube and dried in a vacuum centrifuge tube (Eppendorf style). After 200 μL water was added and mixed by vortexing, the sample was positioned in the auto-sampler and electrokinetically injected. Using the sample parameters given above, the electropherogram and integration report shows the product, the largest peak eluting at 12 min, to constitute 84% of the total integrated area. The integrated percent of the product peak is often cited as a measure of purity by Gel CE and by HPLC. However, to be accurate, it must be clarified that only a single wavelength, typically 260 nm, is monitored. Also, the relative extinction coefficients of all the species in the sample are not corrected. The integration parameters also influence the quantitation, based on threshold reject values, peak identification format, and so on. In practical terms, the purity shown in Fig. 3A would indicate excellent performance for this oligonucleotide in most, if not all, applications, such as PCR, sequencing, probe experiments, and so on. Most of the sample components that elute prior to the product are failure sequences, sometimes referred to as "*N*-" species, whereas the desired, full length product is called the "*N*" peak or band. Later eluting impurities arise from other synthesis imperfections, comprised of either higher mol wt or less charged species. These are termed "*N*+". They may be caused by incomplete deprotection or branching of the oligonucleotide during synthesis. This elution pattern is, of course, consistent with slab gel electrophoresis (PAGE) where the components *N*–, *N*, and *N*+ are viewed in a vertical lane format of bands. This sample, Fig. 3A, shows several discrete *N*– peaks and virtually no *N*+ peaks. Since the detectability limit by UV shadow or staining detection with PAGE is about .05 A_{260} U, these low-level impurities may not be apparent. In addition, the high resolution of gel-capillary CE is required to separate some impurities from the product. Figure 3B shows a different synthesis of the same 18-mer sequence with a greater amount of *N*+ impurities and an integrated product peak area of 65%. By PAGE analysis, the oligonucleotides of Figs. 3A and B cannot be differentiated by purity.

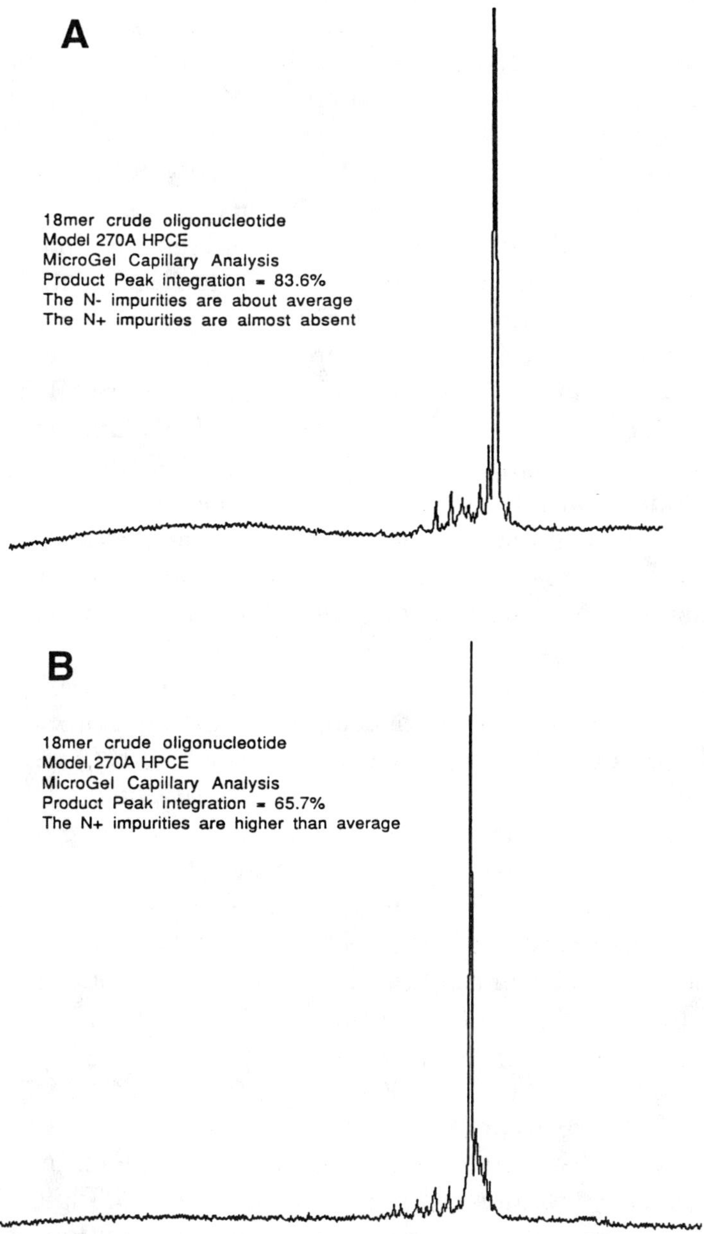

Fig. 3. **A:** Crude 18-mer 5' TCA CAG TCT GAT CTC GAT 3'. **B:** Crude 18-mer 5' TCA CAG TCT GAT CTC GAT 3'.

3.2. Analysis of a 29-mer, 65-mer, 72-mer, and 120-mer

As the length of the oligonucleotides increase, resolution decreases. The size of the product peak (*N*) relative to impurities decreases also. This is simply a function of the imperfect synthesis chemistry. For example, at the typical 98% stepwise efficiency, the yield of a 20-mer will be about 68%. At this same efficiency, the yield of a 40-mer will be 45%. As a consequence, longer oligonucleotides are less pure and the product is less distinct by any analytical method. By injecting more sample, either with a higher concentration, longer injection time, or higher injection voltage, the capillary may be overloaded and result in loss of resolution. Figure 4A (*see* pp. 288 and 289) shows a 29-mer with a relatively high amount of *N*+ impurities. Figure 4B is a 65-mer with a product peak integration of 19%. Note that in this electropherogram, during elution of all the components, the baseline remains high. This pattern is indicative of an overloaded capillary, which may be difficult to avoid when analyzing impure or long oligonucleotides. The 72-mer in Fig. 4C is of higher purity, 51% product peak integration. Here the absorbance returns to the baseline during elution of all the components of the sample.

Even longer oligonucleotides will show single base resolution when critical parameters, such as concentration, injection time, and injection voltage, are optimized, often empirically. The 120-mer shown in Fig. 4D gives a quantitative assessment of purity, unavailable by any other method, for such a long oligonucleotide. The integrated product purity of 10% is consistent with a stepwise efficiency of about 98%, probably the maximum that can realistically be attained. Higher trityl measurements can sometimes be obtained, but they are augmented by side reactions that contribute to the trityl release, but detract from product purity.

3.3. Analysis of a Mixture of 119-mer and 120-mer

As an illustration of the resolving power of Gel CE, the 120-mer in Fig. 4D and a 119-mer (same sequence as the 120-mer minus the third base from the 3' terminus) were mixed and analyzed. Figure 5 (*see* p. 290) is the electropherogram showing useful separation of the two long oligonucleotides, differing in length by only one base. Considering both oligonucleotides were in a crude state of approx 10% purity, the method demonstrates excellent resolution.

3.4. Analysis of Mixture of Two 18-mers

Although separations of oligonucleotides by Gel CE occur primarily on the ratio of charge-to-mass, therefore length, there is also a hydrophobic component to the elution order. The MicroGel matrix is an entangled polymer in which the large oligonucleotide molecules must migrate through pores in response to the electric field. Polyacrylamide also provides a sieving type matrix through which charged molecules pass. The shape and/or hydrophobicity does influence elution patterns. As an example, a mixture of 18-mers, sharing the same base composition ($A_4G_3C_5T_6$) separate in Fig. 6 (*see* p. 291). Even with the same mol wt and net charge, a different sequence order of the three bases at the 5' end allow separation. To achieve separation between oligonucleotides of the same length, let alone the same base composition, is not a typical occurrence. However, this example shows the significance of the chemical interaction between the gel matrix and oligonucleotides.

3.5. Analysis of Trityl-On and Trityl-Off Oligonucleotides

Trityl-on oligonucleotides can separate easily from trityl-off oligonucleotides by Gel CE analysis. The presence of the very hydrophobic 5' dimethoxytrityl (DMT) interacts strongly with the MicroGel matrix, retarding elution, compared to the 5' hydroxyl terminus. The elution difference between trityl-on and trityl-off oligonucleotides by Gel CE is not as pronounced as in reverse-phase HPLC, but more so than in PAGE. The same considerations in preserving the trityl group of the sample for HPLC apply to CE analysis also. Acidic solutions must be avoided and a few microliters of triethylamine, when added to the sample before concentrating under vacuum, may help. Figure 7 (*see* p. 292) shows a mixture of trityl-on and trityl-off oligonucleotides with the same length and sequence. The trityl-on oligonucleotides elute later owing to the hydrophobic interaction with the gel matrix and the increased mass of the molecule.

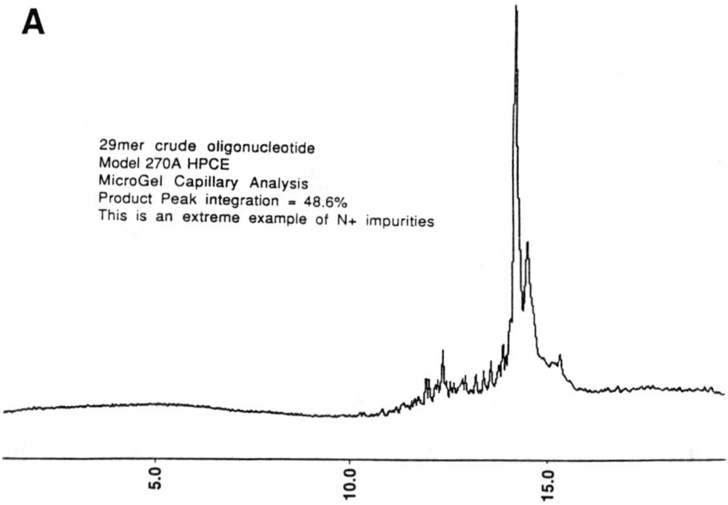

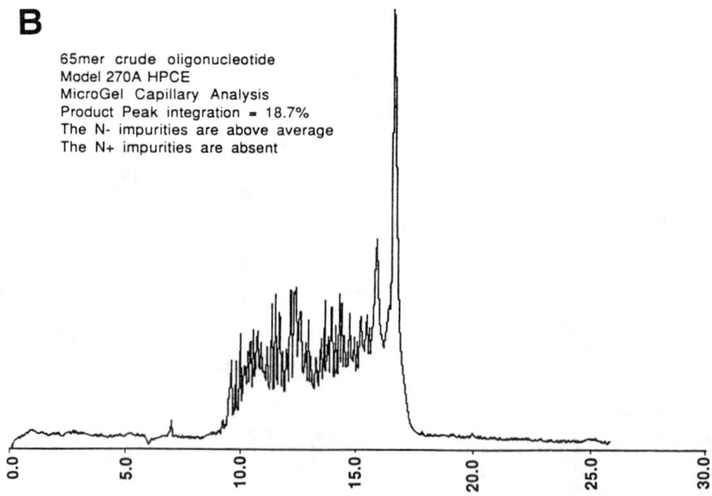

Fig. 4. **A:** 29-mer sequence 5' CCA TGA AGC TTT GAC CAT GAA AAT GGA GA 3'. **B:** 65-mer sequence 5' TCT CTG CGC GAC GTT CGC GGC GGC GTG TTT GTG CAT CCA TCT GGA TTC TCC TGT CAG TTA GCT TT 3'. **C:** 72-mer sequence 5' AGG GCC GAG CGC AGA AGT GGT CCT GCA ACT TTA TCC

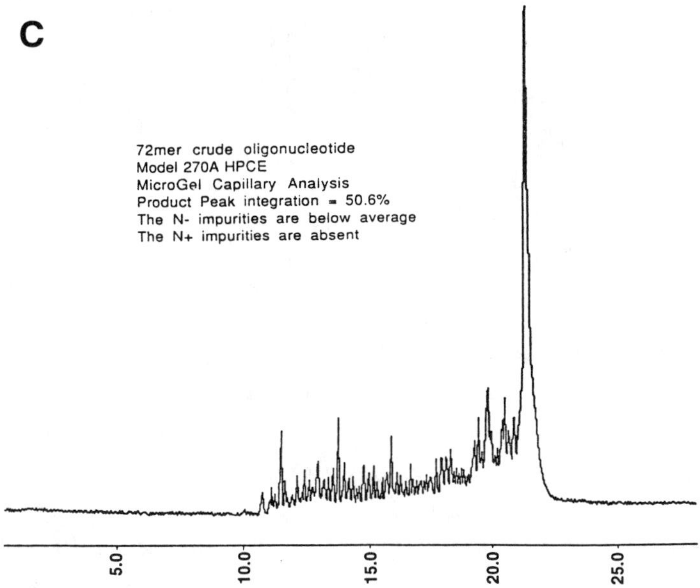

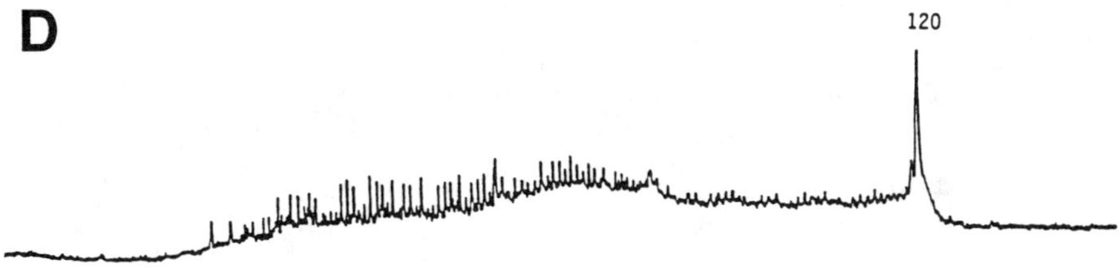

GCC TCC ATC CAG TCT ATT AAT TGT TGC CGG GAA GCT 3'. **D:** 120-mer sequence 5' CAA CAG GGG ATT TGC TGC TTT CCA TTG AGC CTG TTT CTC TGC GCG AGG TTC GCG GCG GCG TGT TTG TGC ATC CAT CTG GAT TCT CCT GTC AGT TAG CTT TTC ACA GTC TGA TCT CGA TAT 3'.

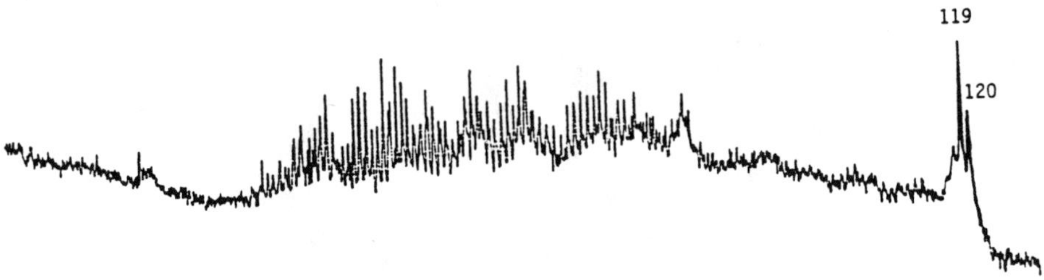

Fig. 5. Mixture of 120-mer and 119-mer.

3.6. Analysis of 5' Amino Oligonucleotides and their Adducts

More and more frequently, oligonucleotides are derivatized post-synthesis and labeled by covalent attachments with other molecules to serve a growing variety of novel applications *(9,10)*. The most common site of attachment is the 5' end. A protected amino phosphoramidite reagent, such as AminoLink 2 (6-*N*-trifluoroacetylaminohexyl-1-diisopropylamino, methyl phosphoramidite), can be used as the final base on the DNA synthesizer, giving a nucleophilic amine group after cleavage and deprotection (*see* Chapter 3 for details). The 5' amino-linked oligonucleotide can be analyzed and purified by the same methods as underivatized, 5' OH oligonucleotides. The aminohexyl phosphate moiety confers a slight retarding effect during electrophoresis, relative to the 5' hydroxyl, because of the slight increase in mass and net zero charge difference. Gel-capillary CE is a convenient method to assess the efficiency of the amino phosphoramidite coupling and to gauge purity of the crude amino-link oligonucleotide product. Figure 8A shows a mixture of 5' hydroxyl and 5' amino-link oligonucleotides. The amino-link oligonucleotides consistently elute later than their 5' hydroxyl counterpart sequences. Further purification at this stage may not be necessary.

Amino-linked oligonucleotides are then reacted further, in solution, usually with an active ester molecule, such as NHS-biotin or NHS-fluorescent dyes, to make biotinylated and fluorescent dye oligonucle-

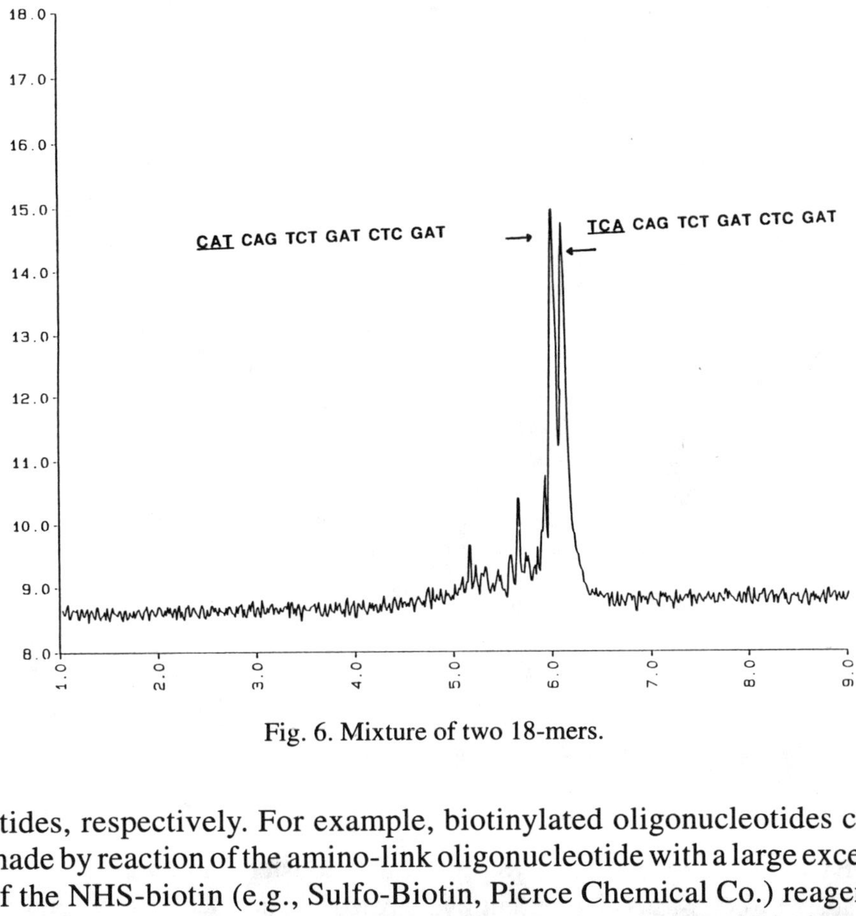

Fig. 6. Mixture of two 18-mers.

otides, respectively. For example, biotinylated oligonucleotides can made by reaction of the amino-link oligonucleotide with a large excess of the NHS-biotin (e.g., Sulfo-Biotin, Pierce Chemical Co.) reagent. Careful purification is necessary to remove the unreacted biotin and the organic solvents, base, and salts present in the coupling reaction. Alternatively, biotin-oligonucleotides may be conveniently prepared on the DNA synthesizer with biotin phosphoramidite reagents, available from many sources. Figure 8B shows an OPC-purified, biotinylated 25-mer. The biotin phosphoramidite used here contained a trityl group, allowing efficient purification by OPC even when coupling efficiency is low. The corresponding unlabeled 25-mer elutes 1 min earlier (data not shown). Some versions of biotin phosphoramidites allow multiple incorporation of biotin throughout the oligonucleotide. Biotin-oligonucleotides elute significantly later than their 5' hydroxy counterparts, owing to the hydrophobic nature and mass addition of the biotin and the linker moieties.

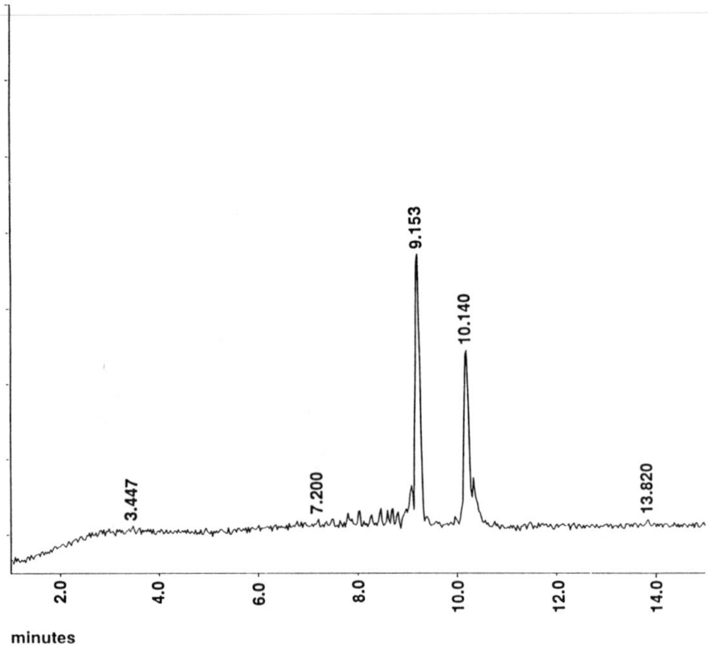

Fig. 7. Mixture of 5' DMT and 5' OH 18-mers 5' TCA CAG TCT GAT CTC GAT 3'.

By similar procedures, a variety of active-ester fluorescent dyes are available to construct fluorescent dye-labeled oligonucleotides. A crude 5' 5-carboxy fluorescein labeled 18-mer is shown in Fig. 8C. Like biotin, fluorescent dyes retard the electrophoretic velocity of oligonucleotides, but otherwise behave normally and exhibit well formed peaks. The tallest peak in Fig. 8C is the unlabeled 5' OH 18-mer that failed to couple with the fluorescent dye compound. The dye-labeled product follows, slightly resolved into two peaks, reflecting a diastereomeric carbon in the linker moiety.

3.7. Analysis of Oligoribonucleotides

Oligoribonucleotides, RNA, are excellent samples for Gel CE. Being more hydrophilic, they elute slightly earlier than their DNA counterparts. RNase degradation does not seem to be significant when the RNA sample is dissolved in purified, filtered water (*see* Note 4). All

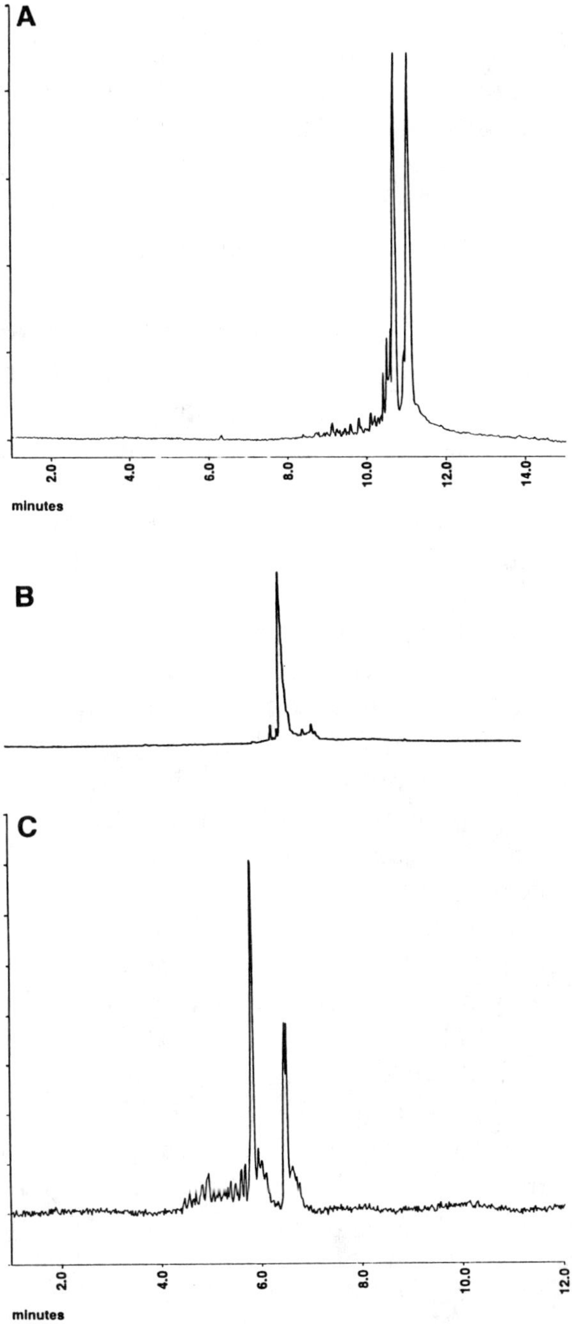

Fig. 8. **A:** Mixture of 5' OH and 5' aminohexylphosphate 5' TGT AAA ACG ACG GCC AGT 3'. **B:** OPC-purified 5' biotin AGG CGA GCA GAA GTG TCC TGC ACT T 3'. **C:** Mixture of 5' OH and 5' (5-FAM-linker) 5' TGT AAA ACG ACG GCC AGT 3'.

the same considerations and separation parameters pertaining to DNA are relevant for RNA. Considering the lower synthesis efficiency, higher costs, and more stringent applications for RNA, it is all the more important to conduct the high resolution, quantitative analysis that Gel CE provides. Figure 9 shows an electopherogram of a crude 22-mer RNA oligonucleotide *(11)*.

3.8. Analysis of Phosphorothioate Oligonucleotides

Internucleotide phosphate analogs are being intensively studied for their inhibition of gene expression with nuclease protection. These experiments, usually referred to as the antisense effect, are most frequently conducted with phosphorothioate oligonucleotides, with one of the nonbridging oxygen atoms of the internucleotide phosphate being replaced with a sulfur atom. The synthesis of the phosphorothioate analogs is very efficiently conducted on the DNA synthesizer with the usual phosphoramidite chemistry, using tetraethylthiuram disulfide (TETD) as a sulfurizing agent *(12)*, instead of the iodine oxidizing reagent (*see* Chapter 8 in vol. 20 of this series). Phosphorothioate oligonucleotides are stable, easily handled compounds. They are significantly more hydrophobic than their phosphodiester, oxygen-containing counterparts. Also, the sulfurizing reaction is not stereospecific at the chiral phosphorous center, yielding a large number of chemically distinct diastereomeric products. The net result on analysis, by any of the common methods, is a slight broadening of the product peak or band. Gel CE is again a very useful technique for assessing the purity of phosphorothioate oligonucleotides, either in the crude or purified state. Oligonucleotides containing both phosphodiester and phosphorothioate linkages may be easily prepared in a single synthesis operation. Figures 10A, B, and C show three electropherograms of a 25-mer sequence consisting of phosphodiester and phosphorothioate in three different arrays. The sequence with a single phosphorothioate linkage, Fig. 10A, shows the narrowest peaks and the earliest elution time of the product. The other two electropherograms, Figs. 10B and C, show the effects of the presence of many phosphorothioate linkages; broadened peaks, lower resolution, and slightly increased elution times.

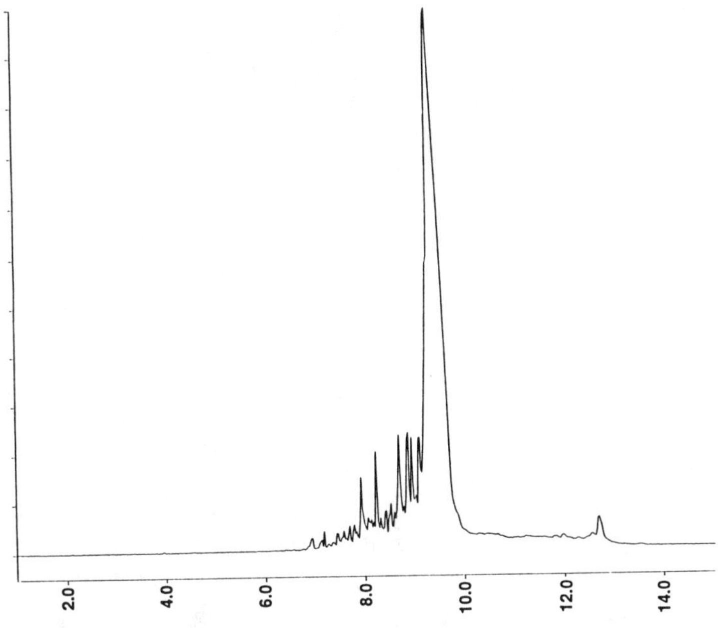

Fig. 9. Crude RNA 22-mer 5' AUA AUG GUU UGU UUG UCU UCG U 3'.

4. Discussion

Gel-capillary electrophoresis offers a direct method to evaluate the purity of synthetic oligonucleotides. The denaturing gel matrix, under high voltage conditions, gives the predictable, familiar elution patterns for oligonucleotides, as well as analogs, RNA, and their labeled conjugates. Coupled with an appropriate data system, integrator, or chart recorder, Gel CE thus attains most of the virtues of both PAGE and HPLC. With the attributes of high resolution, speed, and ease of automation, Gel CE is well suited for oligonucleotide analysis.

Further investigations are needed to understand the factors controlling resolution. It is clear that electrophoretic mobility of an oligonucleotide in a gel-filled capillary is influenced by more than mass and

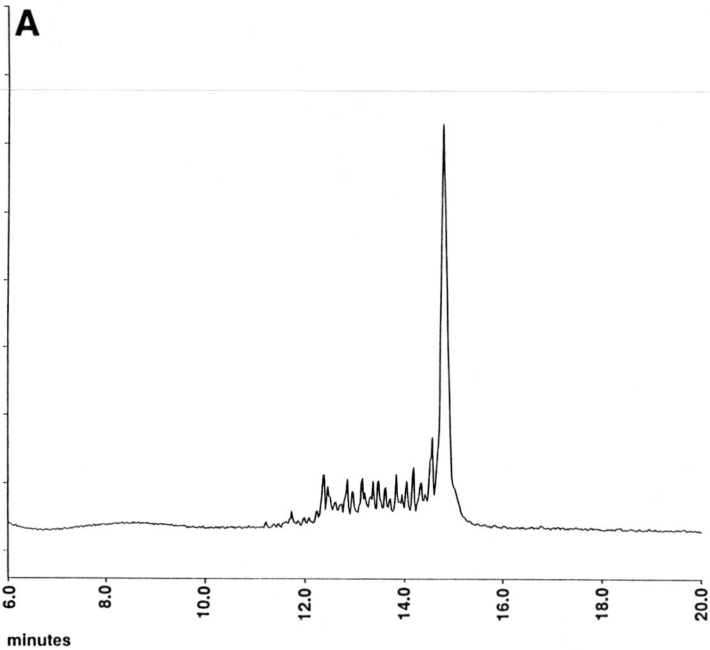
A
6.0
8.0
10.0
12.0
14.0
16.0
18.0
20.0
minutes

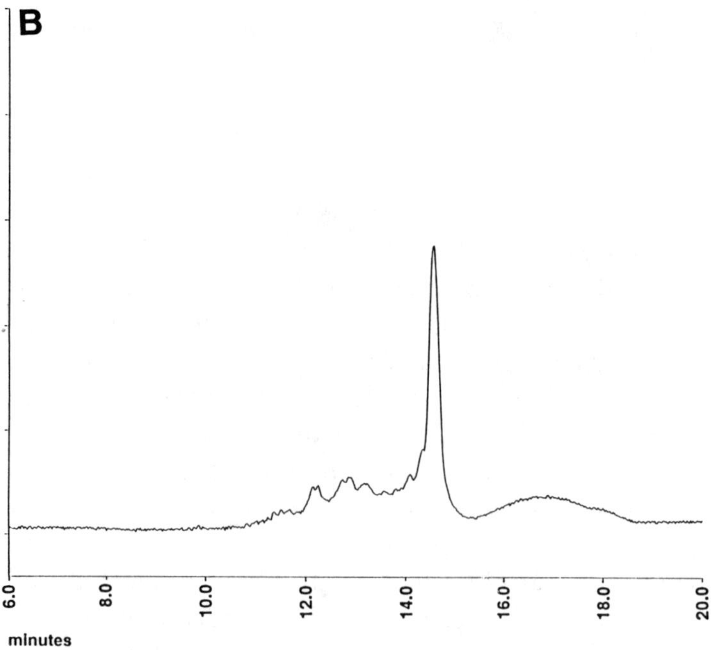
B
6.0
8.0
10.0
12.0
14.0
16.0
18.0
20.0
minutes

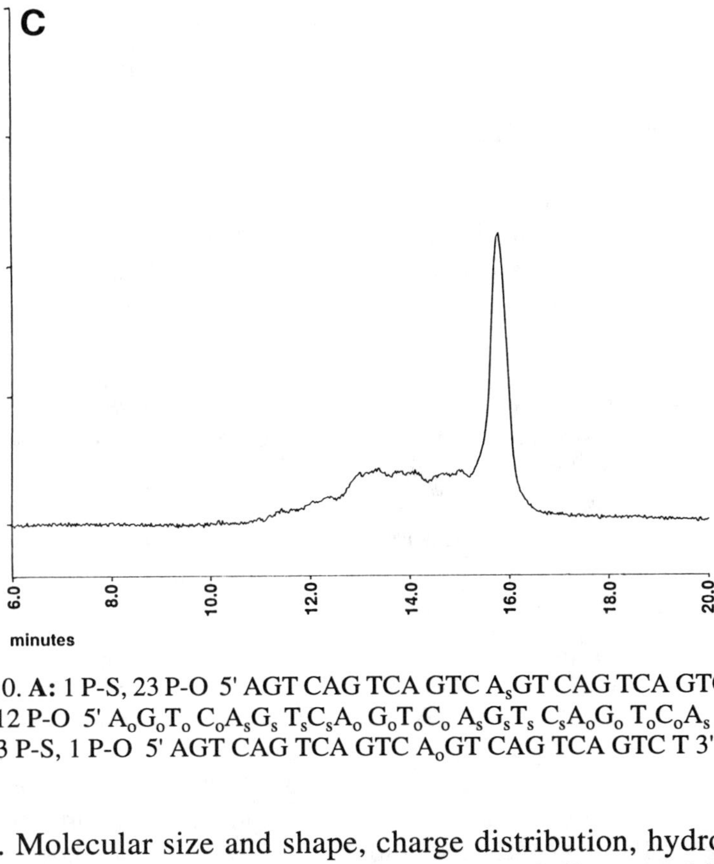

Fig. 10. **A:** 1 P-S, 23 P-O 5' AGT CAG TCA GTC A_sGT CAG TCA GTC T 3'. **B:** 12 P-S, 12 P-O 5' $A_oG_oT_o$ $C_oA_sG_s$ $T_sC_sA_o$ $G_oT_oC_o$ $A_sG_sT_s$ $C_sA_oG_o$ $T_oC_oA_s$ $G_sT_sC_s$ T 3'. **C:** 23 P-S, 1 P-O 5' AGT CAG TCA GTC A_oGT CAG TCA GTC T 3'.

charge. Molecular size and shape, charge distribution, hydrophobic elements, hydrogen bonding, and interactions of all types with the gel matrix seem to have more influence in Gel CE than in PAGE. The highly desirable denaturing ability of the MicroGel matrix is considerable. The gel materials are still in an early development phase and there remain opportunities to further optimize resolution of oligonucleotides. Practical features, such as reliability, durability, and shelf life of the gel capillaries, must also be improved. The variables influencing the mass of oligonucleotide sample entering a capillary during the electrokinetic injection period is not yet well understood. Advances here will better ensure reproducible peak heights, accurate integrations, increased capillary lifetimes, and optimal resolution. Coupled with second generation capillary-electrophoresis instruments, advances in gel matrices and method developments will establish gel-capillary electrophoresis as a powerful technique for oligonucleotide analysis.

5. Notes

1. Caution: Capillary Electrophoresis operates at very high voltages, therefore all recommended safety precautions should always be observed.
2. Oligonucleotide samples prepared by simple evaporation of the crude ammonia solution can be used directly for electrokinetic injection and do not require further sample preparation procedures. Samples in salt containing media or present in low purity, such as crude long oligonucleotides, usually benefit from a desalting operation. Convenient desalting operations include OPC *(13)* or precipitation (*see* Note 3 below).
3. Desalting operations remove inorganic salts, traces of organic compounds, protecting group by-products, and other low-mol-wt impurities that are present in crude oligonucleotide ammonia solutions. Desalting may be achieved with:
   a. Trityl-off or trityl-on oligonucleotides.
   b. RNA oligonucleotides, after 2' desilylation deprotection.
   c. Polyacrylamide gel (PAGE) purification extracts.
   d. Anion-exchange HPLC fractions.

   OPC desalting protocol:
   a. Pass 5 mL of acetonitrile through the OPC to waste.
   b. Pass 5 mL of 2*M* TEAA (triethylammonium acetate) through the OPC to waste.
   c. Dissolve the oligonucleotide in 1–3 mL of aqueous solution (e.g., 0.1*M* TEAA). The loading solution should not contain organic solvents. Pass the diluted solution through the OPC at a rate of about 1 drop/s. Collect the eluate and pass it through a second time.
   d. Pass 15 mL of water through the OPC to waste.
   e. Elute the desalted oligonucleotide with 1 mL of 1:1 acetonitrile:water at the rate of about 1 drop/s.

   Precipitation desalting protocol:
   a. Quantitate the sample to determine concentration.
   b. Aliquot 1 A_{260} U and dry in an Eppendorf-style tube.
   c. Dissolve in 30 µL water and 5 µL 3*M* sodium acetate by vortexing.
   d. Add 100 µL ethanol (isopropanol for <15-mers), vortex.
   e. Chill on ice or in a refrigerator for at least 30 min.
   f. Centrifuge 3 min, remove, and discard supernatant.
   g. Add 100 µL ethanol, vortex, and centrifuge 3 min.
   h. Remove and discard supernatant, being careful not to disturb pellet.

4. Oligoribonucleotide (RNA) samples are susceptible to RNase degradation. RNases are ubiquitous and relatively stable enzymes. To minimize their effects, vessels that come in contact with the RNA samples should be the disposable type and preferably autoclaved. Disposable gloves should be worn when handling RNA samples, to minimize the greatest laboratory source of RNases; human skin. Stringent precautions are typically observed when handling relatively smaller amounts of higher mol-wt, native RNA derived from biological sources. The greater quantities of lower mol-wt RNA made by chemical synthesis are at less risk and do not seem to warrant extreme precautions. Wearing gloves, using freshly prepared solutions from deionized, filtered water, and disposable vessels, will adequately minimize RNase degradation.

Acknowledgments

The author wishes to thank his colleagues, Huynh Vu, Christie McCollum, Pete Theisen, Cindy Lotys, Ravi Vinayak, Peter Wright, Larry DeDionisio, Bob Dubrow, and Junko Stevens for their technical assistance, and to Jenny Andrus and Beth Sanchez for their editorial assistance.

References

1. Cohen, A. S., Najarian, D. R., Paulus, A., Guttman, A., Smith, J. A., and Karger, B. L. (1988) Rapid separation and purification of oligonucleotides by high-performance capillary gel electrophoresis. *Proc. Natl. Acad. Sci. USA* **85,** 9660–9663.
2. Paulus, A. and Ohms, J. I. (1990) Analysis of oligonucleotides by capillary gel electrophoresis. *J. Chromatog.* **507,** 113–123.
3. Demorest, D. and Dubrow, R. (1991) Factors influencing the resolution and quantitation of oligonucleotides separated by capillary electrophoresis on a gel filled capillary. *J. Chromatog.* **559,** 43–56.
4. Dubrow, B., DeDionisio, L., and Andrus, A. (1991) Gel capillary electrophoresis for the analysis of synthetic oligonucleotides, *International Meeting of the Electrophoresis Society*, Washington, DC, March 19–21.
5. Dubrow, R. S. (1991) Analysis of synthetic oligonucleotide purity by capillary gel electrophoresis. *Am. Lab.* March, pp. 64–67.
6. Grossman, P. D., Lauer, H. H., Moring, S. E., Mead, D. E., Oldham, M. F., Nickel, J. H., Goudberg, J. R. P., Krever, A., Ransom, D. H., and Colburn, J. C. (1989) A practical introduction to free solution capillary electrophoresis of proteins and peptides. *Am. Biotech. Lab.* February, pp. 35–45.
7. Moring, S. E., Colburn, J. C., Grossman, P. D., and Lauer, H. H. (1990) Analytical aspects of an automated capillary electrophoresis system. *LC GC* **8,** 34.

8. Karger, A. E., Harris, J. M., and Gesteleand, R. F. (1991) Multiwavelength fluorescence detection for DNA sequencing using capillary electrophoresis. *Nucleic Acids Res.* **19, No. 18,** 4955–4962.
9. Keller, G. H. and Manak, M. M. (1989) Non-radioactive labeling procedures, in *DNA Probes*, Stockton, New York, pp. 105–148.
10. Goodchild, J. (1990) Conjugates of oligonucleotides and modified oligonucleotides: a review of their synthesis and properties. *Bioconj. Chem.* **1, No. 3,** 165–187.
11. Vinayak, R., McCollum, C., and Hampel, A. (1992) Chemical synthesis of RNA using fast oligonucleotide deprotection chemistry. *Nucleic Acids Res.* **20(b),** 1265–1269.
12. Vu, H. and Hirschbein, B. L. (1991) Internucleotide phosphite sulfurization with tetraethylthiuram disulfide. Phosphorothioate oligonucleotide synthesis via phosphoramidite chemistry. *Tetrahedron Lett.* **32, No. 26,** 3005–3008.
13. Applied Biosystems, Foster City, CA (March 1991) New applications for the oligonucleotide purification cartridge, *User Bulletin* Number 59.

CHAPTER 12

Nuclear Magnetic Resonance Studies of Oligonucleotides

Jerzy W. Jaroszewski,
Siddhartha Roy, and Jack S. Cohen

1. Introduction

NMR spectroscopy is an extremely valuable and versatile method for studies of nucleosides, nucleotides, and their oligomers. It can provide information about chemical composition, three-dimensional structures, internal motions, and interactions with other macromolecules or ligands. It can be used to follow outcome and yield of a synthesis, to ascertain identity and purity of the product, to identify and quantify byproducts, and finally, to obtain a largely complete picture of the spatial arrangement of atoms in the molecule. It can be used at all stages of synthesis, development, and investigations of oligonucleotides.

There are three principal NMR parameters of interest in interpretation of NMR data. The first is chemical shift value (δ), which is the resonance frequency of a nucleus, and reflects its electronic environment and hence the chemical identity. The second is J-coupling constant (spin-spin, through-bond, or scalar coupling), which depends, inter alia, on the geometry of the bonding path connecting two nuclei, and which can be used to derive structural (configurational and conformational) information. The third parameter is relaxation, which can be used to obtain information about spatial proximity of nonbonded nuclei (via so-called through-space or dipolar coupling), and about internal motions.

From: *Methods in Molecular Biology, Vol. 26: Protocols for Oligonucleotide Conjugates*
Edited by: S. Agrawal

All NMR active nuclei present in nucleic acids, i.e., those nuclei that have nonzero nuclear spin quantum number **I**, can be, and have been, used to derive some or all of the above-mentioned types of information. Among those, protons (^{1}H) can be detected with the highest sensitivity, and ^{1}H NMR is used by far most extensively. Phosphorus (^{31}P) and carbon (^{13}C) have sensitivity of respectively 6.6% and 0.018% (corrected for the isotopic abundance of ^{13}C, which is only 1.1%) of that of ^{1}H, and these nuclei also provide very important and easily accessible information. ^{2}H, ^{15}N, ^{14}N, and ^{17}O are limited to more special applications. NMR studies of some nuclei are greatly aided by isotopic labeling.

Any NMR investigation starts with recording of normal (one-dimensional or 1D) spectrum. Recent progress in multidimensional NMR (2D, 3D) methods greatly facilitated studies of oligonucleotides, providing structural information faster and with greater reliability, and extended the range of molecular sizes amenable to investigations. In particular, various variants of COSY (correlated spectroscopy), NOESY (nuclear Overhauser effect spectroscopy), and heteronuclear correlation experiments are useful for studies of oligonucleotides. Although a detailed discussion of these techniques, reviewed elsewhere *(1–6)*, lies beyond the scope of this chapter, a few examples will be given to emphasize their impact. Since the application of NMR spectroscopy for studies of oligonucleotides constitutes a vast area, the topic is treated selectively and only some examples from the literature are cited.

2. Nucleosides and Nucleotides

NMR studies of nucleosides and nucleotides go back to the early days of NMR spectroscopy *(7–9)*. Extensive tabulations of ^{1}H chemical shift (δ) and coupling constant (J) data for dinucleotides and deoxydinucleotides are available in the literature *(10–13)* and serve as a reference for oligomer spectra. However, in contrast to peptides, with average ^{1}H chemical shift parameters derived and used for calculation of random coil spectra, the chemical shifts of nucleotides and deoxynucleotides are strongly influenced by the sequence and by intermolecular interactions (stacking), even in single-stranded, low-mol-wt oligomers. The observed ranges of chemical shifts are given in Fig. 1.

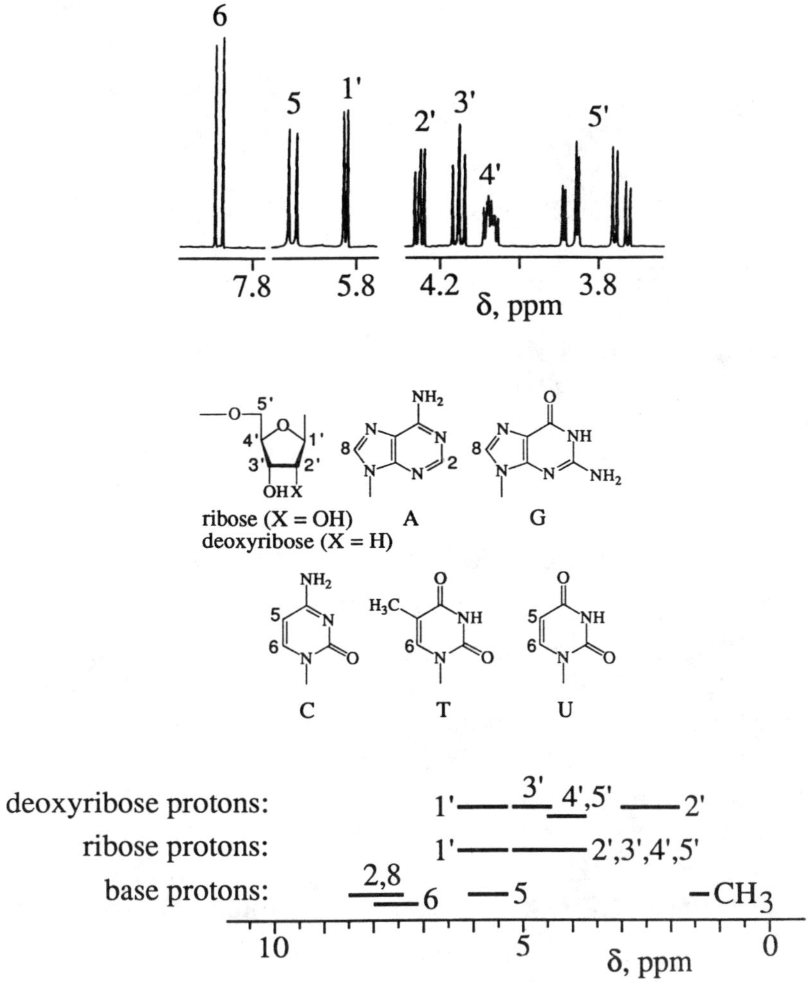

Fig. 1. Chemical shift ranges in nucleotides, deoxynucleotides, and their oligomers; protons are designated as indicated. The amino (NH_2) and imino (NH) protons appear within a broad range of about δ 6.5–9 and 10–15 ppm, respectively (adapted from ref. *1*). Example of a mononucleoside spectrum (cytidine in D_2O, 400 MHz) is shown on top.

The ^1H-^1H coupling constants within the sugar moiety depend strongly on ring conformation and are thus of diagnostic value. Borderline conformations and the corresponding coupling constants are shown in Fig. 2. Intermediate values are observed with mixed conformations. Extensive treatment of this topic is given in the literature *(3,4,14)*.

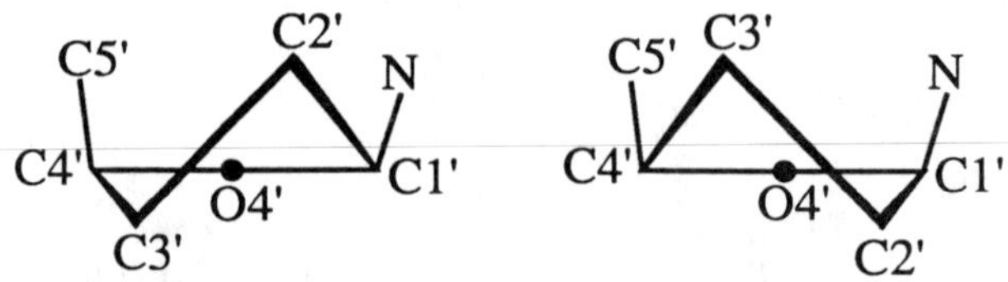

	C2'-endo	C3'-endo
J(H1',H2'):	9.5 Hz	0 Hz
J(H1',H2"):	6 Hz	8 Hz
J(H2',H3'):	6 Hz	6 Hz
J(H2",H3'):	0 Hz	9 Hz
J(H3',H4'):	0 Hz	9 Hz

Fig. 2. Main conformations of the sugar ring and the corresponding ^{1}H-^{1}H coupling constants (adapted from ref. *4*).

3. Unmodified Oligonucleotides

3.1. ^{1}H NMR Spectra

^{1}H NMR spectroscopy can be used for studies of oligonucleotides at various levels. Single stranded oligomers can be analyzed for the presence of particular nucleotides, and the ratio between them, by assessment of the intensity and number of peaks in various regions of the spectra (*see* Fig. 1). Although ^{1}H NMR spectroscopy of single-stranded oligomers is very important as an analytical tool, the most interesting usage of ^{1}H NMR is perhaps analysis of spectra of duplex oligonucleotides *(15–17)*. Two classes of protons are of particular interest: nonexchangeable protons and exchangeable protons.

3.1.1. Nonexchangeable Protons

Protons bonded to carbon atoms (aromatic and sugar protons) are valuable for conformational analysis of short oligonucleotide duplexes. This requires that a sequence-specific assignment of signals of individual nucleotides present is carried out. The sequence-specific assignment usually requires the following steps:

1. Classification of resonances in 1D spectra according to classes (base protons, H1' protons, H2' protons, and so on).
2. Identification of spin systems, i.e., all or most resonances that belong to the same sugar. This is obtained notably from COSY spectra, which

provide connectivities between chemical shifts according to the presence of J couplings. Various variants of COSY experiments as well as J-resolved 2D spectra help to assign overlapping signals.

3. Sequence specific assignment of spin systems of individual nucleotides. This assignment is based on NOE effects (Nuclear Overhauser effects *(18)* originating from dipolar coupling between nuclei separated by <4.5–5 Å), obtained from two-dimensional NOESY spectra. In particular, NOE effects between the anomeric protons (H1') and H6 protons of pyrimidines or H8 protons of purines are useful. Thus, in a right-handed helix, the aromatic proton (H6 or H8) of the base at the 5'-end exhibits NOE to the anomeric proton of the sugar to which the base is attached (intranucleotide NOE). The latter proton also exhibits NOE to the aromatic proton of the base of the second nucleotide, and so on.

In general, the anomeric proton of nucleotide *n* has NOE to its own aromatic proton and to the aromatic proton of the base upstream ($n + 1$). Conversely, each aromatic proton has NOE to its own anomeric proton and to the anomeric proton of the nucleotide downstream ($n - 1$). The anomeric proton at the 3'-end and aromatic proton at the 5'-end have only one (intranucleotide) NOE each. The sequence-specific assignment is aided by identification of NOE effects involving H2' protons. This, in combination with identification of spin patterns from COSY and related spectra, gives the basis for assignment of all resonances in the molecule. In reality, assignment of all H4' and H5' resonances is rarely achievable because of extensive signal overlap. An example of sequence-specific assignment of anomeric and aromatic resonances from a NOESY spectrum is shown in Fig. 3.

Once specific assignment of most of the signals is carried out, one can proceed to identification of other NOE crosspeaks *(1,3)*, and to semiquantitative or quantitative evaluation of cross-relaxation rates from 1D and 2D data, which can lead to three-dimensional structure of the duplex *(1)*. Thus, the NOESY spectra yield volumes of crosspeaks, which can be translated *(19)* to interproton distances in the molecule. The duplex structure can be obtained from distance geometry calculations or restrained molecular dynamics using distance information from the NOE data *(1)*.

In the left-handed Z-DNA the sequence-specific assignment procedure outlined above fails. The intranucleotide H1'-H8 distance in Z-DNA is decreased (because of syn guanosine conformations) lead-

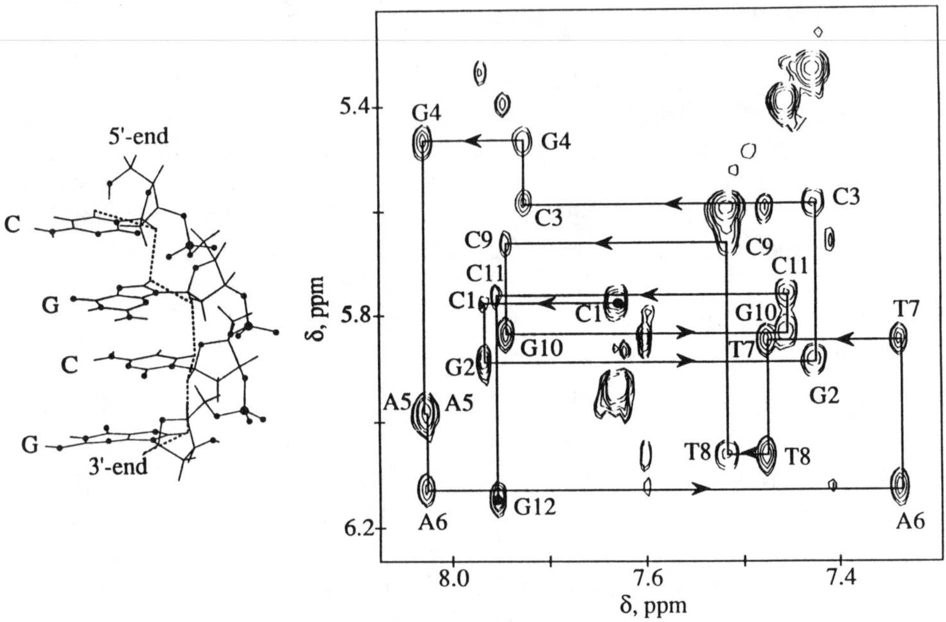

Fig. 3. Left: Schematic illustration of the principle of sequence-specific assignment of nucleotides in a right-handed duplex from NOE effects (stippled lines). Right: Sequence-specific assignment of aromatic and anomeric proton resonances in the Dickerson dodecamer duplex d(CpGpCpGpApApTpTpCpGpCpG) in D_2O at 400 MHz (Jaroszewski et al., unpublished work; *see* ref. *15*).

ing to very strong NOE, whereas the internucleotide connectivities are absent. Z-DNA can also be distinguished by the absence of intranucleotide NOE between base and H2' protons, which are strong in B-DNA *(20,21)*. ^{1}H NMR spectra are also valuable for analysis of unusual conformations, such as hairpins and mismatched duplexes *(22–24)*, and for studies of helix-coil transitions *(25)*.

3.1.2. Exchangeable Protons

There are several classes of exchangeable protons in oligonucleotides, i.e., imino protons, amino protons, and hydroxylic protons. The imino protons are a particularly interesting group, since they resonate far downfield from other protons, and only one imino resonance is contributed by each basepair. In addition, under certain conditions, their exchange rate with the bulk solvent may reflect a basepair opening rate *(26,27)*. Imino protons have been particularly valuable for studies of moderate-sized nucleic acids, such as single-stranded RNA or tRNA, where structural analysis utilizing nonexchangeable protons fails *(28,29)*.

3.2. ^{31}P NMR Spectra

Phosphate groups of nucleotides can be detected by means of ^{31}P NMR spectra *(30,31)*. The chemical shift of ^{31}P is dependent on the conformation of nucleic acid duplex and hence may be used to detect helix transitions. For example, conversion of B-DNA to Z-DNA causes one of the phosphates of dinucleotide repeat to shift about 4 ppm downfield *(30)*.

Heteronuclear J couplings (4–7 Hz) between ^{31}P and H3', H5', and H5" of the sugar moiety cause complications in the ^{1}H NMR spectra of oligonucleotides *(13)*, but can also be used as a diagnostic tool for establishing nucleotide connectivity *(23,32–34)*. A convenient means to obtain ^{1}H-^{31}P spin-spin connectivities is via long-range heteronuclear chemical shift correlations (Fig. 4) *(34)*.

3.3. ^{13}C, ^{15}N, and ^{2}H NMR Spectra

Although ^{1}H and ^{31}P NMR spectra can provide a great deal of information, they are unsuitable for studies of many interesting aspects of protein-nucleic acid interactions and nucleic acid dynamics. Relaxation properties of ^{13}C and ^{2}H nuclei reflect directly the dynamic characteristics of the isotope-bearing molecular region. ^{15}N NMR chemical shifts are sensitive to hydrogen bonding, making them a very valuable probe to study intermolecular interactions. Because of their low sensitivity and low isotopic abundance, some NMR studies involving these nuclei at natural abundance are not feasible. The former problem can be solved by indirect detection via J-coupled protons, whereas the latter by chemical or biochemical enrichment. For these reasons, a resurgence of interest in these nuclei has recently been observed *(35–38)*.

^{13}C NMR spectra of isotopically enriched nucleotides can also provide ^{1}H-^{13}C and ^{13}C-^{13}C coupling constants, which afford a means to assess nucleotide conformation *(14)*. ^{13}C-^{31}P coupling constants data are available from the literature *(39)*.

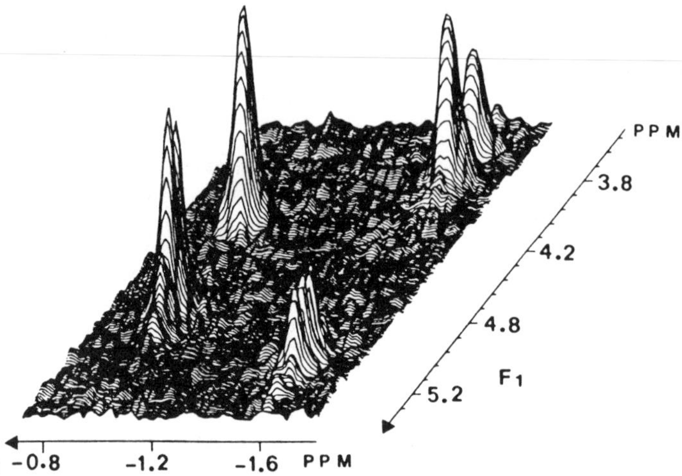

Fig. 4. ^{1}H-^{31}P chemical shift correlation for the branched trimer A[(2'-5'A)(3'-5'A)] (from ref. *34*, reproduced by permission of the copyright owner).

4. Modified Oligonucleotides

4.1. Phosphorothioates and Phosphorodithioates

Initially, substitution of a nonbridging phosphate oxygen for sulfur *(40)* was carried out in order to assign specific phosphorus resonances in the ^{31}P NMR spectra of oligomers, although this experiment was soon replaced by the ^{1}H-^{31}P shift correlation experiment mentioned above. The replacement of one of the nonbridging oxygens of the phosphate group introduces chirality on phosphorus, and phosphorothioates prepared using standard automated synthesis from phosphoramidite precursors are mixtures of isomers. Thus, spectra of such materials have characteristics of nonhomogeneous material (diffuse signals). However, many resonances distant to the phosphorothioate moiety remain to be largely unaffected, and ^{1}H NMR spectra of phosphorothioates can be to a certain degree analyzed and compared to those of the unmodified counterparts. ^{1}H NMR spectra of stereospecifically substituted phosphorothioates have been studied *(41)*.

^{31}P NMR spectra are a very convenient means for analyzing purity of phosphorothioates, since the phosphorus resonance is shifted approx 56 ppm downfield on sulfur substitution *(42)*. The use of ^{31}P NMR

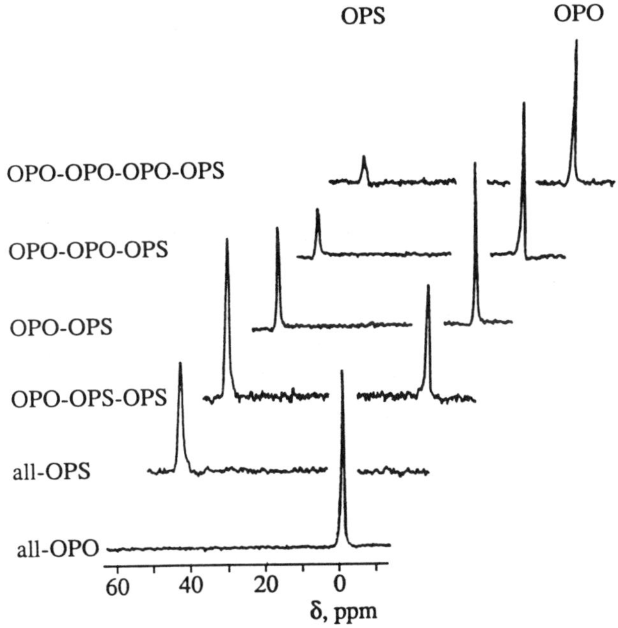

Fig. 5. ^{31}P NMR spectra (162 MHz) of phosphate/phosphorothioate (OPO/OPS) copolymers with repeating sequences as shown. The integration of signals gives ratios between phosphate and phosphorothioate content (Cohen et al., unpublished results).

also allows characterization of copolymers with different proportions of phosphodiesters and phosphorothioates (Fig. 5). In the case of phosphoroselenoates this difference has been used to measure kinetics of their degradation *(43)*.

In phosphorodithioates, the phosphorus signal is shifted downfield by 110–115 ppm relative to phosphodiesters *(44,45)*. If most or all of the phosphate groups in a self-complementary oligomer are replaced by phosphorodithioate groups, the melting point of the duplex is strongly diminished *(45)*, which can lead to the presence of mixtures of species and exchange-broadening of NMR signals at room temperature. ^{1}H NMR spectra of a decamer containing two phosphorodithioate sites in the middle of the sequence have been studied *(44)*. Unlike the parent (unmodified) sequence, which forms stable duplex, it adopted a hairpin conformation, emphasizing destabilizing effect of the phosphorodithioate groups on duplex formation.

4.2. Other Modified Oligodeoxynucleotides

Other modifications of oligonucleotides can be detected by NMR either indirectly or directly. Thus, the methyl groups of methylphosphonates are immediately apparent in ^{1}H and ^{13}C NMR spectra *(46,47)*. Like the phosphorothioates, methylphosphonates contain chiral phosphorus, the chemical shift of which (about 35 ppm) is strongly different from that of the phosphodiesters. Since the chemical shift of phosphorus depends on its chirality, the stereochemical purity of methylphosphonates can be conveniently assessed by ^{31}P NMR (Fig. 6) *(48,49)*. Use of 2D methods for resolution of overlapping proton resonances of a methylphosphonate tetramer is illustrated in Fig. 7.

Phosphate triesters *(50)* and phosphoramidates *(51,52)* have also been studied by NMR spectroscopy. Another synthetic type of analog has the α-configuration of base attached to C1' rather than the natural β-configuration. Morvan et al. carried out an extensive analysis of properties of α-oligodeoxynucleotide hexamers, and showed that they adopt the C2'-endo conformation and form antiparallel duplexes via Watson–Crick base pairing *(53)*, although they form parallel duplexes with complementary β-oligomers *(54)*.

The presence of linked groups will usually be immediately apparent in the NMR spectra *(55)*, and proof of the identity of the linker and the assessment of the amount of the linker present can be obtained by analysis of the spin patterns and integration of the resonances.

5. Practical Considerations

Recording of NMR spectra of oligonucleotides requires that a variety of precautions related to the problem to be solved are taken. In order to obtain reproducible chemical shift data, the oligomer solution has to be standardized in terms of pH, temperature, ionic strength, and presence of metal ions that can strongly affect the results. EDTA is often added to avoid effects of metal ion impurities. These precautions are especially important for studies related to duplex formation and duplex geometry.

To obtain good quality 2D NMR data, one should preferably have an approx 1 m*M* solution of oligomer (about 0.5 μmol in 0.5 mL of solvent in a 5-mm NMR tube), although higher quality data can often

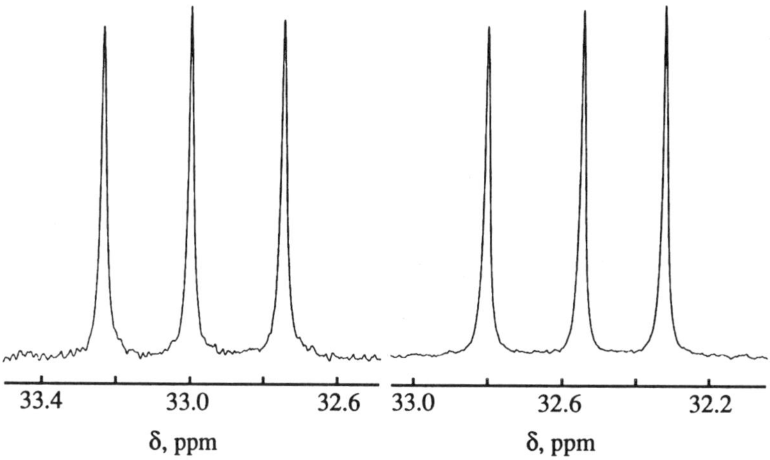

Fig. 6. ^{31}P NMR spectra (162 MHz) of $S_pS_pS_p$-diastereomer (left) and $R_pR_pR_p$-diastereomer (right) of tetra(thymidine methanephosphonate), d[Tp(CH_3)Tp(CH_3)-Tp(CH_3)T], in pyridine-d_5 (Jaroszewski et al., unpublished results).

be obtained in less time if more material is available. 1D NMR spectra can be obtained with much less material. Structural analyses are normally performed in D_2O solution (or H_2O if exchangeable protons have to be studied; this requires use of water signal suppression techniques). Repeated freeze-drying of samples from D_2O prior to the analysis in deuterated solvents is preferable in order to remove water signal caused by residual water and exchangeable protons. Also, particular care should be exercised in the case of samples obtained from purification by HPLC with a buffer containing triethylamine, because the residual triethylammonium ions can obscure proton resonances of interest.

If the aim of the NMR analysis is to characterize or confirm identity of the material, other solvents, such as pyridine-d_5 or methanol-d_4 can be preferable. When the aim is to study changes of spectra as a function of temperature, pH changes with the temperature should be considered. Finally, quantitative analyses based on integration of ^{31}P and ^{13}C signals must be carried out without saturation of the resonances by pulsing too fast.

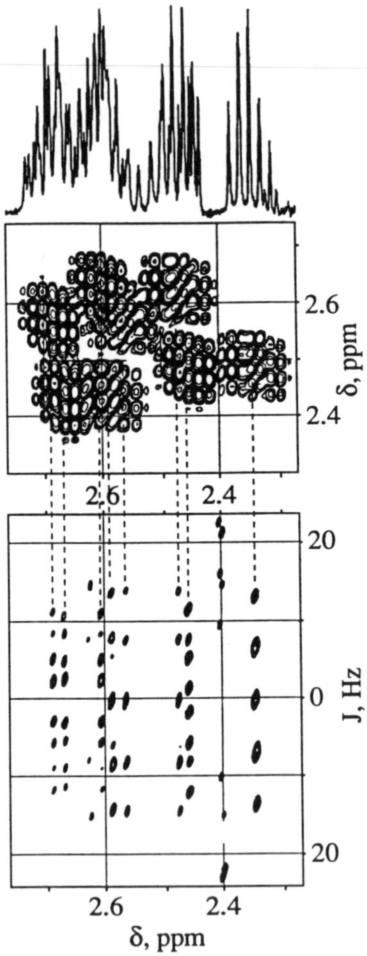

Fig. 7. Normal (1D) ^{1}H NMR spectrum (top), COSY spectrum (center) and J-resolved spectrum (bottom) of $R_pR_pR_p$-diastereomer of tetra(thymidine methanephosphonate), d[Tp(CH_3)Tp(CH_3)Tp(CH_3)T], in pyridine-d_5 (400 MHz), showing the H2' region: 8 overlapping signals are readily resolved in the 2D spectra (Jaroszewski et al., unpublished work).

6. Concluding Remarks

It would be inconceivable to anticipate progress in the area of oligonucleotide chemistry without the direct application of NMR spectroscopy. NMR is utilized at all levels of the subject, and no other technique is comparable to NMR in terms of the scope of applications and usefulness of the information obtained. The only general drawback of NMR is the relative insensitivity of the method, so that much more material is required than for detection by, for example, UV or chromatographic methods.

The latest innovation in the area of high-resolution NMR is use of gradient methods, derived from a magnetic resonance imaging area *(56,57)*. This allows a complete reduction of water signal from the spectrum, so that H_2O solutions can be used for most applications mentioned above, making observation of the important imino protons more readily available than before. Moreover, gradients offer improvements in the 2D methods, by eliminating the need of lengthy phase cycling for quadrature detection in ω_1. Since the future of oligonucleotides as therapeutics looks promising, it can be expected that application of NMR in this field will accelerate.

References

1. Wüthrich, K. (1986) *NMR of Proteins and Nucleic Acids.*Wiley, New York.
2. Kessler, H., Gehrke, M., and Griesinger, C. (1988) Two-dimensional NMR spectroscopy: background and overview of the experiments. *Angew. Chem. Int. Ed. Engl.* **27,** 490–536.
3. Van de Ven, F. J. M. and Hilbers, C. W. (1988) Nucleic acids and nuclear magnetic resonance. *Eur. J. Biochem.* **178,** 1–38.
4. Hosur, R. V., Govil, G., and Miles, H. T. (1988) Application of two-dimensional NMR spectroscopy in the determination of solution conformation of nucleic acids. *Magn. Res. Chem.* **26,** 927–944.
5. Clore, G. M. and Gronenborn, A. M. (1989) Determination of three-dimensional structures of proteins and nucleic acids in solution by nuclear magnetic resonance spectroscopy. *Crit. Rev. Biochem. Mol. Biol.* **24,** 479–564.
6. Clore, G. M. and Gronenborn, A. M. (1991) Two- three-, and four-dimensional NMR methods for obtaining larger and more precise three-dimensional structures of proteins in solution. *Ann. Rev. Biophys. Chem.* **20,** 29–63.
7. Lemieux, R. U. (1961) The configuration and conformation of thymidine. *Can. J. Chem.* **39,** 116–120.
8. Jardetzky, C. D. (1960) Proton magnetic resonance studies on purines, pyrimidines, ribose nucleosides, and nucleotides. Ribose conformation. *J. Am. Chem. Soc.* **82,** 229–233.
9. Jardetzky, C. D. (1962) Proton magnetic resonance of nucleotides. Ribose conformation. *J. Am. Chem. Soc.* **84,** 62–66.
10. Evans, F. E. and Sarma, R. H. (1976) Nucleotide rigidity. *Nature* **263,** 567–572.
11. Lee, C.-H., Ezra, F. S., Kondo, N. S., Sarma, R. H., and Danyluk, S. S. (1976) Conformational properties of dinucleoside monophosphates in solution: dipurines and dipyrimidines. *Biochemistry* **15,** 3627–3639.

12. Ezra, F. S., Lee, C.-H., Kondo, N. S., Danyluk, S. S., and Sarma, R. H. (1977) Conformational properties of purine-pyrimidine and pyrimidine-purine dinucleoside monophosphates. *Biochemistry* **16,** 1977–1987.
13. Cheng, D. M. and Sarma, R. H. (1977) Intimate details of the conformational characteristics of deoxyribodinucleoside monophosphates in aqueous solution. *J. Am. Chem. Soc.* **99,** 7333–7348.
14. Kline, P. C. and Serianni, A. S. (1990) ^{13}C-Enriched ribonucleosides: synthesis and application of ^{13}C-^{1}H and ^{13}C-^{13}C spin-coupling constants to assess furanose and N-glycoside bond conformation. *J. Am. Chem. Soc.* **112,** 7373–7381.
15. Hare, D. R., Wemmer, D. E., Chou, S.-H., Drobny, G., and Reid, B. R. (1983) Assignment of the non-exchangeable proton resonances of d(C-G-C-G-A-A-T-T-C-G-C-G) using two-dimensional nuclear magnetic resonance methods. *J. Mol. Biol.* **171,** 319–336.
16. Haasnoot, C. A. G., Westerink, H. P., van der Marel, G. A., and van Boom, J. H. (1984) Discrimination between A-type and B-type conformations of double helical nucleic acid fragments in solution by means of two-dimensional nuclear Overhauser experiments. *J. Biomol. Struct. Dyn.* **2,** 345–360.
17. Nilsson, L., Clore, G. M., Gronenborn, A. M., Brünger, A. T., and Karplus, M. (1986) Structure refinement of oligonucleotides by molecular dynamics with nuclear Overhauser effect interproton distance restraints: application to 5' d(C-G-T-A-C-G)$_2$. *J. Mol. Biol.* **188,** 455–475.
18. Neuhaus, D. and Williamson, M. (1989) The nuclear Overhauser effect in structural and conformational analysis. VCH, New York.
19. Keepers, J. W. and James, T. L. (1984) A theoretical study of distance determination from NMR. Two-dimensional nuclear Overhauser effect spectra. *J. Magn. Res.* **57,** 404–426.
20. Borah, B., Cohen, J. S., and Bax, A. (1985) Conformation of double-stranded polydeoxynucleotides in solution by proton two-dimensional nuclear Overhauser enhancement spectroscopy. *Biopolymers* **24,** 747–765.
21. Behling, R. W. and Kearns, D. R. (1986) ^{1}H Two-dimensional nuclear Overhauser effect and relaxation studies of poly(dA)·poly(dT). *Biochemistry* **25,** 3335–3346.
22. Roy, S., Weinstein, S., Borah, B., Nickol, J., Appella, E., Sussman, J. L., Miller, M., Shindo, H., and Cohen, J. S. (1986) Mechanism of oligonucleotide loop formation in solution. *Biochemistry* **25,** 7417–7423.
23. Roy, S., Sklenar, V., Appella, E., and Cohen, J. S. (1987) Conformational perturbation due to an extra adenosine in a self-complementary oligodeoxynucleotide duplex. *Biopolymers* **26,** 2041–2052.
24. Woodson, S. A. and Crothers, D. M. (1988) Structural model for an oligonucleotide containing a bulged guanosine by NMR and energy minimization. *Biochemistry* **27,** 3130–3141.
25. Patel, D. J. (1977) d-CpCpGpG and d-GpGpCpC self-complementary duplexes: NMR studies of the helix-coil transition. *Biopolymers* **16,** 1635–1656.
26. Patel, D. J., Kozlowski, S. A., Marky, L. A., Broka, C., Rice, J. A., Itakura, K., and Breslauer, K. J. (1982) Premelting and melting transitions in the d(CGCGAATTCGCG) self-complementary duplex in solution. *Biochemistry* **21,** 428–436.

27. Guéron, M., Kochoyan, M., and Leroy, J.-L. (1987) A single mode of DNA base-pair opening drives imino proton exchange. *Nature* **328,** 89–91.
28. Kime, M. J., Gewirth, D. T., and Moore, P. B. (1984) Assignment of resonances in the downfield proton spectra of *Escherichia coli* 5S RNA and its nucleoprotein complexes of ribonuclease resistant fragments. *Biochemistry* **23,** 3559–3568.
29. Roy, S. and Redfield, A. G. (1981) Nuclear Overhauser effect study and assignment of D stem and reverse-Hoogsteen basepair proton resonances in yeast $tRNA^{Asp}$. *Nucleic Acids Res.* **9,** 7073–7083.
30. Chen, C. W. and Cohen, J. S. (1984) DNA and RNA conformations, in *Phosphorus-31 NMR: principles and applications* (Gorenstein, D. G., ed.), Academic, Orlando, FL, pp. 233–263.
31. Cheng, D. M., Kan, L.-S., and Ts'o, P. O. P. (1987) The studies of ^{31}P NMR of nucleic acids and nucleic acid complexes, in *Phosphorus NMR in Biology* (Tyler, B. C., ed.), CRC Press, Boca Raton, pp. 135–151.
32. Pardi, A., Walker, R., Rapoport, H., Wider, G., and Wüthrich, K. (1983) Sequential assignment for the ^{1}H and ^{31}P atoms in the backbone of oligonucleotides by two-dimensional nuclear magnetic resonance. *J. Am. Chem. Soc.* **105,** 1652–1653.
33. Byrd, R. A., Summers, M. F., Zon, G., Fouts, C. S., and Marzilli, L. G. (1986) A new approach for assigning ^{31}P NMR signals and correlating adjacent nucleotide deoxyribose moieties via ^{1}H-detected multiple quantum NMR. Application to the adduct of d(TGGT) with the anticancer agent (ethylenediamine) dichloroplatinum. *J. Am. Chem. Soc.* **108,** 504–505.
34. Remaud, G. (1988) Practical aspects of 2D NMR for assigning the non-exchangeable protons in DNA-RNA fragments. *J. Biochem. Biophys. Meth.* **17,** 253–276.
35. Roy, S., Hiyama, Y., Torchia, D. A., and Cohen, J. S. (1986) New enzymatic synthesis of 2'-deoxynucleoside-2',2"-d_2 and the determination of sugar ring flexibility by solid state deuterium NMR. *J. Am. Chem. Soc.* **108,** 1675–1678.
36. Bax, A., Griffey, R. H., and Hawkins, B. L. (1983) Sensitivity-enhanced correlation of ^{15}N and ^{1}H chemical shifts in natural-abundance samples via multiple quantum coherence. *J. Am. Chem. Soc.* **105,** 7188–7190.
37. Massefski, Jr., W., Redfield, A., Das Sarma, U., Bannerji, A., and Roy, S. (1990) [7-^{15}N]Guanosine-labeled oligonucleotides as nuclear magnetic resonance probes for protein-nucleic acid interaction in the major groove. *J. Am. Chem. Soc.* **112,** 5350–5351.
38. Goswami, B. and Jones, R. A. (1991) Nitrogen 15 labeled deoxynucleotides. Synthesis of [1-^{15}N]- and [2-^{15}N]-labeled 2'-deoxyguanosines. *J. Am. Chem. Soc.* **113,** 644–647.
39. Lankhorst, P. P., Haasnoot, C. A. G., Erkelens, C. and Altona, C. (1984) Carbon-13 NMR in conformational analysis of nucleic acids fragments. *Nucleic Acids Res.* **12,** 5419–5428.
40. Stec, W. J., Zon, G., Egan, W., and Stec, B. (1984) Automated solid-phase synthesis, separation, and stereochemistry of phosphorothioate analogs of oligodeoxyribonucleotides. *J. Am. Chem. Soc.* **106,** 6077–6079.

41. LaPlanche, L. A., James, T. L., Powell, C., Wilson, W. D., Uznanski, B., Stec, W. J., Summers, M. F., and Zon, G. (1986) Phosphorothioate-modified oligodeoxyribonucleotides. NMR and UV spectroscopic studies of the R_p-R_p, S_p-S_p, and R_p-S_p duplexes [d(GGsAATTCC)]$_2$, derived form diastereomeric O-ethyl phosphorothioates. *Nucleic Acids Res.* **14,** 9081–9091.
42. Stein, C. A., Subasinghe, C., Shinozuka, K., and Cohen, J. S. (1988) Physicochemical properties of phosphorothioate oligodeoxynucleotides. *Nucleic Acids Res.* **16,** 3209–3221.
43. Mori, K., Boizeau, C., Cazenave, C., Matsukura, M., Subasinghe, C., Cohen, J. S., Toulme, J. J., and Stein, C. A. (1989) Phosphoroselenoate oligodeoxynucleotides: synthesis, physico-chemical characterization, anti-sense inhibitory properties and anti-HIV activity. *Nucleic Acids Res.* **17,** 8207–8219.
44. Piotto, M. E., Granger, J. N., Cho, Y., Farschtschi, N., and Gorenstein, D. G. (1991) Synthesis, NMR and structure of oligonucleotide phosphorodithioates. *Tetrahedron* **47,** 2449–2461.
45. Bjergårde, K. and Dahl, O. (1991) Solid phase synthesis of oligodeoxyribonucleoside phosphorodithioates from thiophosphoramidites. *Nucleic Acids Res.* **19,** 5843–5850.
46. Noble, S., Fisher, E. F., and Caruthers, M. H. (1984) Methylphosphonates as probes for protein-nucleic acid interactions. *Nucleic Acids Res.* **12,** 3387–3404.
47. Kan, L. S., Cheng, D. M., Miller, P. S., Yano, J., and Ts'o, P. O. P. (1980) Proton nuclear magnetic resonance studies on dideoxyribonucleoside methylphosphonates. *Biochemistry* **19,** 2122–2132.
48. Lesnikowski, Z. J., Jaworska, M., and Stec, W. J. (1988) Stereoselective synthesis of oligo(thymidine methanophosphonates). *Nucleic Acids Res.* **16,** 11,675–11,689.
49. Lesnikowski, Z. J., Jaworska, M., and Stec, W. J. (1990) Octa(thymidine methanephosphonates) of partially defined stereochemistry: synthesis and effect of chirality at phosphorus on binding to pentadecadeoxyriboadenylic acid. *Nucleic Acids Res.* **18,** 2109–2115.
50. Summers, M. F., Powell, C., Egan, W., Byrd, R. A., Wilson, W. D., and Zon, G. (1986) Alkyl phosphorotriester modified oligodeoxyribonucleotides. NMR and UV spectroscopic studies of ethyl phosphotriester (ET) modified R_pR_p and S_p-S_p duplexes {d[GG AA(ET)TICC]}$_2$. *Nucleic Acids Res.* **14,** 7421–7436.
51. Jäger, A., Levy, M. J., and Hecht, S. M. (1988) Oligonucleotide N-alkylphosphoramidates: synthesis and binding to polynucleotides. *Biochemistry* **27,** 7237–7246.
52. Mag, M. and Engels, J. W. (1988) Phosphoramidate analogs of dinucleotides: synthesis and proton assignment by two-dimensional NMR spectroscopy. *Nucleosides Nucleotides* **7,** 725–728.
53. Morvan, F., Rayner, B., Imbach, J. L., Chang, D. K., and Lown, J. W. (1987) α-DNA. Characterization by high-field ^{1}H-NMR, anti-parallel self-recognition and conformation of the unnatural hexadeoxyribonucleotides α-[d(CpApTpGpCpG)] and α-[d[CpGpCpApTpG)]. *Nucleic Acids Res.* **15,** 4241–4255.

54. Morvan, F., Rayner, B., Imbach, J. L., Lee, M., Hartley, J. A., Chang, D.-K., and Lown, J. W. (1987). α-DNA. Parallel annealing, handedness and conformation of the duplexes of the unnatural α-headeoxyribonucleotide α-[d(CpApTpGpCpG)] with its β-complement β-[d(GpTpApCpGpC)] deduced from high field proton NMR. *Nucleic Acids Res.* **15,** 7027–7044.
55. Mori, K., Subasinghe, C., and Cohen, J. S. (1989) Oligodeoxynucleotide analogs with 5'-linked antraquinone. *FEBS Lett.* **249,** 213–218.
56. van Zijl, P. C. M. and Moonen, C. T. W. (1992) Solvent suppression strategies for in vivo NMR spectroscopy, in *NMR, Basic Principles and Progress*, vol. **22** (Seeling, J. and Rudin, M., eds.), Springer Verlag, Heidelberg, Germany, pp. 67–108.
57. Ruiz-Cabello, J., Vuister, G. W., Moonen, C. T. W., van Gelderen, R., Cohen, J. S., and van Zijl, P. C. M. (1992) Gradient-enhanced heteronuclear correlation spectroscopy. Theory and experimental aspects. *Magn. Reson. Med.* **100,** 282–302.

CHAPTER 13

Mass Spectrometry of Nucleotides and Oligonucleotides

Thomas D. McClure and Karl H. Schram

1. Introduction

The application of mass spectrometry (MS) to the structural analysis of large biopolymers, such as nucleic acids, combines the desirable characteristics of low-level detectability with high selectivity. Current MS techniques are capable of providing mol wt and structurally specific information with submicrogram quantities of sample.

Considerable effort is presently being expended in developing nucleic acid analogs as therapeutic agents, e.g., complementary (antisense) oligonucleotides for the treatment of cancer and acquired immunodeficiency syndrome (AIDS) *(1)*. One difficulty with these oligonucleotides is a short half-life in vivo in the presence of nucleases. Changes to the internucleotide phosphoester linkages (as shown in Fig. 1) confer resistance to nuclease degradation but also prevent characterization of these oligonucleotides by conventional biochemical techniques *(2)*. Fortunately, MS affords an alternative to biochemical methodology for analysis of both the reactive precursors *(3)* and the modified oligonucleotides *(4)*.

Mass spectrometry has undergone major developments in recent years that have resulted in a significant expansion of the number and type of compound classes and the size of the molecules that can be analyzed. Production of gas phase ions from polar, thermally labile and/or high-mol-wt samples is currently possible as a result of the discovery of new desorption ionization methods and other advances in instrumentation.

From: *Methods in Molecular Biology, Vol. 26: Protocols for Oligonucleotide Conjugates*
Edited by: S. Agrawal Copyright ©1994 Humana Press Inc., Totowa, NJ

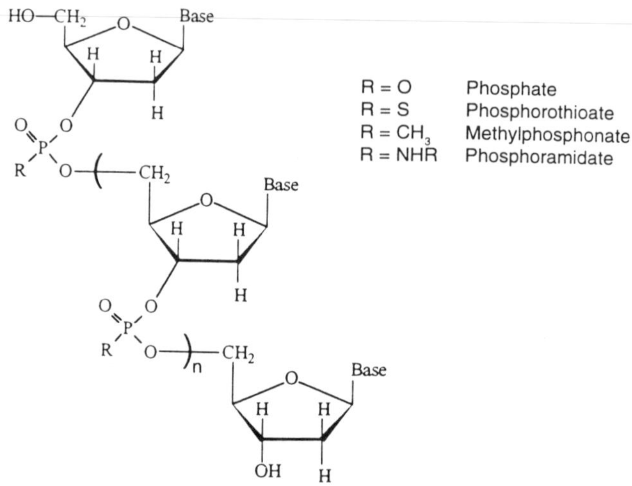

Fig. 1. The general structure of oligonucleotides (adapted from ref. *1* with permission).

The first of the new techniques to be introduced was fast atom bombardment (FAB), also known as liquid secondary ion mass spectrometry (LSIMS) *(5)*, developed by Barber and coworkers *(6)* in 1981. Prior to the introduction of FAB, simple MS analysis of biomolecules was limited, by low sample volatility, to compounds with mol wt of about 1500 daltons. Other important desorption methods currently available for the analysis of nucleic acid components include ^{252}Cf plasma desorption (PD) *(7)*, matrix-assisted laser desorption (MALD) *(8)* and electrospray ionization (ESI) *(9)*.

Developments in instrumentation of significance in the analysis of nucleic acid components include:

1. Advances in magnet technology to permit the analysis of samples with mol wt in excess of 100 kDa.
2. Improvements in time of flight (TOF), ion trap, Fourier transform (FT), and quadrupole mass analyzers *(10)*.
3. The commercial availability of a variety of mass spectrometers capable of performing mass spectrometry-mass spectrometry (MS/MS) analyses.

MS/MS is also termed tandem mass spectrometry since at least two individual mass analyzers are coupled "in tandem" with ionization and mass selection being performed in the first analyzer while the product ion spectrum is generated with the second analyzer. The significance of MS/MS is that a considerable amount of structural information, including mol wt and sequence information, can be obtained by the direct analysis of a mixture containing a compound, or compounds, of interest.

The purpose of this chapter is to provide scientists interested in nucleic acid synthesis and analysis with an overview of the MS techniques available for the identification and characterization of nucleic acids and their components. The chapter is divided into sections according to compound class, starting with nucleotides and cyclic nucleotides and progressing in mol wt through dinucleoside monophosphates and dinucleotides and ending with oligonucleotides. Each section will describe the methods used in the MS or MS/MS analysis of that compound class. A number of book chapters, reviews, and original literature references of importance in the mass spectral analysis of nucleic acid bases *(11–15)* and nucleosides *(11–20)* are included in the references, but are not discussed.

2. Nucleotides and Cyclic-Nucleotides

2.1. Electron and Chemical Ionization

The conventional ionization methods of electron (EI) and chemical ionization (CI) require the sample to be in the vapor phase for mass spectral analysis. Because of the highly polar nature of the phosphate group of nucleotides, application of heat to vaporize the sample results in decomposition. Whereas derivatization, e.g., formation of the per-trimethylsilyl (TMS) *(21,22)* or per-methyl *(23,24)* analogs allows the use of EI or CI MS for simple nucleotides, these classical ionization techniques are not commonly used for the analysis of nucleic acid components above the nucleoside level.

Desorption chemical ionization (DCI) *(25)* and, historically, field desorption (FD) *(26)* can be applied to the analysis of free nucleotides, but the availability of FAB and other, experimentally more simple, techniques has limited the use of these techniques for the analysis of nucleic acid components.

2.2. Fast Atom Bombardment

Since its introduction in 1981, FAB and, more recently, continuous-flow FAB (Cf FAB) *(27)* have been the preferred ionization methods for the mass spectral analysis of nucleotides. Thus, a large number of literature references and reviews are available describing the application of FAB to the analysis of nucleotides *(13,14,17–19,28)*.

In general, the normal positive ion FAB spectrum of a nucleotide *(29)* shows a "cluster" of ions consisting of the protonated molecule ion, MH^+, and ions of the general formula $[M + (Na)_n]^+$ resulting from replacement of exchangeable hydrogens by sodium (or other alkali metal cations). As shown in Fig. 2, the presence of these "cationized" species results in a complex spectrum and, for this reason, most FAB analyses of mononucleotides are performed in the negative ion detection mode. In most cases, the negative ion FAB spectrum is considerably more simple, with the deprotonated molecule ion, $[M - H]^-$, being used to establish mol wt. Confidence in mol wt assignment is, however, best achieved using information from both the positive and negative FAB spectra.

Structurally significant fragment ions are sparse in the normal positive or negative FAB spectrum, with only the $[B + H]^+$ or $[B - H]^-$ ions being prominent. However, the base related ions in these spectra provide important information of value in locating the site of modification following alkylation of nucleotides *(30)* or for identification of modified bases in nucleotides *(31)*.

The absence of structurally significant information in the FAB spectra of free nucleotides can be overcome to some extent by derivatization. For example, the positive ion FAB spectrum of the TMS derivative of AMP contains a number of ions related to the base, sugar, and phosphate portions of the molecule, along with a prominent MH^+ ion for mol wt assignment *(32)*. Since no heating of the FAB probe is required, derivatization followed by FAB analysis can be used to obtain both mol wt and structural information on mononucleotides. Derivatization also appears to provide considerably lower detection limits for nucleotides of biological interest compared to analysis of the free nucleotide, with low nanogram detection of AMP, tricyclic nucleotide, and 2'-deoxy-5-fluorouridine monophosphate being reported *(33)*.

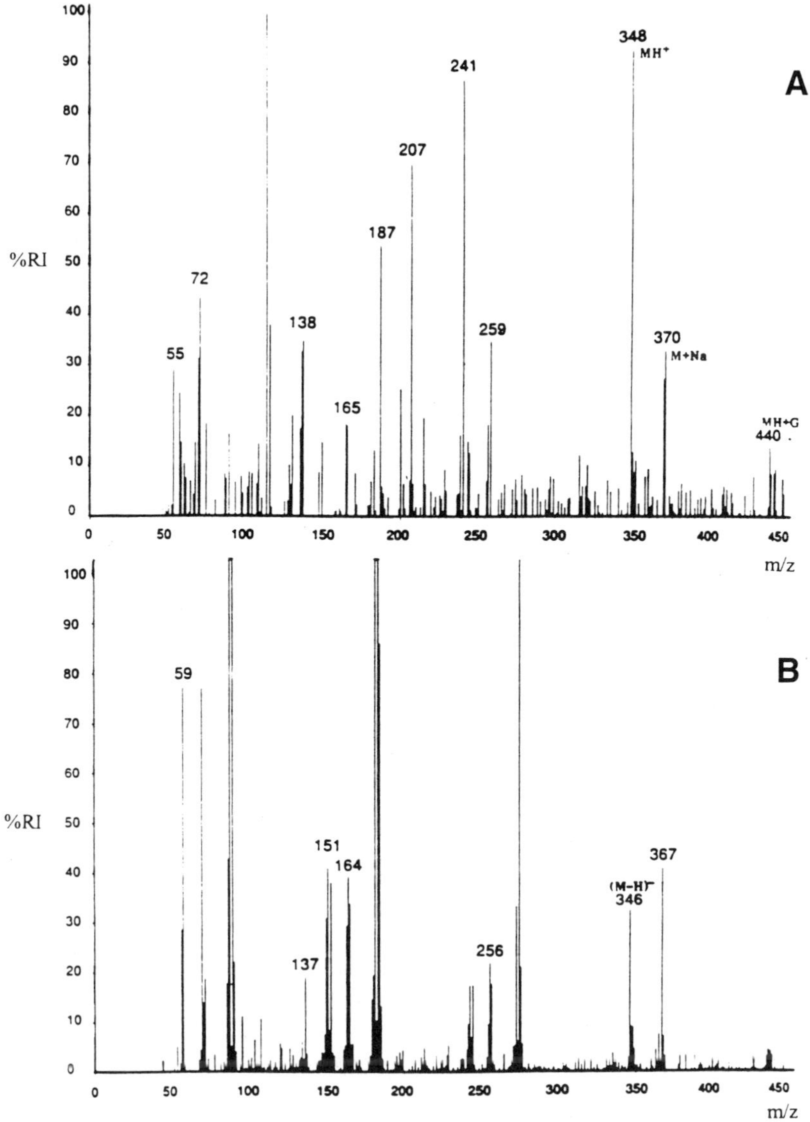

Fig. 2. Positive (**A**) and negative (**B**) ion FAB spectra of adenosine monophosphate using glycerol as the matrix (* matrix peaks). Unless indicated, observed ions are caused by matrix or artifact contribution. Spectrum quality is a function of sample concentration (from ref. *13* with permission).

Tandem mass spectrometry can be used to significantly increase the number and intensity of structurally relevant fragment ions in the FAB MS/MS spectrum compared to the "normal" FAB spectrum of nucleotides. For example, Fig. 3 shows a comparison of the "normal" negative ion FAB spectrum of 5'-GMP in the top panel and the negative ion FAB MS/MS spectrum, recorded following collisional activation, in the lower panel *(34)*. Although both spectra show intense $[M-H]^-$ ions, the normal spectrum lacks any strong ions that are sample related; most of the strong peaks being caused by ions associated with the FAB matrix, glycerol. In contrast, the FAB CAD MS/MS spectrum contains strong ions representing the intact molecule ($[M-H]^-$, m/z 362), the aglycone ($[B-H]^-$, m/z 150), and the sugar phosphate ($[S-H]^-$, m/z 211). Thus, ions indicative of each portion of the nucleotide are present in the MS/MS spectra of nucleotides. Note also that the normal spectrum contains ions from the matrix that "mask" sample related fragment ions, e.g., m/z 211; use of MS/MS very effectively removes these "artifact" ions and provides a clean, easily interpreted spectrum.

Differentiation of 2'-, 3'-, and 5'-monophosphate isomers of adenosine, guanosine, and cytidine is possible *(35)* using FAB in conjunction with mass analyzed ion kinetic energy spectroscopy (MIKES), a simplified form of MS/MS, to accentuate differences in fragmentation pathways and ion stabilities.

MIKES and other MS/MS techniques require a strong precursor ion to produce a useful product ion spectrum and methods have been explored to enhance the formation of preformed ions in the matrix. Approaches that are commonly used to enhance ionization of nucleotides include acidification for positive ion detection, or, for negative ion detection, addition of a weak base, for example diazabicyclo[5. 4. 0]-undec-7-ene (DBU) *(36)*, to the matrix or the use of a basic matrix, such as triethanol amine, to enhance ionization of the phosphate. Ionization in the FAB mode can also be promoted by addition of various substances to alter the surface activity of the matrix. A 1000-fold increase in the intensity of the $[M-H]^-$ ion of ATP has been observed by addition of the cationic surfactant hexadecylpyridinium acetate to the glycerol matrix prior to FAB analysis *(37)*.

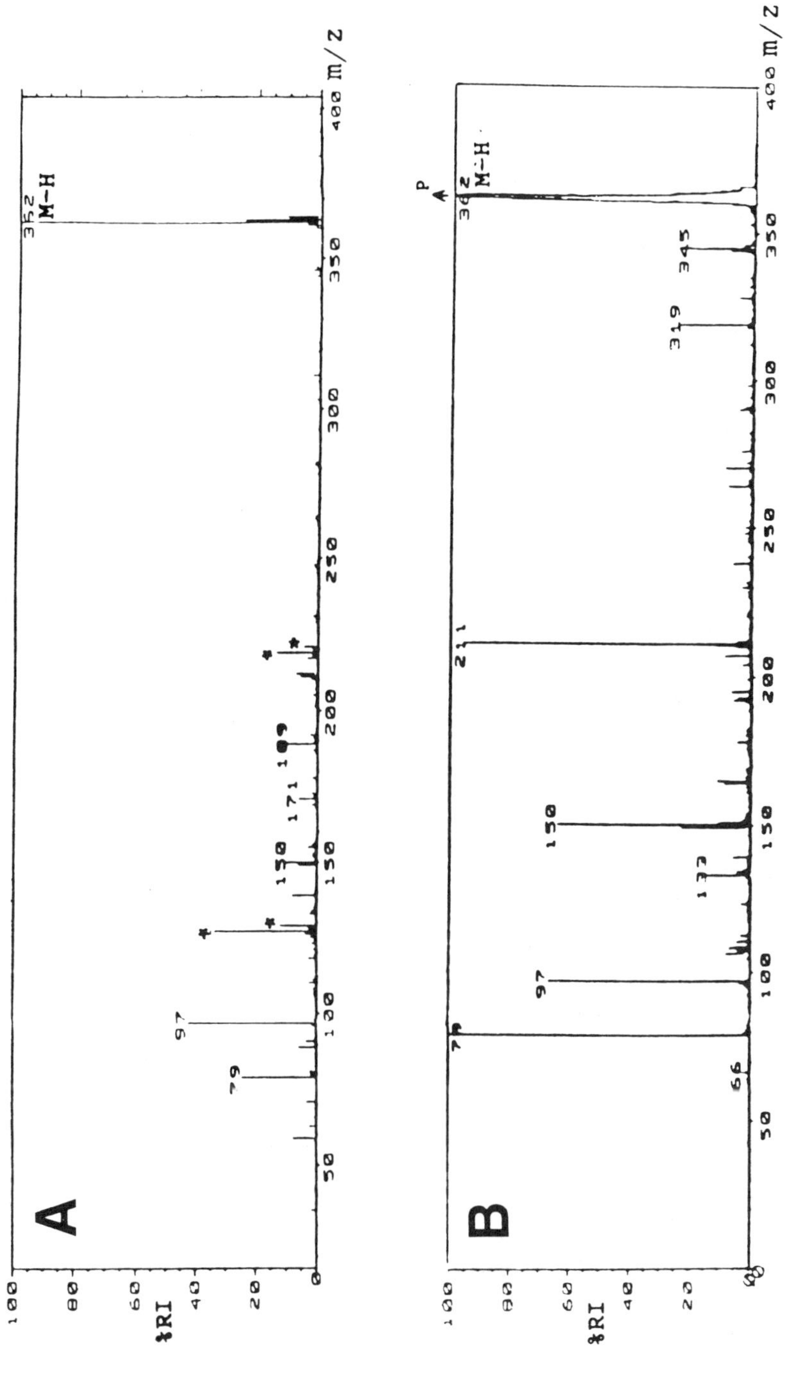

Fig. 3. A comparison of the normal (**A**) negative FAB spectra with the (**B**) negative CID product ion spectra (M + H and M – H selected as precursor icns, respectively) of guanosine 5'-monophosphate (from ref. *34* with permission).

FAB MS/MS techniques have also been applied to the analysis of cyclic nucleotides. Differences in fragmentation between 2',3'- and 3',5'-cyclic nucleotides have been used to differentiate these isomers *(38)* where the FAB MIKES spectrum of the 3',5'-isomer contains a $[BHCH = CHOH]^+$ ion that is absent in the 2',3'-analog. The same approach has been applied to the analysis of a series of amino acid *(38)* and dibutyryl *(39)* derivatives of cyclic nucleotides.

In spite of the wide use of FAB and FAB MS/MS techniques for the analysis of mononucleotides, a number of problems remain to be solved before these techniques will be applicable on a general basis for the analysis of nucleic acid components. The most severe of these problems is the amount of sample required, currently in the mid-nanogram/μL range, to obtain a useful and informative FAB mass spectrum. Thus, analysis of samples containing a limited amount of material, less than mid-nanogram/μL concentrations, will remain a challenge for FAB MS. On the other hand, characterization of synthetic samples, since sample quantity is not a limiting factor, will continue to find FAB MS and FAB MS/MS techniques of considerable value.

2.3. Laser Ionization

Although reports of the analysis of nucleotides and small oligomers by FAB MS have been extensive, the use the two new ionization techniques of matrix assisted laser desorption (MALD) and electrospray ionization (ESI) have been limited, most probably because these new methods find their greatest utility in the analysis of high-mol-wt samples.

Lasers have been used to produce gas phase ions of protected nucleotides for MS analysis in a technique known as multiphoton ionization (MPI). The combined use of an infrared laser for sample evaporation with ionization with a tunable laser has been used in the analysis of protected nucleotides; the tunable laser used for ionization provided the ability to select the hardness or softness of ionization by adjustment of beam intensity. This technique has been applied to the analysis of two series of nucleotides: O-(2-chlorophenyl-2-cyanoethyl)-phosphate-protected and 5'-O-dimethoxytrityl, O-(2-chlorophenyl-2-cyanoethyl)-3'-phosphate-protected deoxynucleotides *(40)*. The spectra of both of these compound classes are dominated by cleavage points directed by the protecting groups, with the type and intensity of the fragmentation being a function of the power used in ionization. Aside from glycosidic bond cleavage, few structurally-informative ions are observed in the laser desorption mass spectra of nucleotides.

2.4. Electrospray

Two reports have described the use of ESI for the analysis of mononucleotides. The positive ion ESI spectra of AMP and ADP in a 1:1 solution of methanol/water show singly charged [MH^+] ions as the base peak (most intense ion) of the spectrum and no fragment ions were observed *(9)*. Both singly- and doubly-charged [M – H] ions are observed in the negative ion ESI spectrum of ATP *(41)* and formation of $[M - H_sPO_3]^-$ and $[M - Na - HPO_3]^-$ ions are observed that arise from the terminal phosphate group.

3. Dinucleoside-Monophosphates and Dinucleotides

The nomenclature used to describe the peaks observed in the mass spectrum of a nucleic acid component more complex than a mononucleotide is not standardized and can lead to confusion. Adaptation of the system developed to describe the peaks in the mass spectra of a peptide, with modifications appropriate to oligonucleotides, has been recommended *(42)* and this system will be used in this chapter. The letters *a*, *b*, *c*, and *d* designate the fragment retaining the charge and possessing a free 5'-position; the letters *w*, *x*, *y*, and *z* refer to the fragment retaining the charge that also contains the terminal free 3'-OH function. The subscripts indicate the position of cleavage along the phosphate backbone beginning with the 5'-end and follow the same convention as is used in numbering the position of the bases as shown in Fig. 4.

3.1. Electron Ionization of TMS Derivatives

Electron ionization has been used to study the mass spectrometry of a series of dinucleoside-monophosphates as their per-TMS derivatives *(43)*. Besides the mol wt indicating ions $M^{+\cdot}$ and M-15, structurally useful fragment ions observed are a_1, z_1, $d_1 + 2H$, $w_1 + 2H$, $d_1 + H{-}CH_3TMSOH$, and $w_1 + H{-}CH_3TMSOH$. The process of differentiating between unknown molecules where the bases are interchanged, such as ApU and UpA, cannot rely on the masses of the

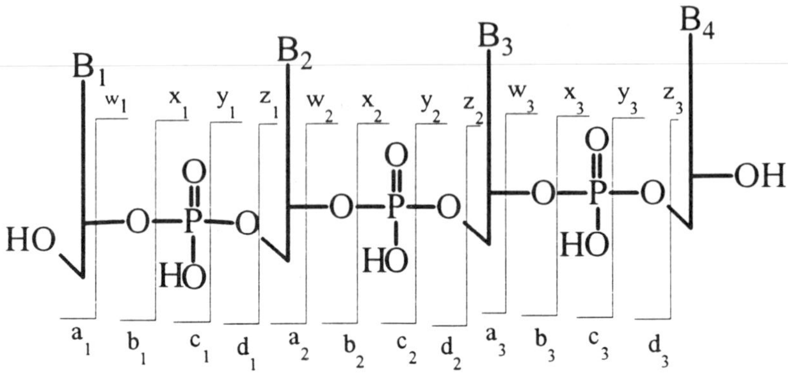

Fig. 4. Fragmentation nomenclature for dinucleoside monophosphates, dinucleoside diphosphates, and oligonucleotides (from ref. *42* with permission).

ions involving fragmentation at a_1 vs z_1 or w_1 vs d_1 as they are isobaric. However, for the compounds studied, fragmentations related to the O–C5' (d_1 and z_1) bond are more facile, producing more intense ions than at the O–C3' (a_1 and w_1) position. Thus, ion intensities were used to distinguish between d_1/z_1 and a_1/w_1 ions, in turn permitting differentiation of positional isomers.

3.2. Desorption Chemical Ionization

Chemical ionization and desorption chemical ionization (DCI) MS and MS/MS were used to determine the structure of the phosphate methylated dinucleoside monophosphate Tp(ME)T *(25)*. Chemical ionization was unsuccessful in providing either mol wt or structural information on the sample, but DCI provided an MH^+ ion for the intact molecule. Ions corresponding to a mononucleotide with the structure Tp(Me) and the nucleoside were also observed in the spectrum. Similar results were obtained in the negative ion detection mode.

3.3. Fast Atom Bombardment

The ionization method of choice for the mass spectral analysis of dinucleoside monophosphates (B_1pB_2) and dinucleoside diphosphates (e.g., pB_1pB_2) is fast atom bombardment in the negative ion detection mode. The structural information in the mass spectra of these compound classes is similar to that observed for nucleoside monophosphates and includes ions corresponding to $[M - H]^-$, $[M - H - B_1$

$_{or\ 2}]^-$ *(17)*. The positive ion spectra contain some useful information, especially when B_1 differs from B_2, but the structurally important ions are less evident, or may be masked, because of formation of cationized species with either the matrix or sample. Thus, the negative ion detection mode is most commonly used in order to eliminate the cationization or masking of significant ions *(29,31)*.

Dissimilarities in the fragmentation pattern of 3',5'-($B_{1,3'}p_{5'}B_2$) and 5',5'-($B_{1,5'}p_{5'}B_2$) linked dinucleoside monophosphates can be used to differentiate these isomers since the 3',5'-bond is less stable than the 5',5'-bond *(44)*. Lower stability of specific bonds resulted in specific cleavages in the 3',5'-series of compounds; loss of one of the two bases, followed by cleavage of the phosphate group and subsequent loss of the second base. Owing to greater bond stability, the 5',5'-compounds did not have these selected fragmentation pathways resulting in a more complicated spectrum.

The use of negative ion FAB has been reported *(45)* as the method of choice for the characterization of deoxy-dinucleotide monophosphate synthons used in oligonucleotide synthesis. With the exception of the 3'-terminus, these synthons were fully protected (e.g., dimethoxytrityl group on the 5'-terminus, cyanoethyl group on the phosphate oxygen, and various base-specific groups on the exocyclic amines in the bases) and were unambiguously characterized by ions from *d*- and *w*- cleavages as well as ions resulting from losses of all or part of the protecting groups. As a consequence of reduced polarity, tetraethylene glycol rather than glycerol was used as the matrix.

Using FAB MS, partially protected -di(deoxyribo)nucleoside diphosphate synthons also have been characterized using two ions: The deprotonated molecule provides mol wt, which is unique for all nonisomeric combinations of A, C, G, and T; and the ion from the w_1 cleavage, which differentiates between base-positional isomers (e.g., d-ApC vs d-CpA). The ions resulting from y_1 and d_1 cleavages are used to verify sequence *(46)*.

Negative ion FAB has also been used in the analysis of protected dinucleoside monophosphate dimers with a carbamate group replacing the phosphodiester linkage *(47)*. Derivatives used in this study included a monomethoxytrityl group in the 5'-position and protection of exocyclic amino groups in the aglycone with benzoyl groups, leaving the 3'-OH function free. If ions resulting from the loss of the

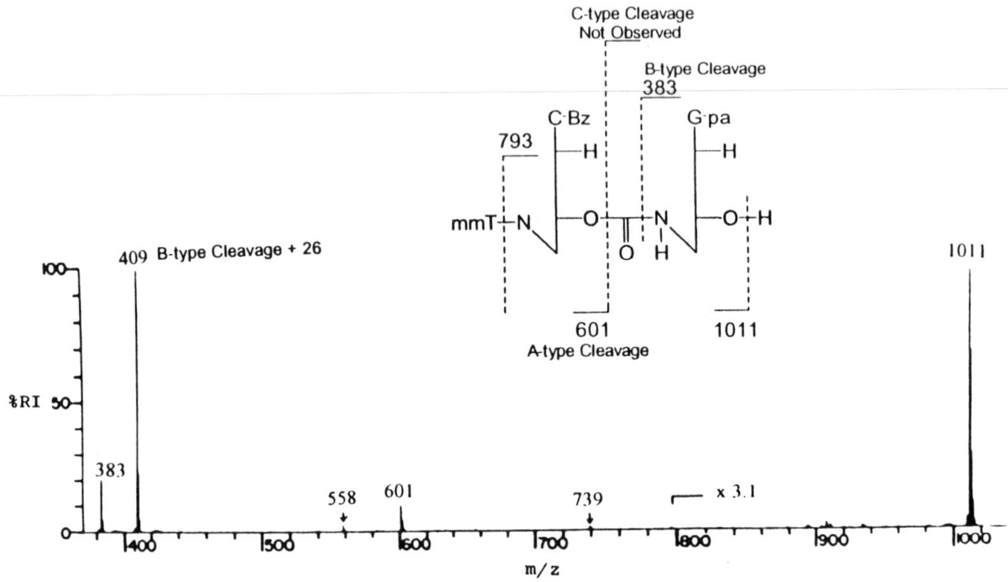

Fig. 5. Negative ion FAB mass spectrum of a partially protected (3'OH left unprotected) modified dimer d(CG). Modification involves replacement of phosphodiester linkage with carbamate linkage. The A-type, B-type, and C-type fragmentations are as indicated (from ref. *47* with permission).

protecting groups are ignored, three types of cleavage, A-, B-, and C-, were observed, as shown in Fig. 5.

An A-type cleavage results from breaking the carbamate carbonyl-3'O bond with the charge being carried by the 3'-OH fragment; a B-type cleavage breaks the carbonyl-amino group bond in the carbamate function, with the charge residing on the 5' end of the fragment; and a C-type cleavage is the same as the A-type cleavage except that the charge is retained by the carbamate group, which rearranges to a more stable structure. Ions resulting from A- and B-type cleavages are prominent in the spectrum and sequence ions may be initiated from either end of the molecule. Figure 5 shows the negative ion FAB spectra of two of these carbamate linked, protected dimers.

Although the normal FAB spectrum of free deoxydinucleoside monophosphates does not allow differentiation of structural isomers, e.g., d[AT] and d[TA], unambiguous identification can be achieved using MS/MS techniques *(17,48)*. Figure 6 shows the normal and

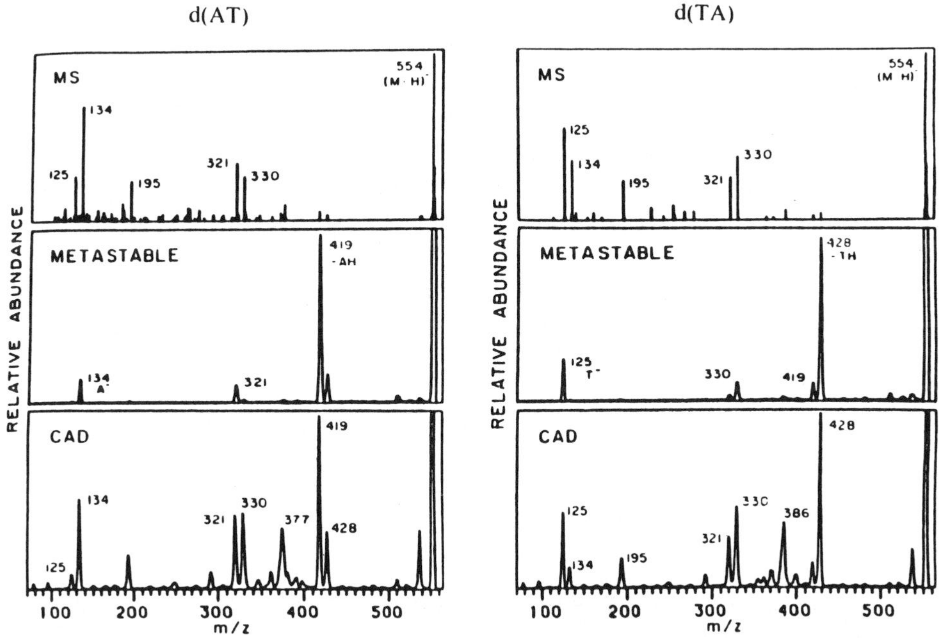

Fig. 6. A comparison of the normal and product (M-H selected as the precursor ion) negative ion FAB spectra for: 1. d(AT) and 2. d(TA) (from ref. *48* with permission).

CID MS/MS FAB spectra of the d[AT] and d[TA] pair. Most significant in distinguishing one isomer from the other is the specific elimination of the base on the 3'-terminal position in the CAD MS/MS spectrum of the $[M - H]^-$ ion, a loss barely evident in the normal spectrum *(48)*. The CID MS/MS spectrum also produces fragmentation of the deoxyribose ring, forming the $[M - H - BH - 42]^-$ ion by elimination of the 1'- and 2'-carbon atoms and the ring oxygen; this ion shifts by 16 daltons to higher mass in the CID MS/MS spectra of ribose containing analogs. The M-17 ion in the CID MS/MS spectrum is also characteristic of a deoxyribose.

3.4. ^{252}Cf-Plasma Desorption

Negative ion ^{252}Cf-plasma desorption (PD) has been used to study the mass spectrometry of fully protected dinucleoside monophosphates *(49)*. Mol wt information is available from the negative ion PD mass spectrum by the appearance of an $[M - H]^-$ ion and sequence information, most especially the d_1 and w_1 type ions. Fragments arising from loss of the various protecting groups are also present.

3.5. Matrix Assisted Laser Desorption

Tandem mass spectrometry with a Fourier transform (FT) mass spectrometer and matrix assisted laser desorption ionization has been used in the analysis of free dinucleoside monophosphates *(50)*. An intense $[M - H]^-$ ion, d_1 and w_1 ions, and ions corresponding to loss of BH^+ ions were characteristic of these molecules. Elimination of the bases was preferential for B_1H over B_2H whereas bond breaking in the phosphate group favored cleavage at w_1 over d_1.

4. Oligonucleotides

An excellent review has been published *(51)* discussing the application of mass spectrometry nucleic acid constituents with an emphasis on high mass compounds.

4.1. Electron Ionization

Electron ionization has, surprisingly, been applied to the analysis of synthetic intermediates and final products of the solid phase synthesis of oligonucleotides. In pyrolysis mass spectrometry (Py-MS) *(42)*, excess heat is applied to the sample to affect decomposition of the sample in a defined manner. Unfortunately the technique suffers from a number of problems, e.g., different samples have different decomposition temperatures, control of the decomposition is difficult and the products of decomposition are not always well defined. Py MS has, however, been used for characterization of oligonucleotides where an increase in the ion current, characteristic of a given base, is monitored; the addition of a given residue to an oligonucleotide chain is considered successful if the ion current characteristic of the added base increases by 20% *(19)*. Py-EI-MS has also been used to confirm the incorporation of the modified nucleoside N^6-Methyladenosine (m^6A) into a synthetic oligonucleotide for studies of cellular repair mechanisms *(52)*. Ions characteristic of the base portion of m^6A in the Py MS spectrum and the ion profile, shown in Fig. 7, were used to confirm incorporation of the modified base into d(GGm^6ATACC).

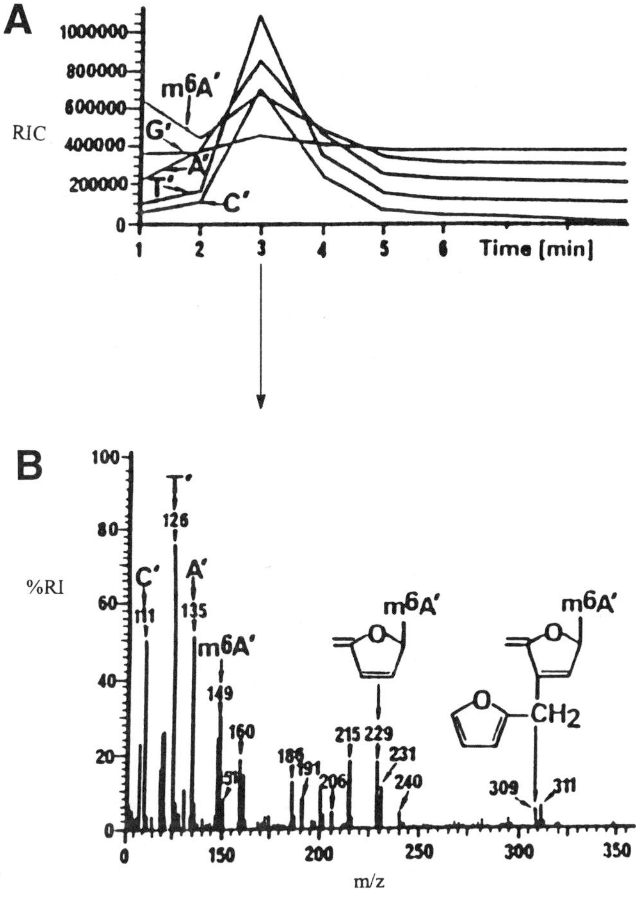

Fig. 7. Pyrolytic MS of the modified octamer d(GpGpm^{6}ApTpApTpCpC). **A.** ion chromatogram during pyrolysis experiment. **B.** EI spectrum after 3 min of pyrolysis (from ref. *52* with permission).

4.2. Fast Atom Bombardment

Negative-ion FAB MS has been applied to the analysis of free oligonucleotides where both mol wt and sequence information are available in the mass spectrum *(53,54)*. Figure 8 shows the mass spectrum and structural assignments for ions in the spectrum of the decamer d(GpApApGpApTpCpTpTpC). Mol wt is indicated by the $[M - H]^-$

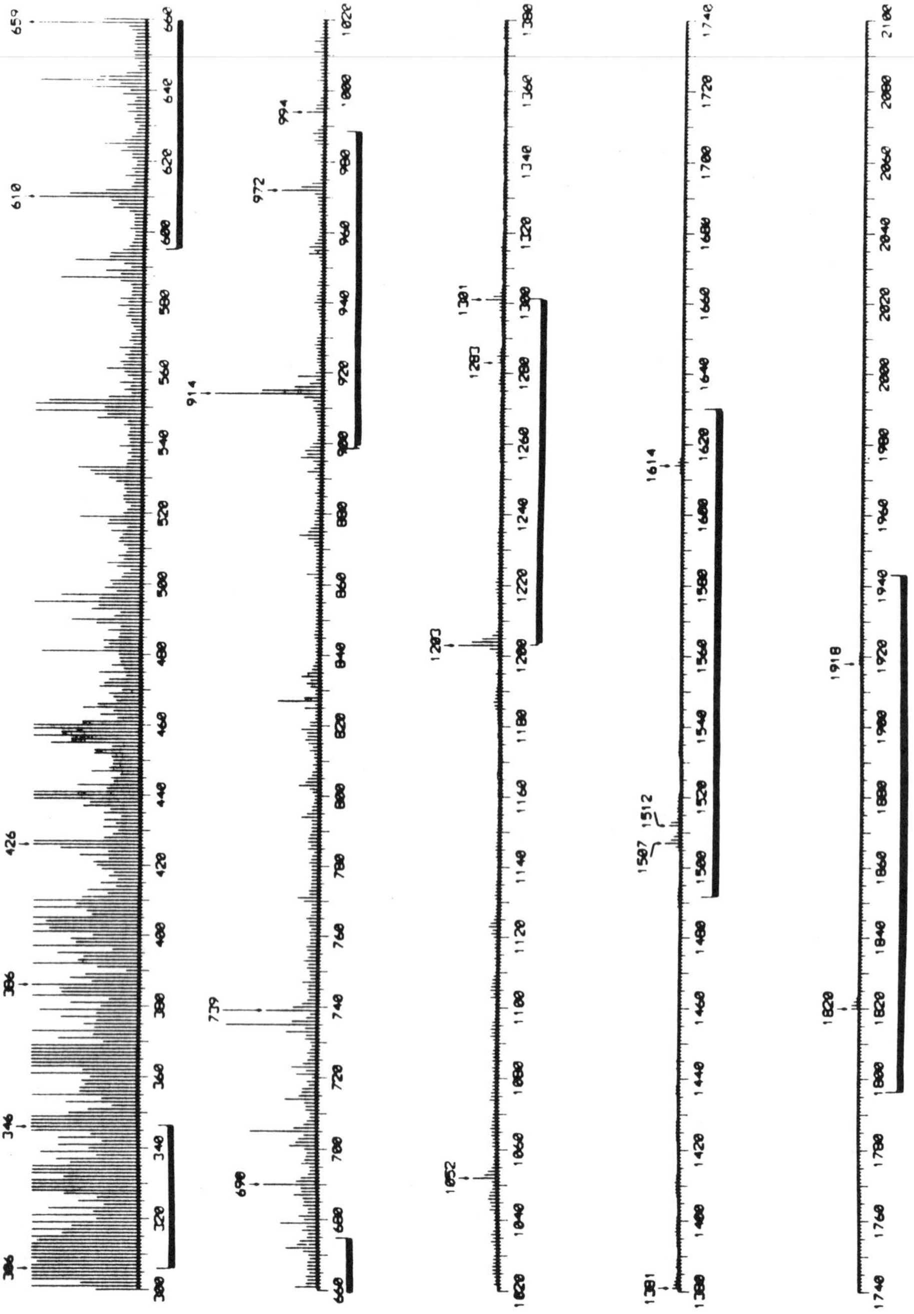
659
610
426
386
346
306
994
972
914
739
690
1301
1283
1203
1052
1614
1512
1507
1381
1918
1820

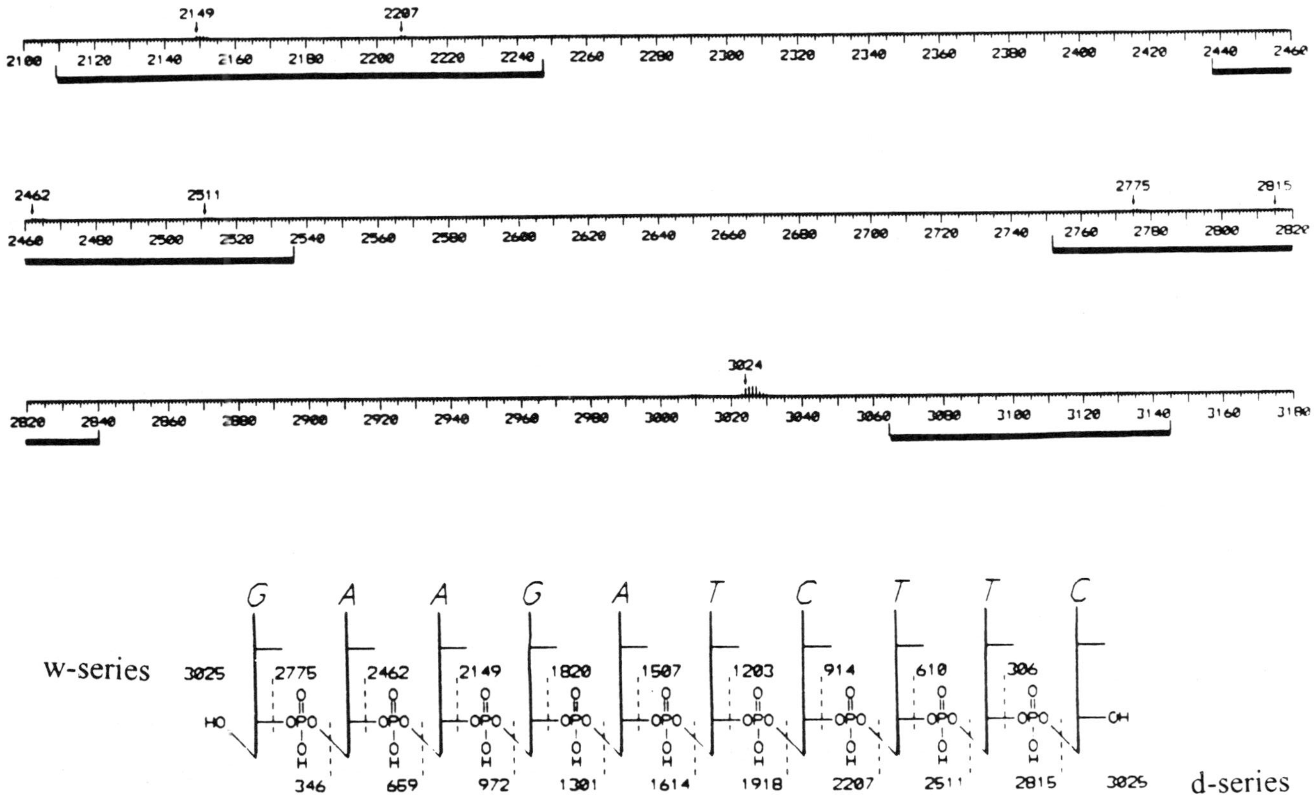

Fig. 8. Negative ion FAB mass spectrum of d(GpApApGpApTpCpTpTpC) using glycerol as the matrix. The underlines indicate sequence ion regions (from ref. *54* with permission).

ion at m/z 3024 and sequence ions, the *w*- and *d*-ion series, arise from both ends of the chain. The *w*-series, often termed the 5'-phosphate sequence ions, are more intense than the *d*-series, the 3'-phosphate sequence ions, which allows direction to be assigned to the sequence. The difference in intensity of these ion series is thought to reflect the greater stability of a charge located at the secondary 3'-carbon atom relative to the less stable location of the charge on the primary 5'-position. Exceptions to the intensity differences have been reported *(31)* and may be a result of the size of the oligonucleotide. Also, addition of a protecting group to the 5'-end changes the relative intensity of the fragments such that the *d*-series become more intense than the *w*-series *(55)*. Ten residues appears to be the limit to which the spectrum contains the complete *d*- and *w*-ion series *(19)*. Oligonucleotides of greater length can still be sequenced by combining partial sequence information from each ion series and noting overlap. This process has been termed the "half-sequence" method and may be applicable to the sequencing of 20mers *(56)*.

The ambiguity in identifying *d*- and *w*-series ions can be removed by using MS/MS techniques. Application of CID MS/MS to the analysis of oligonucleotides ranging in size from trimers to hexamers has been described *(57)*. In each case, the *d*-series ions were more abundant than the *w*-series ions. Furthermore, oligonucleotides from dimers to tetramers showed a facile cleavage of BH^+ ion from the 3' terminus; differences were also observed in the product ion spectra of ribo- and deoxyribo-sugar residues; the deoxynucleotide spectra showed a preference for cleavage of sugar-phosphate bonds to eliminate nucleosides whereas the ribonucleotides tended to fragment by loss of mononucleotides.

A significant limitation to the use of FAB MS for the characterization of free or blocked oligonucleotides is the sensitivity of the technique. Normally, a concentration of sample of greater than 1–10 μg/μL is required to obtain a useful spectrum. An improvement in sensitivity is achieved by introducing the sample into the mass spectrometer using a continuous flow FAB probe (Cf FAB) *(58)*. For example, Cf FAB has been successful in the analysis of 38 pmol of the hexamer d(GTTAAC), which is an order of magnitude improvement in sensitivity relative to static FAB; the mobile phase in this experiment was 80:20:0. 5 methanol/water/triethanolamine *(59)*.

4.3. ^{252}Cf-Plasma Desorption

The analysis of fully protected oligonucleotides has been reported by PD MS using both positive *(60)* and negative *(61)* ion detection. The positive ion spectra were complicated by formation of cationized species and by the presence of structurally uninformative fragments arising from the protecting groups. Less complicated spectra were recorded using negative ion mode. In either case, however, both mol wt and full sequence information were observed on hexamers. Negative ion PD MS has also been used in the analysis of free trimers where $[M - H]^-$, *w*-series, *d*-series and numerous phosphate-related ions were recorded for each of the ten samples examined *(62)*.

4.4. Matrix Assisted Laser Desorption

Matrix assisted laser desorption (MALD) is an exciting new ionization method providing a significant improvement in sensitivity relative to FAB and having particular application in the determination of mol wt for large biopolymers. For example, a singly charged protonated molecule of intact yeast tRNA having a mol wt of approx 24,900 daltons has been recorded (*see* Fig. 9) using an aqueous solution containing 100 ng/µL of sample *(63)*. The ability to obtain mol wt information on an intact tRNA molecule must be considered a landmark in the application of mass spectrometry to biochemical problems, and MALD should be of considerable importance in establishing the mol wt of naturally occurring and synthetic oligonucleotides.

MALD can also be used for establishing the mol wt of smaller, unprotected oligomers, e.g., tetramers, where the sample requirements for analysis are in the 10–100 pmol range *(64)*.

Finally, when MALD is coupled with a Fourier transform (FT) mass spectrometer, both mol wt and sequence may be obtained by using the MS/MS capability of this mass analyzer. Such analyses have been reported for tri- and tetrameric oligonucleotides *(50)*.

4.5. Electrospray Ionization

Electrospray ionization (ESI) is a second new ionization method that hold considerable promise for the analysis of high-mol-wt nucleic acid components. The characteristic feature of ESI is the formation of multiple charges on the molecule being examined, resulting in the appearance of protonated molecule ions at m/z values determined by

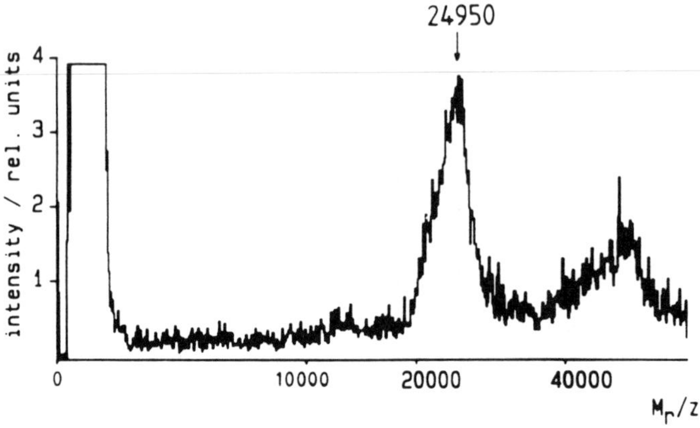

Fig. 9. Negative ion matrix assisted laser desorption (MALD) mass spectrum of yeast tRNA (mol wt = 24,952 daltons). Matrix was nicotinic acid (from ref. *63* with permission).

the number of charges on each of the ions. The practical result of multiple charging is to lower the mass of the protonated molecules, thus, in effect, expanding the mass range of the mass spectrometer by a factor of 10 to 100 times. The coupling of an ESI source to both quadrupole and magnetic instruments has been reported.

As with other desorption methods, the formation of sodium adducts with the molecule ion compromises the quality of the positive ion spectrum in ESI. Purification of the sample using gel electrophoresis followed by formation of the ammonium salt prior to ESI analysis reduces formation of the sodium adducts and results in the recording of a much cleaner spectrum of oligonucleotides *(65)*. The ESI spectrum of a 77-mer, $A_{23}C_{17}G_{29}T_8$ with a mol wt of 24,056 daltons, pretreated in this manner is shown in Fig. 10; the mass spectrum was recorded using a 50 pmol/µL concentration of the oligonucleotide. The ion at m/z 924. 5 corresponds to a single sodium adducted to the $[M-27H]^{26-}$ ion. A mass accuracy of at least 0. 03% in determination of mol wt is possible with the quadrupole mass analyzer used in this study, which should allow ESI MS to check for additions, deletions, depurination, and complete deblocking of synthetic oligonucleotides.

Sensitivity and mass accuracy are important considerations when mass spectrometry is used to confirm the identity of products obtained by automatic synthesis. These criteria are demonstrated in the ESI

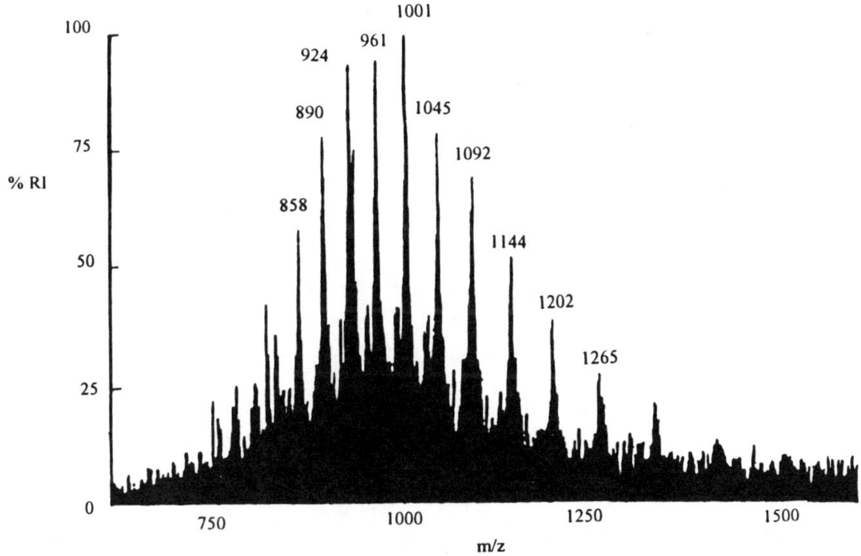

Fig. 10. Negative ion electrospray mass spectrum of a 50 pmol/µL solution of a 77-mer [$A_{23}C_{17}G_{29}T_8$, mol wt = 24,039] (from ref. *65* with permission).

analysis of the synthetic 14-mer d(CpApTpGpCpCpApTpGpGpCp-ApTpG), having a calculated mass for the protonated molecule of 4260.7 daltons. Analysis of a 200 pmol sample of this 14-mer by ESI provided an observed mass for the MH^+ ion of 4261.4 daltons, a difference of 0. 4 daltons and an accuracy of 0. 009% *(66)*.

The analysis of two intact tRNA molecules with mol wt in excess of 24 daltons has also been reported *(67)* and the spectra of these molecules is shown in Fig. 11. Similar to MALD, the ESI can be used for mol wt determination, but no fragmentation has yet been observed with either of these techniques.

5. Concluding Remarks

Mass spectrometry has only recently become an important tool for the analysis of nucleotides, dimers, and oligonucleotides. New ionization methods and improvements in instrument design and capabilities have expanded the application of mass spectrometry to compound classes, e.g., oligonucleotides, almost unimaginable only a short time ago. Along with this expansion in capability, however, has come confusion as to which method is best for the analysis of

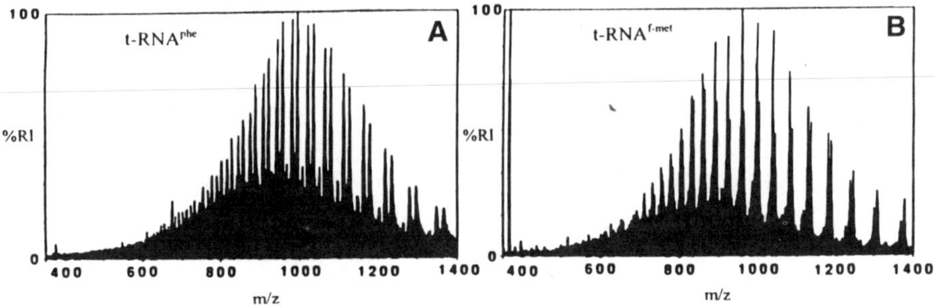

Fig. 11. Negative ion electrospray mass spectrum of **(A)** tRNA^{Phe} from Brewer's yeast [mol wt = 24,926] and **(B)** tRNA$^{f\text{-}met}$ from *E. coli* [mol wt = 24,926. 3] (from ref. *67* with permission).

nucleic acid components. The answer is that no one technique will answer all of the questions asked by a synthetic nucleotide chemist, a biochemist, or a molecular biologist. Access to all of the methods mentioned in this chapter would provide the greatest potential in solving a problem related to the structure of a nucleic acid component. Few individual laboratories are so equipped.

A clear definition of the problem to be solved can, however, simplify the choices. Possibly the most conceptually simple division of problems is by mol wt. Thus, for samples below mol wt 2 kDa, FAB MS or FAB MS/MS would be the method of choice since both mol wt and sequence information may be obtained. Other methods applicable to this mass range include PD or MALD on a time of flight analyzer, ESI on quadrupole or sector instruments, or, in some cases, even DCI or EI.

Mol wt determination of oligonucleotides above the 2 kDa level are best made using MALD or ESI, although PD may also be used in some cases. Observation of protonated molecule ions of intact tRNA molecules is a landmark achievement and illustrates the potential of mass spectrometry for the analysis of high-mol-wt oligonucleotides. Although sequencing of these high-mol-wt oligos is not yet possible, new methods are currently under investigation that may yield such information in the not too distant future. Considering the sensitivity, selectivity, and structural/mol wt information available in a mass spectrum, access to state of the art mass spectrometry should be considered a fundamental requirement for anyone involved in the synthesis, biochemistry, or molecular biology of nucleic acid components.

References

1. Zamecnik, P. C. and Agrawal, S. (1991) The hybridization or antisense, approach to the chemotherapy of AIDS, in *Aids Research Reviews*, vol. 1 (Koff, W. C., Wong-Staal, F., and Kennedy, R. C., eds.), Marcel Dekker, New York, pp. 301–313.
2. Agrawal, S., Temsamani, J., and Tang, J. Y. (1991) Pharmacokinetics, biodistribution, and stability of oligodeoxynucleotide phosphorthioates in mice. *Proc. Natl. Acad. Sci. USA* **88,** 7595–7599.
3. Toren, P. C., Betsch, D. F., Weith, H. L., and Coul, J. M. (1986) Determination of impurities in nucleoside 3'-phosphoradmidites by fast atom bombardment mass spectrometry. *Anal. Biochem.* **152,** 291–294.
4. Arlandini, E., Gioia, B., Brasca, M. G., and Fustinoni, S. (1990) Comparison of FAB and FD mass spectrometry in the analysis of unusually linked nucleotides. *Nucleosides Nucleotides* **2,** 424–431.
5. Burlingame, A. L., Baillie, T. A., and Derrick, P. J. (1986) Mass spectrometry. *Anal. Chem.* **58,** 165R–211R.
6. Barber M., Bordoli, R. S., Sedgwick, R. D., and Tyler, A. N. (1981) Fast atom bombardment of solids (F. A. B.): A new ion source for mass spectrometry. *J. Chem. Soc. Chem. Commun.* 325–327.
7. Roepstorff, P. and Sundqvist, B. (1986) Plasma desorption mass spectrometry of high-molecular weight biomolecules, in *Mass Spectrometry in Biomedical Research* (Gaskell, S. J., ed), Wiley, New York, pp. 269–285.
8. Hillenkamp, F., Karas, M., Ingendoh, A., and Stahl, B. (1990) Matrix-assisted UV laser desportion/ionization: a new approach to mass spectrometry of large biomolecules, in *Biological Mass Spectrometry* (Burlingame, A. L. and McCloskey, J. A., eds.), Elsevier, Amsterdam, pp. 49–60.
9. Whitehouse, C. M., Dreyer, R. N., Yamashita, M., and Fenn, J. B. (1985) Electrospray interface for liquid chromatographs and mass spectrometers. *Anal. Chem.* **57,** 675–679.
10. Brunnee, K. (1987) The ideal mass analyzer: fact or fiction? *Int. J. Mass Spectrom. Ion Proc.* **76,** 121–237.
11. McCloskey, J. A. (1974) Mass spectrometry, in *Basic Principles in Nucleic Acid Chemistry*, vol. 1 (Ts'o, P. O. P., ed.), Academic, New York, pp. 209–309.
12. Hignite, C. (1980) Nucleic acids and derivatives, in *Biochemical Applications of Mass Spectrometry, First Supplementary Volume* (Waller, G. R. and Dermer, O. C., eds.), Wiley, New York, pp. 527–566.
13. Schram, K. H. (1989) Purines and pyrimidines, in *Mass Spectrometry* (Lawson, A. M., ed.), Walter de Gruyter, Berlin, pp. 508–570.
14. Schram, K. H. (1990) Mass spectrometry of nucleic acid components, in *Biomedical Applications of Mass Spectrometry*, vol. 34. (Suelter, C. H. and Watson, J. T., eds.), Wiley, New York, pp. 203–287.
15. Jankowski, K., Jocelyn-Parf, J. R., and Wightman, R. H. (1986) Mass spectrometry of nucleic acids. *Adv. Heterocyclic Chem.* **39,** 79–115.
16. Pang, H., Schram, K. H., Smith, D. L., Gupta, S. P., Townsend, L. B., and McCloskey, J. A. (1982) Mass spectrometry of nucleic acid constituents. Trimethylsilyl derivatives of nucleosides. *J. Org. Chem.* **47,** 3923–3932.

17. Crow, F. W., Tomer, K. B., Gross, M. L., McCloskey, J. A., and Berstrom, D. E. (1984) Fast atom bombardment combined with tandem mass spectrometry for the determination of nucleosides. *Anal. Biochem.* **139,** 243–262.
18. Schram, K. H. (1988) Analysis of nucleosides, nucleotides and oligonucleotides using fast atom bombardment mass spectrometry. *Trends Anal. Chem.* **7,** 28–32.
19. Crain, P. F. (1990) Mass spectrometric techniques in nucleic acid research. *Mass Spectrom. Rev.* **2,** 505–554.
20. Slowikowski, D. and Schram, K. H., (1985) Fast atom bombardment mass spectrometry of nucleosides. Comparison with electron impact and chemical ionization mass spectra. *Nucleosides and Nucleotides* **4,** 347–376.
21. Lawson, A. M., Stillwell, R. N., Tacker, M. M., Tsuboyama, K., and McCloskey, J. A. (1971) Mass spectrometry of nucleic acid components. Trimethylsilyl derivatives of nucleotides. *J. Am. Chem. Soc.* **2,** 1014–1023.
22. Schram, K. H. (1990) Preparation of trimethylsilyl derivatives of nucleic acid components, in *Methods in Enzymology*, vol. **193** (McCloskey, J. A., ed.), Academic, Orlando, pp. 791–795.
23. Teece, R. G. and Schram, K. H. (1986) Preparation and gas phase analysis of permethylated nucleosides, in *Nucleic Acid Chemistry,* vol. **3** (Townsend, L. B. and Tipson, R. S., eds.), Wiley-Interscience, New York, pp. 311–328.
24. Petit, G. R., Einck, J. J., and Brown, P. (1978) Structural biochemistry 15. Mass spectrometry of permethylated nucleotides. *Biomed. Mass Spectrom.* **5,** 153–160.
25. Isern-Flecha, I., Jiang, X. -Y., Cooks, R. G., Pfleiderer, W., Chae, W. -G., and Chang, C. -J. (1987) Characterization of an alkylated dinucleotide by desorption chemical ionization and tandem mass spectrometry. *Biomed. Environ. Mass Spectrom.* **14,** 17–22.
26. Schulten, H. -R. and Schiebel, H. M. (1985) Field desorption and fast atom bombardment mass spectrometry of pyridine nucleotides and nucleoside triphosphates. *Fres. Z. Anal. Chem.* **321,** 531–537.
27. Bertrand, M. J., Benham, V., St-louis, R., and Evans, M. J. (1989) Continuous flow fast atom bombardment mass spectrometry of mononucleotides and their metal complexes. *Can. J. Chem.* **67,** 911–920.
28. Slowikowski, D., and Schram, K. H. (1985) Fast atom bombardment mass spectrometry of nucleosides, nucleotides and oligonucleotides. *Nucleosides Nucleotides* **4,** 309–345.
29. Eagles, J., Javanaud, C., and Self, R. (1984) Fast atom bombardment mass spectrometry of nucleosides and nucleotides. *Biomed. Mass Spectrom.* **11,** 41–47.
30. Fenselau, C. (1984) Fast atom bombardment and middle molecule mass spectrometry. *J. Nat. Prod.* **47,** 215–225.
31. Hogg, A. M., Kelland, J. G., and Vederas, J. C. (1986) Investigation of ribo- and deoxyribonucleosides and -nucleotides by fast atom bombardment mass spectrometry. *Helv. Chim. Acta* **69,** 908–917.

32. Schram, K. H., and Slowikowski, D. L. (1986) Fast atom bombardment of trimethylsilyl derivatives of nucleosides and nucleotides. *Biomed. Environ. Mass Spectrom.* **13,** 263–264.
33. Weng, C. -M., Hammargren, W. M., Slowikowski, D., Schram, K. H., Borysko, C., Wotring, L., and Townsend, L. B. (1989) Low nanogram detection of nucleotides using fast atom bombardment mass spectrometry. *Anal. Biochem.* **178,** 102–106.
34. Sakurai, T., Matsuo, T., Kusai, A., and Nojima, K. (1989) Collisionally activated decomposition spectra of normal nucleosides and nucleotides using a four-sector tandem mass spectrometer. *Rapid Commun. Mass Spectrom.* **3,** 212–216.
35. Walton, T. J., Ghosh, D., Newton, R. P., Brenton, A. G., and Harris, F. M. (1990) Differentiation of isomeric purine and pyrimidine mononucleotides by fast atom bombardment tandem mass spectrometry. *Nucleosides and Nucleotides* **2,** 967–983.
36. Vigny, P. and Viari, A. (1989) Mass spectrometry applied to natural products: nucleosides, nucleotides and nucleic acids. *Mass Spectrometry, Specialist Periodical Reports* **10,** 253.
37. Ligon, W. V. and Dorn, S. B. (1986) Improved secondary ion mass spectral sensitivity for adenosine triphosphate disodium salt. *Fres. Z. Anal. Chem.* **325,** 625–626.
38. Newton, R. P., Brenton, A. G., Walton, T. J., Harris, F. M., Ghosh, D., Jenkins, A. M., and Kingston, E. E. (1990) Application of fast atom bombardment mass spectrometry and mass-analyzed ion kinetic energy spectrum scanning to studies of cyclic nucleotide biochemistry. *Nucleosides and Nucleotides* **2,** 365–368.
39. Newton, R. P., Walton, T. J., Basaif, S. A., Jenkins, A. M., Brenton, A. G., Ghosh, D., and Harris, F. M. (1989) Identification of butyryl derivatives of cyclic nucleotides by positive ion fast atom bombardment mass spectrometry and mass-analyzed ion kinetic energy spectrometry. *Org. Mass Spectrom.* **24,** 679–688.
40. Lindner, J., Grotemeyer, J., and Schlag, E. W. (1990) Applications of multiphoton ionization mass spectrometry: small protected nucleosides and nucleotides. *Int. J. Mass Spectrom. Ion Proc.* **100,** 267–285.
41. Hiraoka, K. and Kudaka, I. (1990) Electrospray interface for liquid chromatography/mass spectrometry. *Rapid Commun. Mass Spectrom.* **4,** 519–526.
42. McLuckey, S. A., Van Berkel, G. J., and Glish, G. L. (1992) Tandem mass spectrometry of small, multiply charged oligonucleotides. *J. Am. Soc. Mass Spectrom.* **3,** 60–70.
43. Hunt, D. F., Hignite, C. E., and Bieman, K. (1968) Structure elucidation of dinucleotides by mass spectrometry. *Biochem. Biophys. Res. Commun.* **33,** 378–383.
44. Sindona, G., Uccella, N., and Weclawek, K. (1982) Structure determination of isomeric oligodeoxynucleotide salts by fast atom bombardment mass spectrometry. *J. Chem. Res. (S)* 184–185.

45. Wolter, A., Mohringer, C., Koster, H., and Konig, W. A. (1987) Negative ion FAB mass spectrometric analysis of non-charged key intermediates in oligonucleotide synthesis: rapid identification of partially protected dinucleoside monophosphates. *Biomed. Environ. Mass Spectrom.* **14,** 111–116.
46. Grotjahn, L., Frank, R., Heisterberg-Moutsis, G., and Blocker, H. (1984) Fast identification by FAB mass spectrometry of building blocks for oligonucleotide synthesis. *Tet. Lett.* **25,** 5373–5376.
47. Griffin, D., Laramee, J., Deinzer, M., Stirchak, E., and Weller, D. (1988) Negative ion fast atom bombardment mass spectrometry of oligodeoxynucleotide carbamate analogs. *Biomed. Environ. Mass Spectrom.* **17,** 105–111.
48. Cerny, R. L., Gross, M. L., and Grotjahn, L. (1986) Fast atom bombardment combined with tandem mass spectrometry for the study of dinucleotides. *Anal. Biochem.* **156,** 424–435.
49. McNeal, C. J., Ogilvie, K. K., Theriault, N. Y., and Nemer, M. J. (1982) A new method for sequencing fully protected oligonucleotides using ^{252}Cf-plasma desorption mass spectrometry. 1. Negative ions of dinucleoside monophosphates. *J. Am. Chem. Soc.* **104,** 972–975.
50. Hettich, R. and Buchanan, M. (1991) Structural characterization of normal and modified oligonucleotides by matrix assisted laser desorption Fourier transform mass spectrometry. *J. Am. Soc. Mass Spectrom.* **2,** 402–412.
51. McCloskey, J. A. and Crain, P. F. (1992) Progress in mass spectrometry of nucleic acid constituents: analysis of xenobiotic modifications and measurements at high mass. *Int. J. Mass Spectrom. Ion Proc.* **118/119,** 593–615.
52. Guy, A., Molko, D., Wagrez, L., and Teoule, R. (1986) Chemical synthesis of oligonucleotides containing N^6-methyladenine residues in the GATC site. *Helv. Chim. Acta* **69,** 1034–1040.
53. Grotjahn, L., Frank, R., and Blocker, H. (1982) Ultrafast sequencing of oligodeoxyribonucleotides by FAB-mass spectrometry. *Nucleic Acids Res.* **10,** 4671–4678.
54. Grotjahn, L. and Steinert, H. (1985) Mass spectrometry in molecular design, in *Mass Spectrometry in the Health and Life Sciences* (Burlingame, A. L. and Castagnoli, Jr., N., eds.), Elsevier, Amsterdam, pp. 597–614.
55. Grotjahn, L., Blocker, H., and Frank, R. (1985) Mass spectroscopic sequence analysis of oligonucleotides. *Biomed. Mass Spectrom.* **12,** 514–524.
56. Jankowski, K. and Soler, F. (1984) Sequencing of polynucleotides via FAB mass spectrometry: a half-sequence method (part IX). *J. Bioelectricity* **3,** 299–304.
57. Cerny, R. L., Tomer, K. B., Gross, M. L., and Grotjahn, L. (1987) Fast atom bombardment combined with tandem mass spectrometry for determining structures of small oligonucleotides. *Anal. Biochem.* **165,** 175–182.
58. Caprioli, R. M. (1990) Design and operation, in *Continuous-Flow Fast Atom Bombardment Mass Spectrometry* (Caprioli, R. M., ed.), Wiley, New York, pp. 1–27.
59. Iden, C. R. and Rieger, R. A. (1989) Structure analysis of modified oligodeoxyribonucleotides by negative ion fast atom bombardment mass spectrometry. *Biomed. Environ. Mass Spectrom.* **18,** 617–619.

60. McNeal, C. J., Ogilvie, K. K., Theriault, N. Y., and Nemer, M. J. (1982) A new method for the analysis of fully protected oligonucleotides by ^{252}Cf-plasma desorption mass spectrometry. 3. Positive ions. *J. Am. Chem. Soc.* **104,** 981–984.
61. McNeal, C. J., Ogilvie, K. K., Theriault, N. Y., and Nemer, M. J. (1982) A new method for the analysis of fully protected oligonucleotides by ^{252}Cf-plasma desorption mass spectrometry. 3. Negative ions of subunits in the stepwise synthesis of a heptaribonucleotide. *J. Am. Chem. Soc.* **104,** 976–980.
62. Viari, A., Ballini, J. P., Vigny, P., Shire, D., and Dousset, P. (1987) Sequence analysis of unprotected trideoxyribonucleoside diphosphates by ^{252}Cf-plasma desorption mass spectrometry. *Biomed. Environ. Mass Spectrom.* **14,** 83–90.
63. Hillenkamp, F., Karas, M., Ingendoh, A., and Stahl, B. (1990) Matrix assisted UV-laser desorption/ionization: a new approach to mass spectrometry of large biomolecules, in *Biological Mass Spectrometry* (Burlingame, A. L. and McCloskey, J. A., eds.), Elsevier, Amsterdam, pp. 49–60.
64. Spengler, B., Pan, Y., Cotter, R. J., and Kan, L. S. (1990) Molecular weight determination of underivatized oligodeoxyribonucleotides by laser-desorption mass spectrometry. *Rapid Commun. Mass Spectrom.* **4,** 99–102.
65. Stults, J. T. and Marsters, J. C. (1991) Improved electrospray ionization of synthetic oligodeoxynucleotides. *Rapid Commun. Mass Spectrom.* **5,** 359–363.
66. Covey, T. R., Bonner, R. F., and Shushan, B. I. (1988) The determination of protein, oligonucleotide and peptide molecular weights by ion-spray mass spectrometry. *Rapid Commun. Mass Spectrom.* **2,** 249–256.
67. Smith, R. D., Loo, J. A., Edmonds, C. G., Barinaga, C. J., and Udseth, H. R. (1990) New developments in biochemical mass spectrometry: electrospray ionization. *Anal. Chem.* **62,** 882–899.

CHAPTER 14

Extracting Thermodynamic Data From Equilibrium Melting Curves for Oligonucleotide Order-Disorder Transitions

Kenneth J. Breslauer

1. Preface

A large number of research groups have focused considerable effort toward thermodynamically characterizing melting transitions in nucleic acid molecules. This interest reflects the fact that the resulting thermodynamic data permit one to define the nature of the molecular forces that control nucleic acid structures, while also providing a data base for predicting the stability and the temperature-dependent melting behavior of these structures.

In the past, as well as now, the overwhelming majority of nucleic acid systems investigated have been formally either monomolecular (e.g., single-stranded) or bimolecular (e.g., duplex). The relevant thermodynamic equations required to analyze melting profiles for such equilibria are well established, and even have found their way into textbooks. More recently, however, higher-order nucleic acid structures (e.g., junctions, triplexes, tetraplexes) have received intense attention. To extract thermodynamic data from melting curves of these "new" structural forms, it became necessary to extend the existing equations so as to be applicable to equilibrium events involving molecularities greater than two. To this end, several years ago, my colleague, Luis Marky, and I generalized the existing mono- and bimolecular equations so as to permit their application to studies on higher-order nucleic acid struc-

From: *Methods in Molecular Biology, Vol. 26: Protocols for Oligonucleotide Conjugates*
Edited by: S. Agrawal Copyright ©1994 Humana Press Inc., Totowa, NJ

tures of any molecularity. We presented these equations and their derivations in an article published in *Biopolymers* (Calculating Thermodynamic Data for Transitions of Any Molecularity from Equilibrium Melting Curves, L. A. Marky and K. J. Breslauer, *Biopolymers* **26,** 1601–1620, © 1987, John Wiley and Sons, Inc.). What follows, in large part, is a reproduction of that article, with a new section added in which I describe explicitly the calculation of melting temperatures (T_m), and qualitatively discuss the influence of salt concentration on the T_m of oligomeric nucleic acid structures.

2. Synopsis

In this chapter, we derive the general forms of the equations required to extract thermodynamic data from equilibrium transition curves on oligomeric and polymeric nucleic acids of any molecularity. Significantly, since the equations and protocols are general, they also can be used to characterize thermodynamically equilibrium processes in systems other than nucleic acids. We briefly review how the reduced forms of the general equations have been used by many investigators to evaluate mono- and bimolecular transitions and then explain how these equations can be generalized to calculate thermodynamic parameters from common experimental observables for transitions of higher molecularities. We emphasize the strengths and weaknesses of each method of data analysis so that investigators can select the approach most appropriate for their experimental circumstances. We also describe how to analyze calorimetric heat capacity curves and noncalorimetric differentiated melting curves so as to extract both model-independent and model-dependent thermodynamic data for transitions of any molecularity. The general equations and methods of analysis described in this chapter should be of particular interest to laboratories that currently are investigating association and dissociation processes in nucleic acids that exhibit molecularities greater than two.

3. Introduction

We have learned a great deal about the sequence-dependent conformational states present in naturally-occurring DNA and RNA polymers from studies on specially designed and synthesized oligonucleotides *(1–5)*. Particular attention has been focused on thermodynamically characterizing the secondary structures formed by these oligomeric model systems through investigations of their temperature-dependent "melting" behavior. In fact, the results obtained from these model systems have provided data bases from which thermodynamic libraries have been established that characterize all ten Watson–Crick nearest-neighbor interactions in both DNA *(6)* and RNA *(7–9)*. These thermodynamic data now provide an empirical basis for predicting the stability ($\Delta G°$) and temperature-dependent melting behavior ($\Delta H°$) of any DNA or RNA duplex region by inspection of its primary sequence *(6–9)*.

To date, the overwhelming majority of oligomer systems studied have modeled secondary structures formed by monomolecular (e.g., hairpins) and bimolecular (e.g., duplexes) associations of oligomer strands *(10–36)*. *(See also* other work cited in refs. *4* and *5.)* The equations required to extract thermodynamic data from melting studies on oligomer systems exhibiting these two molecularities already are published and are reasonably well known *(37–43)*. (*See* ref. *37* for an excellent anthology and discussion of the original papers describing the early development of the theory of helix-coil transitions in biopolymers.) Undoubtedly, future modeling of biologically important structures will involve oligomer associations well beyond the bimolecular level. In fact, recent efforts to model cruciform formation by immobile junction structures have required oligomer association processes at the tetramolecular level *(44–49)*. Consequently, it would be useful to derive the general forms of the equations required to extract thermodynamic data from melting studies on oligomer systems of any molecularity. In this chapter, we derive the relevant equations and describe how they can be used to calculate thermodynamic parameters from experimental data. We also describe a useful method for extracting complete thermodynamic profiles from calorimetric measurements in the absence of optical data. As part of an independent and concurrent effort, Privalov and Potekhin also have derived general equations that they use to analyze calorimetric heat capacity curves for protein transitions *(50)*. Although their and our formulations for analyzing calorimetric data differ significantly, the information content of the two sets of equations is similar. Significantly, however, in contrast to Privalov and Potekhin, we explicitly include a concentration dependent term in several of our formulations. Such a term is important when using noncalorimetric techniques to characterize thermodynamically association or dissociation processes for relatively short oligomers.

4. Analyzing the Shape of an Equilibrium Melting Curve to Calculate ΔH_{VH}

A thermally-induced order-disorder transition in any nucleic acid can be monitored by following at an appropriate wavelength the increase in UV absorption with increasing temperature. The resulting absorption vs temperature profile commonly is called a UV melting curve. Figure 1A shows a typical experimental curve. The shape of this curve as well as other equilibrium melting profiles (e.g., viscosity vs temperature, heat capacity vs temperature) can be analyzed to yield a value for the van't Hoff transition enthalpy. We describe below how this analysis can be accomplished for a UV absorption vs temperature curve. However, keep in mind that the protocol is general and therefore can be used to analyze the temperature dependence of any equilibrium property that is directly related to the concentrations of the two equilibrating species.

If we define α as equal to the fraction of single strands in the duplex state, then any experimental absorbance vs temperature curve can be converted into an α vs temperature profile by assuming that the fractional change in absorbance at any temperature monitors the extent of reaction. This conversion is accomplished graphically as illustrated in Fig. 1 by taking the ratio at each temperature of the height between the upper baseline and the experimental curve (x) and the height between the lower and upper baselines ($x + y$). We now have constructed the curve in Fig. 1B which expresses how an equilibrium property, α, varies with temperature. Alternatively, we could have plotted α vs the reciprocal of the temperature, $1/T$, to obtain the curve shown in Fig. 1C. Since the equilibrium constant, K, for any transition can be expressed in terms of α, the curves shown in Figs. 1B and 1C reveal how K varies with T and $1/T$. This knowledge of the temperature dependence of K allows us to calculate the transition enthalpy using either form of the van't Hoff equation shown below.

$$\Delta H_{VH} = RT^2 \frac{d\ln K}{dT} \quad \text{or} \quad \Delta H_{VH} = -R \left[\frac{d\ln K}{d(1/T)}\right] \tag{1}$$

Obviously, to complete the calculation of ΔH_{VH}, K must be expressed in terms of α. For this purpose, one generally assumes that the transition proceeds in a two-state (all-or-none) manner. When this condition prevails, the general expression for the equilibrium constant, K, and its value at the melting temperature, $K(T_m)$, will depend on the molecularity

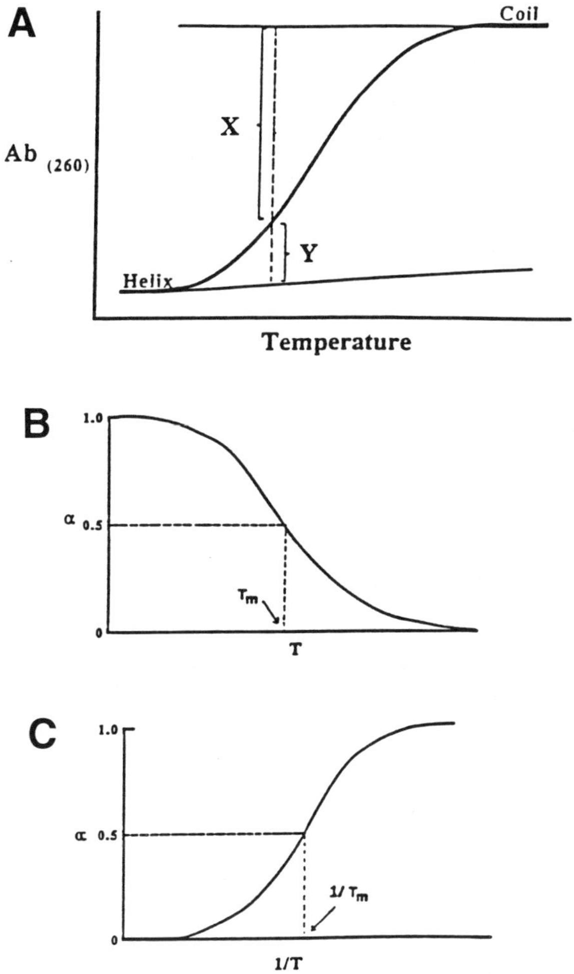

Fig. 1. **(A)** A typical absorbance vs temperature melting curve. *X* corresponds to the distance between an upper baseline and the curve whereas *Y* represents the distance between the lower baseline and the curve. **(B)** A plot of α vs T, where α equals the fraction of single strands in the duplex state. This curve is derived from Fig. 1A as described in the text. **(C)** A plot of α vs 1/T where α equals the fraction of single strands in the duplex state.

of the transition and the nature of the associating sequences (self-complementary vs nonself-complementary). In the sections that follow, we derive the general expressions for the equilibrium constant and explain how they can be used to calculate thermodynamic data.

4.1. Equilibria Involving Nonself-Complementary Sequences

Consider the general equilibrium shown below for the association of nonsclf-complementary sequences to form an n-mer structure

$$A_1 + A_2 + A_3 + \cdots + A_n \leftrightarrow A_1A_2A_3 \cdots A_n$$

where n is the molecularity of the reaction which equals the number of strands that associate to form the final n-mer complex. The general expression for the corresponding equilibrium constant, K, in terms of α and n is

$$K = \frac{[A_1A_2A_3A_n]}{[A_1][A_2][A_3]\cdots[A_n]} = \frac{\alpha(C_T/n)}{[(1-\alpha)C_T/n]^n} = \frac{\alpha}{(C_T/n)^{n-1}\,(1-\alpha)^n} \quad \left[\begin{array}{c}\text{for nonself-}\\ \text{complementary}\\ \text{associations}\end{array}\right] \tag{2}$$

where C_T equals the total strand concentration and each strand is present in equal concentration; namely, C_T/n. This expression is applicable to the general case in which the associating strands all are nonself-complementary. The special case in which the associating sequences all are self-complementary will be described in the next section.

We now have an expression for the equilibrium constant in terms of experimentally accessible parameters; namely, n, C_T, and α. If we define the melting temperature, T_m, as the temperature at which $\alpha = 1/2$, then the general expression for the equilibrium constant shown in Eq. (2) reduces to

$$K_{T_m} = \frac{1/2}{(C_T/n)^{n-1}\,(1/2)^n} = \frac{1}{(C_T/n)^{n-1}(1/2)^{n-1}} = \frac{1}{(C_T/2n)^{n-1}} \tag{3}$$

This expression allows calculation of K at the T_m for an association reaction of any molecularity between nonself-complementary sequences.

We also can derive a general expression for calculating the transition enthalpy. To accomplish this, we substitute Eq. (2) into the first form of the van't Hoff expression (Eq. [1]), differentiate, and solve for ΔH_{VH} at the T_m where $\alpha = 1/2$. These manipulations yield

$$\Delta H_{VH} = (2 + 2n)\, RT_m^{\,2} \left(\frac{\partial\alpha}{\partial T}\right)_{T=T_m} \tag{4a}$$

Since $\partial(1/T)/\partial T = -1/T^2$, Eq. (4a) also can be written as

$$\Delta H_{VH} = -(2 + 2n)R \left(\frac{\partial\alpha}{\partial(1/T)}\right)_{T=T_m} \tag{4b}$$

We could have obtained this expression directly by substituting Eq. (2) into the second form of the van't Hoff expression shown in Eq. (1). Many of the equations presented below can be reformulated in a similar manner.

Either general form of the van't Hoff expression derived above (Eqs. [4a] or [4b]) allows calculation of ΔH_{VH} for an association reaction of any molecularity simply by evaluating the slope of an α vs T or an α vs $1/T$ melting curve at T_m. The resulting value of $(\delta\alpha/\delta T)_{T=T_m}$ or $[\partial\alpha/\partial(1/T)]_{T=T_m}$ and the known molecularity, n, of the reaction are then plugged into Eqs. (4a) or (4b) to calculate ΔH_{VH}. (Eqs. [4a] or [4b] is derived for an *association* reaction. The same expression with the *opposite sign* can be used for *dissociation* processes.) These expressions [Eqs. (4a) or (4b)] allow calculation of ΔH_{VH} for any equilibria involving nonself-complementary sequences. In the next section, we demonstrate that the same expressions (Eqs. [4a] or [4b]) also can be used to calculate ΔH_{VH} for equilibria that involve structures formed from self-complementary sequences.

4.2. Equilibria Involving Self-Complementary Sequences

For an association reaction involving structures formed from self-complementary sequences, the general equilibrium can be written as

$$nA \leftrightarrow A_n$$

For this special case, the general expression for the equilibrium constant, K, in terms of α and n is

$$K = \frac{[A^n]}{[A]^n} = \frac{\alpha(C_T/n)}{[(1-\alpha)C_T]^n} = \frac{\alpha}{nC_T^{\,n-1}(1-\alpha)^n} \tag{5}$$

Note that this expression for the equilibrium constant for an association reaction between self-complementary sequences is *not* identical to the corresponding expression for nonself-complementary sequences. (Compare Eqs. [2] and [5].) This disparity reflects the statistical differences between these two classes of equilibria.

If we define the melting temperature, T_m, as the temperature at which $\alpha = 1/2$, the general expression for K shown above reduces to:

$$K_{T_m} = \frac{1/2}{nC_T^{\,n-1}\,(1/2)^n} = \frac{1}{n(C_T/2)^{n-1}} \tag{6}$$

This expression allows calculation of K at the T_m for an association reaction of any molecularity between self-complementary sequences.

We also can derive a general expression for calculating the transition enthalpy for self-complementary associations. To accomplish this, we substitute Eq. (5) into the van't Hoff expression (Eq. [1]), differentiate, and solve for ΔH_{VH} at the T_m where $\alpha = 1/2$. These manipulations yield equations for calculating ΔH_{VH} that are identical to those derived above for the association of nonself-complementary sequences (Eqs. [4a] and [4b]). This identity reflects the fact that the statistical differences between processes involving these two classes of sequences do not influence the expression used to calculate ΔH_{VH}.

4.3. Calculating ΔH_{VH} from Melting Curves

Equation (4a) can be used to calculate a van't Hoff transition enthalpy for a process of any molecularity. To date, bimolecular (e.g., duplex to single strands) and monomolecular (e.g., hairpin to single strand) processes represent the two most commonly studied classes of nucleic acid transitions. For a monomolecular process ($n = 1$), the leading coefficient in Eq. (4a) is 4 whereas for a bimolecular process ($n = 2$) the corresponding coefficient is 6. These two reduced forms of Eq. (4a) are well known and have been used by many investigators to extract van't Hoff transition enthalpies from the temperature dependence of various equilibrium properties for mono- and bimolecular processes. By contrast, the use of Eq. (4a) to extract van't Hoff transition enthalpies from processes exhibiting molecularities greater than two is much less common in the literature.

However, the recent use of multistrand oligomer systems to model more complex biological structures has resulted in studies on processes that exhibit molecularities greater than two. Consequently, it is useful to examine the reduced forms of Eq. (4a) required to extract van't Hoff transition enthalpies from such studies. One example of a thermodynamic analysis on a higher molecularity process is described in ref. *49*. In this work, a cruciform is modeled by an immobile junction structure that forms through association between four 16-mer sequences *(49)*. To extract thermodynamic data from the melting curves associated with this tetramolecular process (where $n = 4$), the reduced form of Eq. (4a) shown below was used.

$$\Delta H_{VH} = 10RT_m^{\ 2}\left(\frac{\delta\alpha}{\delta T}\right)_{T=T_m}$$

In ref. *49*, a critical appraisal of the ΔH_{VH} data obtained from this equation is provided by comparison with directly measured calorimetric data.

In summary, Eqs. (4a) or (4b) allows one to calculate ΔH_{VH} for any association reaction for which the temperature-dependence of an equilibrium property has been monitored. One simply needs to know the molecularity, *n*, of the reaction and then graphically determine the value of $(\delta\alpha/\delta T)_{T=T_m}$ from the experimental melting curve. Note that the same equation with the opposite sign can be used to evaluate dissociation or "melting" processes.

5. Calculating ΔH_{VH} from the Concentration Dependence of the Melting Temperature

The formation or dissociation of complexes of molecularity greater than one will result in a concentration-dependent equilibrium. Such equilibria therefore can be characterized by determining the concentration dependence of the melting temperature. This approach represents a second method for extracting thermodynamic data from experimental melting profiles. The relevant equations in general form are derived below.

5.1. Concentration Dependent Equilibria Involving Nonself-Complementary Sequences

Consider the general equilibrium reaction shown below for the association of nonself-complementary sequences to form an *n*-mer structure

$$A_1 + A_2 + A_3 + \cdots + A_n \leftrightarrow A_1A_2A_3 \cdots A_n$$

Since for any process at equilibrium, $\Delta G^\circ = -RT\ln K_{eq}$ and $\Delta G^\circ = \Delta H^\circ - T\Delta S^\circ$, we can derive an expression for K_{eq} in terms of ΔH° and ΔS° by equating these two expressions for ΔG° to yield

$$-RT\ln K = \Delta H^\circ - T\Delta S^\circ \tag{7}$$

Eq. (3) provides us with an expression for K at T_m in terms of *n* and C_T when nonself-complementary strands associate. Plugging this expression into the equality given above yields

$$+RT_m\ln (C_T/2n)^{n-1} = \Delta H^\circ - T_m\Delta S^\circ$$

On rearrangement, this expression becomes

$$(n-1)RT_m\ln C_T - (n-1)RT_m\ln 2n = \Delta H^\circ - T_m\Delta S^\circ$$

Dividing by $T_m\Delta H^\circ$ and rearranging the terms yields

$$\underset{(y)}{\frac{1}{T_m}} = \underset{(m)}{\frac{(n-1)R}{\Delta H^\circ}} \underset{(x)}{\ln C_T} + \underset{(b)}{\frac{[\Delta S^\circ - (n-1)\,R\ln 2n]}{\Delta H^\circ}} \quad \left(\begin{array}{c}\text{for associations}\\ \text{of nonself-complementary}\\ \text{strands}\end{array}\right) \tag{8}$$

As emphasized by the symbols in parenthesis, this equation corresponds to a straight line when the reciprocal of the melting temperature ($1/T_m$) is plotted against the natural logarithm of the strand concentration ($\ln C_T$). The slope (m) of such a plot is equal to $[(n-1)R]/\Delta H^\circ$ and the intercept (b) is equal to $[(1/\Delta H^\circ)/(\Delta S^\circ - (n-1)R\ln 2n]$. Figure 2 shows a typical $1/T_m$ vs $\ln C_T$ plot in which these features are emphasized.

5.2. Concentration-Dependent Equilibria Involving Self-Complementary Sequences

For the case in which the associating sequences all are self-complementary rather than nonself-complementary, the equilibrium constant expression shown in Eq. (5) rather than Eq. (2) is used to derive the appropriate relationship between T_m and $\ln C_T$. Paralleling the procedure outlined above, one obtains the expression shown below which is similar to, but different from Eq. (8).

$$\frac{1}{T_m} = \frac{(n-1)R}{\Delta H^\circ} \ln C_T + \frac{[\Delta S^\circ - (n-1)\,R\ln 2 + R\ln n n]}{\Delta H^\circ} \quad \left(\begin{array}{c}\text{for associations}\\ \text{of self-complementary}\\ \text{strands}\end{array}\right) \tag{9}$$

Note that Eqs. (8) and (9) have the same expression for their slopes, but different expressions for their intercepts. This disparity reflects the statistical difference between these two classes of equilibria. Significantly, it should be noted that this statistical factor only influences the intercept and not the slope. Thus, the statistical differences between self-complementary and nonself-complementary equilibria do not influence the enthalpy term. As with Eq. (4), our derivation of Eq. (8) or (9) assumes a two-state transition and a temperature-independent enthalpy. However, this method for determining ΔH_{VH} is much less sensitive to the choice of baselines when analyzing experimental melting curves. This feature results from the fact that the slope of a $1/T_m$ vs $\ln C_T$ line results from *differences* in the T_m values obtained from several melting curves. Consequently, as long as the baselines for a family of concentration-dependent melting curves are selected in a *consistent* manner, the slope of the resulting $1/T_m$ vs $\ln C_T$ line provides a good

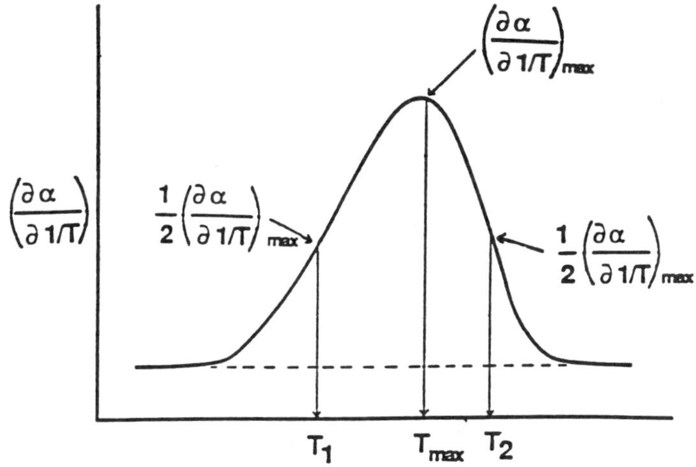

Fig. 2. A plot of $1/T_m$ vs $\ln C_T$. This plot reveals the concentration dependence of the melting temperature. The slope is proportional to the transition enthalpy whereas the *X*-axis intercept is proportional to the transition entropy. The equations describing the specific relationships are shown in the plot as well as described in the text.

measure of the van't Hoff enthalpy for transitions that proceed in a two-state manner. Any deviations of the selected baselines relative to the "true" baselines simply produce a parallel line that has a different intercept but the same slope, thereby altering $\Delta S°$ but not ΔH_{VH}.

The method of data analysis described above is uniquely applicable to relatively short oligomer transitions where the concentration dependence of the melting temperature can be observed. For long oligomers and polymers, this approach cannot be applied, since the monomolecular helix growth steps dominate the bimolecular helix initiation step, thereby producing an artificially reduced concentration dependence or a "pseudo monomolecular" equilibrium for which the melting temperature is concentration independent.

5.3. Calculating ΔH_{VH} from the Concentration Dependence of T_m for Some Common Transitions

Clearly, the melting temperature of a monomolecular process (e.g., hairpin formation) will not exhibit a concentration dependence. This feature is illustrated by the fact that when $n = 1$, Eqs. (8) or (9) reduces to $1/T_m = \Delta S^\circ/\Delta H^\circ$. This simplified form of Eqs. (8) or (9) for the $n = 1$ case shows that for a monomolecular process the melting temperature, T_m, depends only on ΔS° and ΔH° and is independent of C_T. Consequently, to obtain a value for ΔH_{VH} from a melting curve of a monomolecular process requires application of Eq. (4).

Equations (8) or (9) can be used to evaluate any process that exhibits a molecularity of two or above. For example, for the bimolecular duplex formed between two strands of a self-complementary sequence, Eq. (9) reduces to the well known expression

$$\frac{1}{T_m} = \frac{R}{\Delta H^\circ}\ln C_T + \frac{\Delta S^\circ}{\Delta H^\circ} \quad \left(\begin{array}{c}\text{for a bimolecular asssociation}\\ \text{of two self-complementary strands}\end{array}\right)$$

However, if the two strands that associate are nonself-complementary rather than self-complementary, then the intercept (b term) includes an additional factor to account for the entropic differences between these two association reactions. Consequently, when analyzing a bimolecular association between nonself-complementary rather than self-complementary strands Eq. (8) reduces to the expression

$$\frac{1}{T_m} = \frac{R}{\Delta H^\circ}\ln C_T + \frac{\Delta S^\circ - R\ln 4}{\Delta H^\circ} \quad \left(\begin{array}{c}\text{for a bimolecular association}\\ \text{of two nonself-complementary strands}\end{array}\right)$$

5.4. Analyzing the Shape of a Differentiated Equilibrium Melting Curve

An approach very similar to the one just described can be used to obtain a van't Hoff transition enthalpy from a *differentiated* melting curve. Figure 3 shows a tracing of a typical differentiated melting curve where $\partial\alpha/\partial(1/T)$ is plotted against T. Such a curve can be derived from the temperature dependence of any equilibrium property or obtained directly from techniques such as temperature-jump measurements. As originally shown by Gralla and Crothers for mono- and bimolecular transitions *(15)*, the full width or the half-width of a differentiated melting curve at the half height is inversely proportional to the van't Hoff transition enthalpy. For a general equilibrium of the form

$$nA \leftrightarrow A_n$$

the relevant general forms of the van't Hoff equation are

$$\Delta H_{VH} = \frac{B}{(1/T_1) - (1/T_2)} \text{ (for the full width at the half-height)} \quad (10)$$

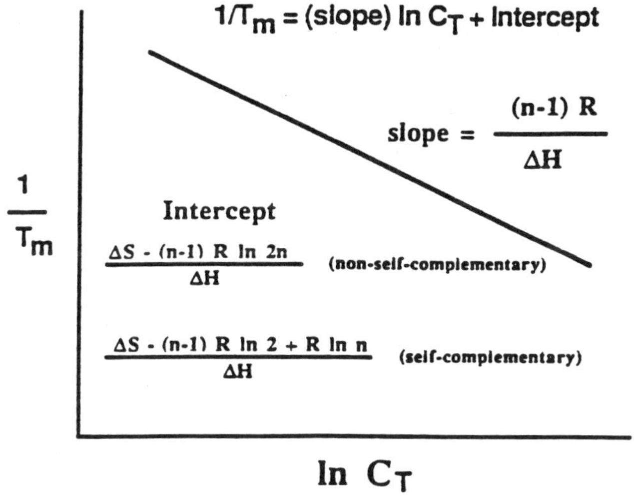

Fig. 3. A typical differentiated melting curve. Several reference temperature points are specifically defined in the figure and described in the text.

$$\Delta H_{VH} = \frac{B'}{(1/T_{max}) - (1/T_2)} \text{(for the upper half width at the half-height)} \quad (11)$$

where T_{max} is the temperature at the maximum and T_1 and T_2 correspond to the lower and upper temperatures, respectively, at which the change in the observable width temperature is equal to one-half of $[\partial\alpha/\partial(1/T)]_{max}$ (*see* Fig. 3). Both B and B' are constants that depend on the molecularity of the process under investigation. Specific values of B and B' are given in Table 1 for processes involving several different molecularities. The detailed derivations of Eqs. (10) and (11) are given in the Appendix of ref. *52*. Examination of these derivations reveals that this method of analyzing differentiated melting curves still incorporates the assumption of a two-state transition and a temperature-independent enthalpy. However, when Eq. (11) is employed, only the high temperature half of the transition need be obtained experimentally (between T_{max} and T_2). Consequently, this approach provides a means of circumventing the lower baseline problem which one frequently encounters in the analysis of the overall shapes of integral absorbance vs temperature curves. Specifically, for transitions with low T_ms where it is difficult to define the lower baseline (or even to obtain the lower half of the melting curve), this method of data analy-

Table 1
Values of the Constants, B and B', in Eqs. (10) and (11)
for Association Reactions
Exhibiting Molecularities (*n*) Between 1 and 5

n	B	–B'
1	7.00	3.50
2	10.14	4.38
3	12.88	5.06
4	15.40	5.63
5	17.79	6.14

B and B' in cal/K-mol.

sis permits a transition enthalpy to be calculated from just the upper half of a melting profile (*see* Fig. 3).

When a differentiated melting curve is obtained directly from temperature-jump experiments, the problem of baseline selection may be reduced further. If the molecular events that give rise to the sloping baselines correspond to processes that are very fast, then the temperature-jump experiment provides a means of kinetically resolving the relatively slow, cooperative helix melting from these fast baseline effects *(15,51)*. In such a case, one obtains a differentiated melting curve that is kinetically filtered of the baseline problems encountered with integral melting curves.

6. Calorimetric Determination of Transition Enthalpies

Differential scanning calorimetry (DSC) also can be used to detect and follow thermally-induced order-disorder transitions in oligonucleotides and other thermally-labile molecules. However, with DSC the excess heat capacity (ΔC_p) rather than the change in absorbance of the solution is monitored. Experimentally, one obtains a ΔC_p vs T melting profile as illustrated in Fig. 4A. Since $\Delta H^\circ = \int \Delta C_p dT$, the area under such a calorimetric transition curve is equal to the transition enthalpy.

In contrast to the model-dependent (usually two-state) ΔH_{VH} values indirectly derived from the temperature-dependence of an equilibrium property, the calorimetrically determined transition enthalpy does *not* depend on the nature of the transition. As with optical studies, with calorimetry one first must define an experimental baseline to analyze for the

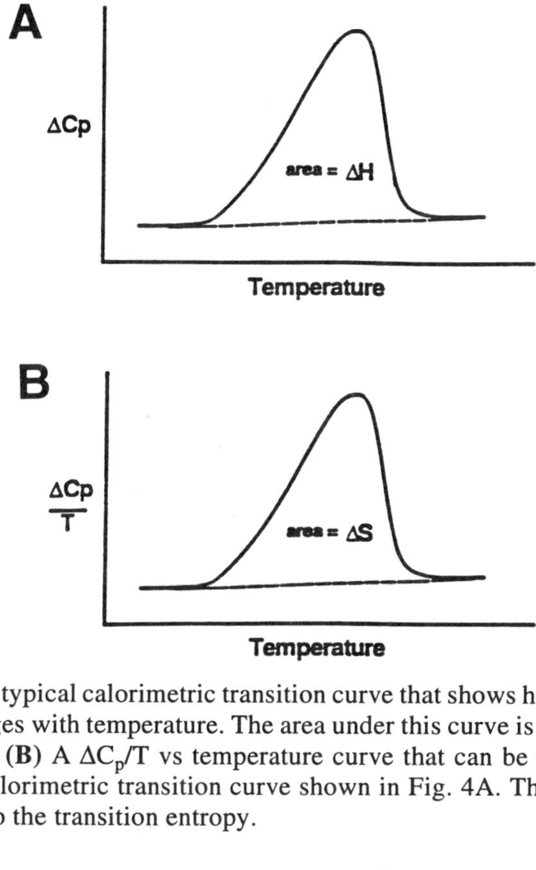

Fig. 4. (**A**) A typical calorimetric transition curve that shows how the heat capacity, ΔC_p, changes with temperature. The area under this curve is equal to the transition enthalpy. (**B**) A $\Delta C_p/T$ vs temperature curve that can be derived from the experimental calorimetric transition curve shown in Fig. 4A. The area under this curve is equal to the transition entropy.

transition of interest. One then measures the total energy required to go from the initial to the final state from the area under the curve and above the baseline. The shape of the curve, which does depend on the nature of the transition, need not be analyzed as is done with optical and other noncalorimetric data. Thus, the calorimetric measurement provides a direct, model-independent determination of the transition enthalpy. For this reason, a comparison of the model-dependent van't Hoff transition enthalpy and the model-independent calorimetric transition enthalpy provides insight into the nature of the transition as will be explained in the next section. Calorimetry also provides a direct measure of the heat capacity change, ΔC^o_p, accompanying the transition. Consequently, one need not assume that ΔC^o_p is zero, as usually is done when analyzing optical data. Interestingly, in contrast to proteins, our DSC studies do not detect significant heat capacity changes accompanying thermally-induced transitions in DNA molecules. In an independent and concurrent effort, Privalov and Potekhin recently also have derived equations that express the dependence of calorimetric enthalpies on n, the reaction molecularity *(50)*. However, their formulation is quite different from the one presented here.

It is of interest to note that a van't Hoff transition enthalpy also can be calculated from the calorimetric data. In a manner paralleling the analysis of a differentiated melting curve, the shape of a calorimetric heat capacity curve can be analyzed using Eqs. (10) or (11) to yield a van't Hoff transition enthalpy that we designate as ΔH_{VH}^{cal}. Values obtained in this manner do not always agree with the corresponding optical ΔH_{VH} derived from Eqs. (4), (8), or (9). For bimolecular transitions we have observed the largest differences between ΔH_{VH}^{cal} and ΔH_{VH} for longer oligomers *(5)*. This result suggests that the thermal and the optical windows may not always equivalently monitor the extent of reaction. For the tetramolecular transition of an immobile junction, we also observe disparities between the ΔH_{VH} data derived by the different methods described above. A critical discussion of the potential origins of these disparities is presented in ref. *49*.

7. Nature of the Transition

The methods described above for extracting thermodynamic data from the temperature dependence of an equilibrium property only can be applied rigorously to transitions that proceed in a two-state (all-or-none) manner. For transitions in which intermediate states are significantly populated, any integral or differentiated equilibrium melting curve will be broadened. According to Eqs. (4), (10), and (11), this broadening will lead to a reduced ΔH_{VH} relative to the true calorimetric value. By contrast, as noted earlier, the calorimetrically determined transition enthalpy is derived directly from the area under (rather than the shape of) a heat capacity curve. Thus, ΔH_{cal} is independent of the nature of the transition as one would expect for a state function.

Comparison of the model-dependent ΔH_{VH} and the model-independent ΔH_{cal} allows one to conclude if the transition proceeds in an all-or-none fashion, thereby providing a test for the applicability of the two-state model to a given transition. If $\Delta H_{VH} < \Delta H_{cal}$, then the transition involves a significant population of intermediate states. However, if $\Delta H_{VH} = \Delta H_{cal}$, then the transition proceeds in a two-state manner and meaningful thermodynamic data can be obtained by monitoring the temperature dependence of an equilibrium and using the equations presented in this paper.

A quantitative comparison of the van't Hoff and calorimetric transition enthalpies provides further insight into the nature of a transition. Specifically, the ratio $\Delta H_{VH}/\Delta H_{cal}$ provides a measure of the fraction of the structure that melts cooperatively; in other words, the size of the cooperative unit. This ability to define the size of the cooperative unit represents an important and unique advantage of the calorimetric measurement. In ref. *49*, we employ this approach and discuss its potential limitations when applied to complex structures such as immobile junctions.

8. Calculating ΔG° and ΔS° From Melting Curve Data

As described above, a van't Hoff transition enthalpy (ΔH_{VH}) can be determined by analyzing the shape of an integral or differentiated melting curve using Eqs. (4), (10), or (11) which each assume a two-state model. By contrast, a model-independent calorimetric transition enthalpy (ΔH^{o}_{cal}) can be determined directly by evaluating the area under an experimental heat capacity curve. The free energy and entropy changes (ΔG° and ΔS°) that correspond to these transition enthalpies (ΔH_{VH} and ΔH^{o}_{cal}) then can be determined by one of the methods described below.

8.1. Calculating ΔG° and ΔS° From Noncalorimetric Melting Curves

8.1.1. Method I

We have shown by Eqs. 2 and 5 that the equilibrium constant, K, for any two-state association reaction can be expressed in terms of the molecularity of the process, *n*, and the extent of association, α (which corresponds to the fraction of single-stranded molecules in the complexed state). Consequently, we can evaluate K for a process of any molecularity at any value of α. Usually, a value of K is determined at the melting temperature, T_m, where $\alpha = 0.5$ (*see* Eqs. [3] and [6]). This K(T_m) value then is extrapolated to some reference temperature, T (usually 298 K) using the experimentally-measured melting temperature, T_m, the calculated van't Hoff transition enthalpy ΔH_{VH} (assumed to be temperature-independent), and the integrated form of the van't Hoff equation shown below:

$$\ln[K(T_m)/K(T)] = \frac{\Delta H^\circ}{R}\left(\frac{1}{T} - \frac{1}{T_m}\right) \tag{12}$$

Using the calculated value of K(T), one can determine the Gibbs free energy change for the transition using the standard thermodynamic relationship $\Delta G^\circ = -RT\ln K(T)$. The corresponding entropy change then can be calculated from the standard equation

$$\Delta G^\circ = \Delta H^\circ - T\Delta S^\circ \tag{13}$$

It should be emphasized that if ΔH° is not known exactly and $(T_m - T)$ is large, the required temperature extrapolation can introduce serious errors.

For a monomolecular process, such as hairpin formation, $\ln K(T_m) = 0$ since K = 1 at the melting temperature. Consequently, for monomolecular associations, the integrated form of the van't Hoff equation reduces to:

$$\ln K(T) = \frac{\Delta H^\circ}{R}\left(\frac{1}{T} - \frac{1}{T_m}\right)$$

Multiplying both sides by RT yields

$$-RT\ln K(T) = \Delta H^\circ\left(1 - \frac{T}{T_m}\right) = \Delta G^\circ \tag{14}$$

This simplified expression for monomolecular processes can be used to calculate the transition-free energy, ΔG°, at any temperature of interest, T, from the experimentally measured values of T_m and ΔH_{VH}. The corresponding ΔS° value then can be calculated using Eq. (13). Significantly, this simplified expression for monomolecular processes also can be used to evaluate melting curves of polymer complexes that formally have molecularities greater than one. This possibility exists because polymer complexes melt in a pseudo monomolecular manner since the growth steps that are monomolecular dominate the bimolecular initiation steps. This pseudo monomolecular behavior is reflected by the fact that the melting temperatures of polymer complexes are concentration independent.

8.1.2. Method II

If a van't Hoff transition enthalpy is determined from concentration-dependent melting studies, a more direct procedure can be employed to determine ΔG° and ΔS°. Inspection of Eqs. (8) or (9) and Fig. 2 reveals that although the slope of a $1/T_m$ vs $\log C_T$ plot yields ΔH_{VH}, the intercept permits calculation of ΔS°. These two thermodynamic parameters then can be used to calculate ΔG° by application of the standard thermodynamic relationship, shown in Eq. (13). In principle, for a two-state transition this method should be equivalent to Method I described above. However, since transition enthalpies calculated by Method I are more sensitive to baseline assignments, the thermodynamic profiles calculated by Method II probably are more reliable.

For a monomolecular association or dissociation process $n = 1$. Thus, at the melting temperature, Eqs. (8) or (9) reduces to:

$$\Delta S^\circ = \frac{\Delta H^\circ}{T_m}$$

Thus, the free energy change, ΔG°, at the T_m can be calculated from ΔS° and ΔH° and extrapolated to any temperature of interest using Eq. (14).

Significantly, all of the methods outlined above for calculating thermodynamic parameters from equilibrium melting curves are applicable only to transitions that proceed in a two-state manner. Such a characterization of the nature of a transition is not possible based exclusively on the temperature dependence of an equilibrium property. Consequently, additional independent information concerning the nature of a transition is required before the approaches described above can be applied to a given oligomer.

8.2. Calculating ΔG° and ΔS° From Calorimetric Melting Curves

8.2.1. Method III

As noted earlier, a single calorimetric transition curve gives a direct, model-independent measure of both the heat capacity and enthalpy changes accompanying a thermally-induced conformational change. Frequently, these calorimetric data have been used in conjunction with independent equilibrium measurements of ΔG° to calculate complete thermodynamic profiles. Significantly, however, as described below, this dependence on noncalorimetric data is not necessary.

In a given DSC experiment, one obtains directly a heat capacity (ΔC_p) vs temperature (T) curve (*see* Fig. 4A). This ΔC_p vs T curve can be converted into a $\Delta C_p/T$ vs T curve (*see* Fig. 4B). Since $\Delta S^\circ = \int(\Delta C_p/T)dT]$, the area under such a curve provides a "direct" measure of the entropy change. Thus, from a single calorimetric transition curve one can obtain ΔC_p, ΔH°, and ΔS°. Using these data, the corresponding value of ΔG° can be calculated at any temperature. This calorimetric approach has two significant advantages relative to optical methods. First, one obtains a model-independent measure of the transition enthalpy and entropy rather than the "two-state values" obtained from analysis of optical data. Second, the calorimetric experiment provides a direct measure of ΔC_p so that the temperature-dependent stability can be assessed. We have used this calorimetric method to characterize thermodynamically the transitions of many oligomers and have compared these results with the corresponding data obtained indirectly from optical melting curves. Significantly, we find that for transitions that proceed through multiple states the optically derived values can seriously deviate from the calorimetrically determined values. Only for two-state transitions do the optical methods of analysis described above yield meaningful thermodynamic data. By contrast, the calorimetric method yields meaningful thermodynamic data regardless of the nature of the transition.

8.3. Calculating Melting Temperatures

One of the primary practical applications of thermodynamic data is its use to calculate melting temperatures. This generalization particularly applies to molecular biologists who frequently need to predict the thermal stabilities (T_m) of probe-gene complexes or other local DNA duplex domains. A melting temperature can be calculated for any transition by application of Eqs. (8) or (9). One simply needs to know the molecularity of the transition, *n,* the total strand concentration, C_T (when $n > 1$), and the transition enthalpy and entropy. The latter two thermodynamic parameters can be obtained experimentally or calculated using published nearest-neighbor data *(6,9).* Several general examples of T_m calculations are given below.

8.3.1. The Monomolecular Case

By definition, $n = 1$ for any monomolecular equilibrium. Consequently, for this case, Eq. (8) reduces to $1/T_m = \Delta S/\Delta H$. Note that the concentration term in Eq. (8) is abolished when $n = 1$, as one would expect for a monomolecular process. Solving this equality for T_m yields

$$T_m = \Delta H/\Delta S \tag{15}$$

This relationship allows one to calculate the melting temperature for any structure formed by a monomolecular equilibrium simply by taking the ratio of ΔH° and ΔS°. These thermodynamic parameters can be obtained experimentally as described above or calculated using published nearest-neighbor thermodynamic data *(6,9)*.

8.3.2. The Bimolecular Case

The majority of equilibria of current interest to molecular biologists are bimolecular in nature (e.g., probe-gene hybridizations). To derive an expression for calculating the T_m of complexes formed by such equilibria, one simply evaluates Eqs. (8) or (9) when $n = 2$. As previously noted, the specific equation used will depend on whether the associating strands are self-complementary or nonself-complementary (9). We will begin with the nonself-complementary case since this reflects the nature of most probe-gene equilibria. Substituting $n = 2$ into Eq. (8) and solving for T_m yields

$$T_m = \frac{\Delta H^\circ}{R\ln C_T + \Delta S^\circ - R\ln 4} \text{ (for nonself-complementary strands)} \quad (16)$$

As expected for such a bimolecular event, the value of T_m will depend on the total concentration of the associating strands (C_T). The corresponding equation for complex formation by self-complementary strands is obtained by substituting $n = 2$ into Eq. (9) to obtain

$$T_m = \frac{\Delta H^\circ}{R\ln C_T + \Delta S^\circ} \text{ (for self-complementary strands)} \quad (17)$$

Once again, as expected for a bimolecular event, the value of T_m depends on C_T as well as ΔH° and ΔS°. Eqs. (16) and (17) allow one to calculate the melting temperature for any bimolecular equilibrium process from knowledge of C_T, ΔH, and ΔS. The latter two thermodynamic parameters can be obtained experimentally using the protocols described in this paper or calculated using published nearest-neighbor thermodynamic data *(6,9)*. Clearly, one can derive corresponding equations for reactions of any molecularity by substituting into Eq. (8) or (9) the appropriate value of n and solving for T_m.

It is important to recognize that the value of T_m also will depend on the salt concentration. For oligomeric duplexes shorter than 14 to 16 base pairs, the value of $\partial T_m/\partial \log[Na^+]$ appears to depend primarily on the chainlength, n, up to a sodium concentration of about 0.4*M* (Erie and Breslauer, unpublished results). In fact, for such duplexes, $\partial T_m/\partial \log[Na^+]$ crudely can be approximated by assigning the derivative a value of $n + 2$. Thus, for a 12-mer duplex ($n = 12$), one can estimate a $\partial T_m/\partial \log[Na^+]$ value of about 14 (12 + 2). For duplexes above 16 base pairs, $\partial T_m/\partial \log[Na^+]$ generally approaches the polymer values, which range from 18 to 22. In addition to the primary influence of duplex chainlength, the exact value of $\partial T_m/\partial \log[Na^+]$ also appears to depend secondarily on base composition, with base sequence exerting a tertiary impact. In the absence of knowledge of $\partial T_m/\partial \log[Na^+]$, one still can use the protocols described above to calculate relative T_m values (ΔT_m) rather than absolute T_ms. Thus, one can predict T_m differences, even when a lack of knowledge of $\partial T_m/\partial \log[Na^+]$ prevents one from accurately predicting absolute T_m values.

9. Concluding Remarks

In this chapter we have presented the general forms of equations that can be used to extract thermodynamic data from experimental results obtained by either calorimetric or noncalorimetric techniques. The detailed derivations of these equations are described in the Appendix of ref. *52*. We believe that these equations will be of widespread interest considering the increase in efforts designed to characterize thermodynamically the molecular forces that dictate and control the structural preferences of nucleic acids in solution.

Significantly, the protocols described here can be used to evaluate association and dissociation reactions of any molecularity. This feature is particularly important since the modeling of more complex biological structures will involve equilibria well beyond the simple bimolecular level. The formation of a DNA junction structure and a DNA tetraplex via tetramolecular association reactions provide two such examples *(49,53)*. The latter study *(53)* also illustrates how one can use the equations presented here to calculate the molecularity of a reaction of known enthalpy from the concentration dependence of T_m. This application is particularly timely since researchers are beginning to discover that certain nucleic acid sequences form higher order complexes that may be biologically significant. Defining the molecularity of these complexes represents an important and often elusive component of characterizing these higher order structures, a definition that can be accomplished by using the equations and protocols presented in this chapter.

In summary, an examination of the current literature reveals that numerous researchers are focusing on oligomeric modeling of biologically relevant structures. The equations and protocols described in this chapter will allow these researchers to characterize thermodynamically these new structural forms by analyzing equilibrium melting curves.

Acknowledgments

This work was supported by National Institutes of Health Grants GM23509, GM34469, and CA 47795, the Charles and Johanna Busch Memorial Fund, the Research Corporation, and the Rutgers Research Council.

References

1. Cantor, C. R. and Schimmel, P. R. (1980) in *Biophysical Chemistry, part I. The Conformation of Biological Macromolecules*, (Freeman, W. H., ed.), San Francisco, Chap. 6, pp. 311–314.
2. Bloomfield, V. A., Crothers, D. M., and Tinoco, Jr., I. (1974) *Physical Chemistry of Nucleic Acids*, Harper and Row, New York, Chap. 6, pp. 293–371.
3. Ts'o, P. O. P. (1974) *Basic Principles in Nucleic Acid Chemistry,* vol. II, Academic, New York.
4. Hinz, H. (1974) in *Biochemical Thermodynamics*, Elsevier Scientific, The Netherlands, pp. 116–168.
5. Breslauer, K. J. (1985) Methods for obtaining thermodynamic data on oligonucleotide transitions, in *Thermodynamic Data for Biochemistry and Biotechnology*, Academic, New York, pp. 377–394.
6. Breslauer, K. J., Frank, R., Blocker, H., and Marky, L. A. (1986) Predicting DNA duplex stability from the base sequence, *Proc. Natl. Acad. Sci. USA* **83,** 3746–3750.
7. Tinoco, Jr., I., Uhlenbeck, O., and Levine, M. D. (1971) Estimation of secondary structure in ribonucleic acids. *Nature* **230,** 363–367.
8. Tinoco, Jr., I., Borer, P. N., Dengler, B., Levine, M. D., Uhlenbeck, O. C., Crothers, D. M., and Gralla, J. (1973) Improved estimation of secondary structure of ribonucleic acids. *Nature New Biol.* **246,** 40–41.
9. Freier, S. M., Kierzek, R., Jaeger, J. A., Sugimoto, N., Caruthers, M. H., Neilson, T., and Turner, D. H. (1986) Improved free energy parameters for prediction of RNA duplex stability. *Proc. Natl. Acad. Sci. USA* **83,** 9373–9377.
10. Martin, F. H., Uhlenbeck, O. C., and Doty, P. (1971) Self-complimentary oligoribo-nucleotides: adenylic-acid-uridylic acid block copolymers. *J. Mol. Biol.* **57,** 201–215.
11. Uhlenbeck, O. C., Martin, F. H., and Doty, P. (1971) Self-complimentary oligoribo-nucleotides: effects of helix defects and guanylic acid-cytidylic acid base pairs. *J. Mol. Biol.* **57,** 217–229.

12. Craig, M. E., Crothers, D. M., and Doty, P. M. (1971) Relaxation kinetics of dimer formation by self-complementary oligonucleotides. *J. Mol. Biol.* **62,** 383–401.
13. Appleby, D. W. and Kallenbach, N. R. (1973) Theory of oligonucleotide stability. I. The effect of single-strand stacking. *Biopolymers* **12,** 2093–2120.
14. Porschke, D., Uhlenbeck, O. C., and Martin, F. H. (1973) Thermodynamics and kinetics of the helix-coil transition of oligomers containing GC base pairs. *Biopolymers* **12,** 1313–1335.
15. Gralla, J. and Crothers, D. M. (1973) Free energy of imperfect nucleic acid helices. II. Small hairpin loops. *J. Mol. Biol.* **73,** 497–511.
16. Gralla, J. and Crothers, D. M. (1973) Free energy of imperfect nucleic acid helices. III. Small internal loops. *J. Mol. Biol.* **78,** 301–319.
17. Borer, P. N., Dengler, B., Tinoco, Jr., I., and Uhlenbeck, O. C. (1974) Stability of RNA double stranded helices. *J. Mol. Biol.* **86,** 843–853.
18. Pohl, R. M. (1974) Thermodynamics of the helix-coil transition of (dG-dC) oligomers. *Eur. J. Biochem.* **42,** 495–504.
19. Breslauer, K. J., Sturtevant, J. M., and Tinoco, Jr., I. (1975) Calorimetric and spectroscopic investigation of the helix-to-coil transition of a ribo-oligonucleotide: rA_7U_7. *J. Mol. Biol.* **99,** 549–565.
20. Felsenfeld, G. and Miles, H. T. (1967) The physical and chemical properties of nucleic acids. *Ann. Rev. Biochem.* **36,** 407–448.
21. Marky, L. A., Canuel, L., Jones, R. A., and Breslauer, K. J. (1981) Calorimetric and spectroscopic investigation of the helix-to-coil transition of the self-complementary deoxyribonucleotide ATGCAT. *Biophys. Chem.* **13,** 141–149.
22. Albergo, D., Marky, L. A., Breslauer, K. J., and Turner, D. H. (1981) Thermodynamics of (dG-dC)3 double helix formation in H_2O and D_2O. *Biochemistry* **20,** 1409–1413.
23. Breslauer, K. J. and Bodnar, C. M. (1979) The effect of salt on the stability of the pH 4.2 rA8 double helix in a series of organic/aqueous mixed solvents: a test of oligoelectrolyte theory. *Biopolymers* **18,** 2167–2174.
24. Martin, F. H. and Tinoco, Jr., I. (1980) DNA–RNA hybrid duplexes containing oligo (dA:rU) sequences are exceptionally unstable, and may facilitate termination of transcription. *Nucleic Acids Res.* **8,** 2295–2299.
25. Breslauer, K. J. and Sturtevant, J. M. (1977) A calorimetric investigation of single stranded base stacking in the ribooligonucleotide A_7. *Biophys. Chem.* **7,** 205–209.
26. Marky, L. A., Blumenfeld, K. S., Kozlowski, S. A., and Breslauer, K. J. (1983) Salt-dependent transitions in the self-complementary deoxydodecanucleotide d(CGCGAATTCGCG): evidence for hairpin formation. *Biopolymers* **22,** 1247–1257.
27. Freier, S. M., Burger, B. J., Alkema, D., Neilson, T., and Turner, D. H. (1983) Effects of 3' dangling end stacking on the stability of GGCC and CCGG double helices. *Biochemistry* **22,** 6198–6206.

28. Freier, S. M., Alkema, D., Sinclair, A., Neilson, T., and Turner, D. H. (1985) Contributions of dangling end stacking and terminal base pair formation to the stabilities of XGGCCp, XCCGGp, XGGCCYp and XCCGGYp helixes. *Biochemistry* **24,** 4533–4539.
29. Freier, S. M., Petersheim, M., Hickey, D. R., and Turner, D. H. (1984) Thermodynamic studies of RNA stability. *J. Biomol. Struct. Dynam.* **1,** 1229–1242.
30. Nelson, J. W., Martin, F. H., and Tinoco, Jr., I. (1981) DNA and RNA oligomer thermodynamics: the effect of mismatched bases on double-helix stability. *Biopolymers* **20,** 2509–2531.
31. Uhlenbeck, O. C., Borer, P. N., Dengler, B., and Tinoco, Jr., I. (1973) The stability of RNA hairpin loops: $A_6C_mU_6$. *J. Mol. Biol.* **73,** 483–496.
32. Patel, D. J., Kozlowski, S. A., Marky, L. A., Broka, C., Rice, J. A., Itakura, K., and Breslauer, K. J. (1982) Premelting and melting transitions in the d(CGCGAATTCGCG) self-complementary duplex in solution. *Biochemistry* **21,** 428–436.
33. Ikuta, S., Chattopadhyaya, R., Ito, H., Dickerson, R. E., and Kearns, D. R. (1986) NMR study of a synthetic DNA hairpin. *Biochemistry* **25,** 4840–4849.
34. Haasnoot, C. A. G., de Bruin, S. H. Y., Berendsen, R. G., Janssen, H. G. J. M., Binnendijik, T. J. J., Hilbers, C. W., van der Marel, G. A., and van Boom, J. H. (1983) Structure, kinetics, and thermodynamics of DNA hairpin fragments in solution. *J. Biomol. Struct. Dyn.* **1,** 115–129.
35. Scheffler, I. E., Elson, E., and Baldwin, R. L. (1970) Helix formation by d(TA) oligomers. II. Analysis of the helix-coil transitions of linear and circular oligomers. *J. Mol. Biol.* **48,** 145–171.
36. Elson, E., Scheffler, I. E., and Baldwin, R. L. (1970) Helix formation by d(TA) oligomers. III. Electrostatic effects. *J. Mol. Biol.* **54,** 401.
37. Poland, D. and Scheraga, H. A. (1970) in *Theory of Helix-Coil Transitions in Biopolymers,* Adenine, New York.
38. Zimm, B. H. and Kallenbach, N. R. (1962) Selected aspects of the physical chemistry of polynucleotides and nucleic acids. *Ann. Rev. Phys. Chem.* **13,** 171.
39. Wang, A. C. and Kallenbach, N. R. (1971) Helical complexes of polyriboinosinic acid with copolymers of polyribocytidylic acid containing inosine, adenosine and uridine residues. *J. Mol. Biol.* **62,** 591–607.
40. Kallenbach, N. R. (1974) Stability of helical nucleic acids, in *Quantum Statistical Mechanics in the Natural Sciences* (Kursunoglu, B., Mintz, S. L., and Widmayer, S. M., eds.), Plenum, New York, pp. 95–118.
41. Applequist, J. and Damle, V. (1965) Thermodynamics of the helix coil equi librium in oligoadenylic acid. *J. Am. Chem. Soc.* **87,** 1450.
42. Poland, D., Vournakis, J. N., and Scheraga, H. A. (1966) Cooperative interactions in single-strand oligomers of deoxyriboadenylic acid. *Biopolymers* **4,** 223.
43. Vournakis, J. N., Poland, D., and Scheraga, H. A. (1967) Anti-cooperative interactions in single-strand oligomers of adenylic acid. *Biopolymers* **5,** 403.
44. Seeman, N. C. (1982) Nucleic acid junctions and lattices. *J. Theor. Biol.* **99,** 237–247.

45. Seeman, N. C. and Kallenbach, N. R. (1983) Design of immobile nucleic acid junctions. *Biophys. J.* **44,** 201–209.
46. Kallenbach, N. R., Ma, R.-I., and Seeman, N. C. (1983) An immobile nucleic acid junction constructed from oligonucleotides. *Nature* **305,** 829–831.
47. Seeman, N. C., Maestre, M. F., Ma, R.-I., and Kallenbach, N. R. (1985) Physical characterization of a nucleic acid junction, in *The Molecular Basis of Cancer* (Rein, R., ed.), Liss, New York, pp. 99–108.
48. Wemmer, D. E., Wand, A. J., Seeman, N. C., and Kallenbach, N. R. (1985) NMR analysis of DNA junctions: imino proton NMR studies of individual arms and intact junctions. *Biochemistry* **24,** 5745–5749.
49. Marky, L. A., Kallenbach, N., McDonough, K. A., Seeman, N. C., and Breslauer, K. J. (1987) The melting behavior of a DNA junction structure: a calorimetric and spectroscopic study. *Biopolymers*, **26,** 1621–1634.
50. Privalov, P. L. and Potekhin, S. A. (1986) in *Methods in Enzymology.* Adenine, New York, pp. 4–51.
51. Breslauer, K. J. and Bina-Stein, M. (1977) Relaxation kinetics of the helix-to-coil transition of a self-complementary ribooligonucleotide: rA_7U_7. *Biophys. Chem.* **7,** 211–216.
52. Marky, L. A. and Breslauer, K. J. (1987) Calculating thermodynamic data for transitions of any molecularity from equilibrium melting curves. *Biopolymers* **26,** 1601–1620.
53. Jin, R., Breslauer, K. J., Jones, R. A., and Gaffney, B. L. (1990) Tetraplex formation of a guanine-containing nonameric DNA fragment. *Science* **250,** 543–546.

Index

O

P